Lecture Notes in Artificial Intell

Edited by J. G. Carbonell and J. Siekmann

T0238194

Subseries of Lecture Notes in Computer Science

Springer
Berlin
Heidelberg
New York
Hong Kong
London
Milan
Paris
Tokyo

Marta Cialdea Mayer   Fiora Pirri (Eds.)

Automated Reasoning with Analytic Tableaux and Related Methods

International Conference, TABLEAUX 2003
Rome, Italy, September 9-12, 2003
Proceedings

 Springer

Series Editors

Jaime G. Carbonell, Carnegie Mellon University, Pittsburgh, PA, USA
Jörg Siekmann, University of Saarland, Saarbrücken, Germany

Volume Editors

Marta Cialdea Mayer
Università di Roma Tre, Dipartimento di Informatica e Automazione
via della Vasca Navale 79, 00146 Rome, Italy
E-mail: cialdea@dia.uniroma3.it

Fiora Pirri
Università di Roma "La Sapienza", Dipartimento di Informatica e Sistemistica
via Salaria 113, 00198 Rome, Italy
E-mail: pirri@dis.uniroma1.it

Cataloging-in-Publication Data applied for

A catalog record for this book is available from the Library of Congress

Bibliographic information published by Die Deutsche Bibliothek
Die Deutsche Bibliothek lists this publication in the Deutsche Nationalbibliographie;
detailed bibliographic data is available in the Internet at <http://dnd.ddb.de>.

CR Subject Classification (1998): I.2.3, F.4.1, I.2, D.1.6, D.2.4

ISSN 0302-9743
ISBN 3-540-40787-1 Springer-Verlag Berlin Heidelberg New York

Springer-Verlag Berlin Heidelberg New York,
a member of BertelsmannSpringer Science+Business Media GmbH

http://www.springer.de

© Springer-Verlag Berlin Heidelberg 2003
Printed in Germany

Typesetting: Camera-ready by author, data conversion by Da-TeX Gerd Blumenstein
Printed on acid-free paper   SPIN: 10931899   06/3142   5 4 3 2 1 0

Foreword

This volume contains the main papers presented at the International Conference on Analytic Tableaux and Related Methods (TABLEAUX 2003) held on September 9–12, 2003 in Rome, Italy. This conference was the continuation of international meetings on the same topic held in Lautenbach near Karlsruhe (1992), Marseille (1993), Abingdon near Oxford (1994), St. Goar near Koblenz (1995), Terrasini near Palermo (1996), Pont-à-Mousson near Nancy (1997), Oisterwijk near Tilburg (1998), Saratoga Springs near Albany NY (1999), St Andrews (2000), and Copenhagen (2002). In 2001 TABLEAUX was part of IJCAR 2001 in Siena.

Tableaux and related methods, such as Gentzen calculi, are a convenient and effective formalism for automating deduction not only in classical logic, but also in various non-standard logics. Examples taken from the papers collected in this volume alone include modal, temporal, intuitionistic, non-monotonic, conditional, paraconsistent, many-valued, intermediate, and description logics. Areas of application include verification of software and computer systems, deductive databases, knowledge representation and its required inference engines, and system diagnosis. The conference brought together researchers interested in all aspects – theoretical foundations, implementation techniques, systems development and applications – of the mechanization of reasoning with tableaux and related methods. Applications and implementations played a quite relevant role, as witnessed by the numerous system descriptions presented at the conference.

TABLEAUX 2003 in Rome was co-located with the International Conference on Theorem Proving in Higher Order Logics (TPHOLs 2003) and the 11th Symposium on the Integration of Symbolic Computation and Mechanized Reasoning (Calculemus 2003). The three events run in parallel provided the opportunity for contacts with a broader community, corroborated by the joint panel discussion on "Open Challenges for Computerized Mathematics," focusing on both deduction and calculus, and the talk by Thierry Coquand, jointly invited by Calculemus 2003 and TABLEAUX 2003.

Acknowledgements. We are grateful to all the people who contributed to the success of the conference TABLEAUX 2003. We thank all the members of the program committee and the other referees for their rigorous work in paper reviewing, the authors and the invited speakers for their contributions, and the tutorial organizers. We also give thanks to our colleague Carla Limongelli, for her invaluable help in many practical matters; Andrea Cecchetti, for installing and helping to maintain the software for our web-based reviewing procedure; and Consulta Umbria, for their professional support for local arrangements, particularly hard in a big city like Rome. A special thanks, finally, goes to our sponsors and the Faculty of Engineering of Università "La Sapienza," in whose seat the conference took place.

June 2003

Marta Cialdea Mayer
Fiora Pirri

Previous Tableaux Workshops/Conferences

1992	Lautenbach, Germany	1993	Marseille, France
1994	Abingdon, UK	1995	St. Goar, Germany
1996	Terrasini, Italy	1997	Pont-à-Mousson, France
1998	Oisterwijk, The Netherlands	1999	Saratoga Springs, USA
2000	St Andrews, UK	2001	part of IJCAR, Siena, Italy
2002	Copenhagen, Denmark		

Invited Speakers

Vito Michele Abrusci	Università di Roma Tre, Italy
Thierry Coquand	Chalmers University of Technology, Sweden (invited jointly with Calculemus 2003)
Johann Schumann	NASA Ames Research Center, USA

Program Chairing

Marta Cialdea Mayer (Chair)	Università di Roma Tre, Italy
Fiora Pirri (Vice-chair)	Università di Roma "La Sapienza," Italy

Program Committee

Peter Baumgartner	Universität Koblenz-Landau, Germany
Bernhard Beckert	Universität Karlsruhe, Germany
Serenella Cerrito	Université d'Evry Val d'Essonne, France
Marta Cialdea Mayer	Università di Roma Tre, Italy
Marcello D'Agostino	Università di Ferrara, Italy
Roy Dyckhoff	University of St Andrews, UK
Uwe Egly	Technische Universität Wien, Austria
Christian Fermüller	Technische Universität Wien, Austria
Melvin Fitting	City University of New York, USA
Ulrich Furbach	Universität Koblenz-Landau, Germany
Didier Galmiche	LORIA Nancy, France
Rajeev P. Goré	Australian National University, Australia
Jean Goubault-Larrecq	ENS Cachan, France
Reiner Hähnle	Chalmers University of Technology, Sweden
Christoph Kreitz	Cornell University, USA
Reinhold Letz	Technische Universität München, Germany
Fabio Massacci	Università di Trento, Italy
Neil V. Murray	University at Albany, SUNY, USA
Fiora Pirri	Università di Roma "La Sapienza," Italy
Nicola Olivetti	Università di Torino, Italy
Peter H. Schmitt	Universität Karlsruhe, Germany

Referees

Each submitted paper was refereed by three members of the program committee. In some cases, they consulted specialists who were not on the committee. We gratefully mention them:

Pietro Abate
Marc Aiguier
Agata Ciabattoni
Rainer Feldmann
Martin Giese
Bernhard Gramlich
Dirk Hoffmann
Dominique Larchey-Wendling

Alexander Leitsch
Daniel Méry
Alex Polleres
Zbigniew Stachniak
Gernot Stenz
Virginie Thion
Simon Thompson
Stefan Woltran

Tutorials

The conference program included the presentation of three tutorials, whose details appear in the same volume as the "Position Papers" (see below).

Hypersequent Calculi for Non-classical Logics
    Agata Ciabattoni (Technische Universität Wien)

The Situation Calculus
    Gerhard Lakemeyer (Rheinisch-Westfälische Technische Hochschule Aachen)

Proofs and Types in Constructive Geometry
    Jan von Plato (University of Helsinki)

Sponsors

Università degli Studi di Roma Tre
AKROS Group
CoLogNet (European Network of Excellence in Computational Logic)
Autonomous Laboratory for Cognitive Robotics (ALCOR)
CISIR (Consorzio Interuniversitario per i Servizi Innovativi in Rete)
RobotItaly
Banca di Roma

Position Papers

The conference program included the presentation of eight position papers. These papers are collected in the proceedings *TABLEAUX 2003 – Position Papers and Tutorials*, Technical Report RT-DIA-80-2003 of the Dipartimento di Informatica e Automazione (Università di Roma Tre), published by ARACNE Editrice, Italy (2003).

A Tableau Formulation of Annotated Logics
  Seiki Akama, Jairo Minoro Abe, Tetsuya Murai

A Parallel Implementation of a Decision Procedure for Propositional Intuitionistic Logic
  Alessandro Avellone, Guido Fiorino, Ugo Moscato

Extending Stålmarck's Method to First Order Logic
  Magnus Björk

Tableaux + Constraints
  Martin Giese, Reiner Hähnle

A Parallel Computation Technique for Linear Time Logic Tableaux
  Carla Limongelli, Andrea Orlandini, Valentina Poggioni

Decision Procedure for a Fragment of FTL with Equality
  Regimantas Pliuskevicius, Aida Pliuskeviciene

A New Form of the Semantic Tableaux Version of the Second Incompleteness Theorem
  Dan E. Willard

Combining Theories Sharing Dense Orders
  Calogero G. Zarba, Zohar Manna, Henny B. Sipma

Table of Contents

System Description

Non Commutative Logic: A Survey (Abstract)

V. Michele Abrusci

Dipartimento di Filosofia
Università degli Studi di Roma Tre, Italy
abrusci@uniroma3.it

Noncommutative Logic (NL) has been introduced by Abrusci and Ruet (Non-Commutative Logic I, Annals of Mathematical Logic, 2001). NL is a refinement of Linear Logic (LL) and a conservative extension of Lambek Calculus (LC). Therefore, NL is a constructive logic (i.e. proofs are programs).

Noncommutative logic allows to deal with commutative and non-commutative conjunctions and disjunctions.

In Noncommutative Logic sequents are order varieties on finite sets of occurrences of formulas.

The talk surveys the main results obtained in Noncommutative logic in 2001-2003, by several authors:

1. proof-nets,
2. sequent calculus, and sequentialization theorem,
3. proof search,
4. phase semantics, and completeness theorem,
5. reduction of proof-nets, and semantics of proofs,
6. modules.

M. Cialdea Mayer and F. Pirri (Eds.): TABLEAUX 2003, LNAI 2796, p. 1, 2003.

Dynamical Method in Algebra: A Survey (Abstract)

Thierry Coquand

Computing Science Department
Göteborg University, Sweden
coquand@cs.chalmers.se

The system D5, of J. Della Dora, C. Dicrescenzo, and D. Duval, allows computations in the algebraic closure of a field, though it is known that such a closure may fail to exist constructively. This mystery has been recently analysed in the work of Coste-Lombardi-Roy. A survey of this work and its connections with the system D5 is presented in this talk. In particular I present in detail the notion of "geometric logic" which allows a quite suggestive notion of proofs which may be of independent interest.

M. Cialdea Mayer and F. Pirri (Eds.): TABLEAUX 2003, LNAI 2796, p. 2, 2003.

Automated Theorem Proving in Generation, Verification, and Certification of Safety Critical Code (Abstract)

Johann Schumann

RIACS / NASA Ames
Moffett Field, CA 94035
schumann@email.arc.nasa.gov

With increased complexity of missions, the need for high quality, mission- and safety-critical software has increased dramatically over the past few years. Several incidents (e.g., Mars Polar Lander) have shown that software errors can cause total loss of a mission. Together with tight budgets and schedules, software development and certification has become a serious bottleneck.

In this talk, I report on work done in the Automated Software Engineering group at NASA Ames. We are developing automatic program synthesis systems for state estimation and navigation, AUTOFILTER, and data analysis (AUTOBAYES). These tools automatically generate documented C/C++ code from compact, high-level specifications.

For safety-critical applications, the generated code must be certified, since the alternative of verifying the code generator is not feasible. We support the certification process by combining code certification with automatic program synthesis. Code certification is a lightweight approach for formally demonstrating important aspects of software quality. Its basic idea is to require the code producer to provide formal proofs that the code satisfies certain safety properties. These proofs serve as certificates that can be checked independently. Full automation can be accomplished by having AUTOBAYES/FILTER synthesize all required detailed annotations (e.g., loop invariants) together with the code. A flexible VCG generates proof obligations for the individual safety policies; the proof tasks are then processed by e-SETHEO.

Whenever a tool is used during certification, the tool must be trusted to produce reliable results. In this talk, I discuss approaches on how to increase trust, like use of small trusted components, traceability between code and proofs, and automatic proof checking, as well as a number of important theoretical (like "is the domain theory consistent?") and practical (e.g., to get the proof out of the prover) issues.

In this talk, I also summarize an application of e-SETHEO to automatically prove safety and effectiveness properties of a flight control system.

M. Cialdea Mayer and F. Pirri (Eds.): TABLEAUX 2003, LNAI 2796, p. 3, 2003.

Tableaux with Four Signs
as a Unified Framework

Arnon Avron

School of Computer Science
Tel-Aviv University
Ramat Aviv 69978, Israel
aa@math.tau.ac.il

Abstract. We show that the use of tableaux with four types of signed formulas (the signs intuitively corresponding to positive/negative information concerning truth/falsity) provides a framework in which a diversity of logics can be handled in a uniform way. The logics for which we provide sound and complete tableau systems of this type are classical logic, the most important three-valued logics, the four-valued logic of logical bilattices (an extension of Belnap's four-valued logic), Nelson's logics for constructive negation, and da Costa's paraconsistent logic C_ω (together with some of its extensions). For the latter we provide new, simple semantics for which our tableau systems are sound and complete.

1 Introduction

There are two main variants of tableau systems for classical logic. One employs two sorts of signed formulas: $\mathbf{T}\varphi$ and $\mathbf{F}\varphi$ (intuitively meaning "φ is true" and "φ is false", respectively). The other employs ordinary formulas, replacing $\mathbf{T}\varphi$ simply by φ and using $\neg\varphi$ as a substitute for $\mathbf{F}\varphi$ (where $\neg$ is the negation connective of the language). This alternative for the use of signs works well for classical logic, but a combination of the two methods is frequently needed for handling weaker logics, since usually the refutation of $\mathbf{F}\varphi$ is not equivalent to the validity of φ. Our goal here is to present what we believe to be a better approach, one which allows for a unified treatment of negation (and other standard connectives!) in a diversity of logics. The idea is to use *four* sorts of signed formulas: $\mathbf{T}^+\varphi$, $\mathbf{T}^-\varphi$, $\mathbf{F}^+\varphi$, and $\mathbf{F}^-\varphi$. The intuitive meaning of these signs can best be explained in terms of positive and negative information (see e.g. [38]). $\mathbf{T}^+\varphi$ intuitively means that there is a positive information for the truth of φ, $\mathbf{T}^-\varphi$ means that there is a negative information for the truth of φ, $\mathbf{F}^+\varphi$ means that there is a positive information for the falsity of φ, $\mathbf{F}^-\varphi$ means that there is a negative information for the falsity of φ.[1] In the rest of this paper we demonstrate the usefulness of the four-signs framework by providing within it sound and complete tableau systems for several well known logics. In the next section we consider logics in

---

[1] A similar idea has motivated the introduction and use of bilattices. See [27, 28, 23, 22, 24, 25].

M. Cialdea Mayer and F. Pirri (Eds.): TABLEAUX 2003, LNAI 2796, pp. 4–16, 2003.

which negation is added to positive classical logic. This includes classical logic itself, the most important three-valued logics, and the four-valued logic of logical bilattices (which is an extension of Belnap's famous four-valued logic). In the last section we treat logics in which (true) negation is conservatively added to positive intuitionistic logic. The systems we consider there are Nelson's two logics for constructive negation ($\mathbf{N}^-$ and $\mathbf{N}$), and da Costa's paraconsistent logic C_ω (together with some of its extensions). For the latter we provide new, simple semantics for which our tableau system(s) are sound and complete.

One point should be noted before we proceed. Tableaux with more than two signs have already been used in the framework of many-valued logics (see the survey papers [29, 13] for the idea and for an extensive list of references). The signs which are employed there correspond to the truth-values of the logic in question (so tableau systems with exactly n signs are used for any n-valued logic). Here, in contrast, the four signs do *not* correspond to truth-values and we use them even for logics which do not have finite characteristic matrix. On the contrary, our goal is to provide a general framework which (as far as possible) is not essentially connected to any specific type of semantics.

2 Many-Valued Extensions of Positive Classical Logic

2.1 Four-Valued Logics

In [12, 11] Belnap suggested the use of logics based on the four truth-values $t, f, \top$, and $\bot$, where t and f are the classical values, $\top$ ("both true and false") represents the truth-value of formulas about which there is inconsistent data, while $\bot$ ("neither true nor false") is the truth-value of formulas on which no data is available. Belnap's structure is nowadays known also as the basic (distributive) *bilattice*, and its logic — as the basic logic of (distributive) bilattices (see [27, 28, 23, 22, 24, 25, 1, 2]). The following is an extension (from [1]) of Belnap's logic with an appropriate implication connective:

Definition 1. *The matrix* $\mathcal{M}_4 = \langle M_4, D_4, O_4 \rangle$: [2]

- $M_4 = \{t, f, \top, \bot\}$
- $D_4 = \{t, \top\}$
- *The operations in* O_4 *are defined by:*
  1. $\neg t = f, \ \neg f = t, \ \neg\top = \top, \ \neg\bot = \bot$
  2. $a \vee b = sup_{\leq_t}(a,b), \quad a \wedge b = inf_{\leq_t}(a,b)$, *where the partial order* $\leq_t$ *is defined by:* $f \leq_t \top \leq_t t$ *and* $f \leq_t \bot \leq_t t$.
  3. $a \supset b = \begin{cases} b & \text{if } a \in D_4 \\ t & \text{if } a \notin D_4 \end{cases}$

As usual, a function v from the set $\mathcal{F}$ of formulas of $\{\neg, \vee, \wedge, \supset\}$ into M_4 is called a valuation in $\mathcal{M}_4$ if it respects the operations in O_4. A valuation v is an $\mathcal{M}_4$-model of a formula φ of $\mathcal{F}$ if $v(\varphi) \in D_4$. v is an $\mathcal{M}_4$-model of a set T of

---

[2] The names of the various matrices discussed in this section are taken from [9].

formulas if it is an $\mathcal{M}_4$-model of each element of T. A formula φ follows in $\mathcal{M}_4$ from T ($T \vdash_{\mathcal{M}_4} \varphi$) if every $\mathcal{M}_4$-model of T is also a $\mathcal{M}_4$-model of φ.

The concept of an $\mathcal{M}_4$-model can be extended to signed formulas as follows:

- v is an $\mathcal{M}_4$-model of $\mathbf{T}^+\varphi$ if $v(\varphi) \in \{t, \top\}$
- v is an $\mathcal{M}_4$-model of $\mathbf{T}^-\varphi$ if $v(\varphi) \in \{f, \top\}$
- v is an $\mathcal{M}_4$-model of $\mathbf{F}^+\varphi$ if $v(\varphi) \in \{f, \bot\}$
- v is an $\mathcal{M}_4$-model of $\mathbf{F}^-\varphi$ if $v(\varphi) \in \{t, \bot\}$

Note that $T \vdash_{\mathcal{M}_4} \varphi$ iff the set $\{\mathbf{T}^+\psi \mid \psi \in T\} \cup \{\mathbf{F}^+\varphi\}$ is not satisfiable.

We present now a tableau system which is sound and complete w.r.t $\vdash_{\mathcal{M}_4}$:

Definition 2. *The Tableau System* $Tab(\mathcal{M}_4)$: [3]

Expansion Rules:

$$(\mathbf{T}^+\neg) \quad \frac{\mathbf{T}^+\neg\varphi}{\mathbf{T}^-\varphi} \qquad\qquad \frac{\mathbf{T}^-\neg\varphi}{\mathbf{T}^+\varphi} \quad (\mathbf{T}^-\neg)$$

$$(\mathbf{F}^+\neg) \quad \frac{\mathbf{F}^+\neg\varphi}{\mathbf{F}^-\varphi} \qquad\qquad \frac{\mathbf{F}^-\neg\varphi}{\mathbf{F}^+\varphi} \quad (\mathbf{F}^-\neg)$$

$$(\mathbf{T}^+\wedge) \quad \frac{\mathbf{T}^+\varphi\wedge\psi}{\mathbf{T}^+\varphi, \mathbf{T}^+\psi} \qquad\qquad \frac{\mathbf{T}^-\varphi\wedge\psi}{\mathbf{T}^-\varphi \mid \mathbf{T}^-\psi} \quad (\mathbf{T}^-\wedge)$$

$$(\mathbf{F}^+\wedge) \quad \frac{\mathbf{F}^+\varphi\wedge\psi}{\mathbf{F}^+\varphi \mid \mathbf{F}^+\psi} \qquad\qquad \frac{\mathbf{F}^-\varphi\wedge\psi}{\mathbf{F}^-\varphi, \mathbf{F}^-\psi} \quad (\mathbf{F}^-\wedge)$$

$$(\mathbf{T}^+\vee) \quad \frac{\mathbf{T}^+\varphi\vee\psi}{\mathbf{T}^+\varphi \mid \mathbf{T}^+\psi} \qquad\qquad \frac{\mathbf{T}^-\varphi\vee\psi}{\mathbf{T}^-\varphi, \mathbf{T}^-\psi} \quad (\mathbf{T}^-\vee)$$

$$(\mathbf{F}^+\vee) \quad \frac{\mathbf{F}^+\varphi\vee\psi}{\mathbf{F}^+\varphi, \mathbf{F}^+\psi} \qquad\qquad \frac{\mathbf{F}^-\varphi\vee\psi}{\mathbf{F}^-\varphi \mid \mathbf{F}^-\psi} \quad (\mathbf{F}^-\vee)$$

$$(\mathbf{T}^+\supset) \quad \frac{\mathbf{T}^+\varphi\supset\psi}{\mathbf{F}^+\varphi \mid \mathbf{T}^+\psi} \qquad\qquad \frac{\mathbf{T}^-\varphi\supset\psi}{\mathbf{T}^+\varphi, \mathbf{T}^-\psi} \quad (\mathbf{T}^-\supset)$$

$$(\mathbf{F}^+\supset) \quad \frac{\mathbf{F}^+\varphi\supset\psi}{\mathbf{T}^+\varphi, \mathbf{F}^+\psi} \qquad\qquad \frac{\mathbf{F}^-\varphi\supset\psi}{\mathbf{F}^+\varphi \mid \mathbf{F}^-\psi} \quad (\mathbf{F}^-\supset)$$

Closure Conditions: *A branch is closed iff for some formula φ it contains either $\{\mathbf{T}^+\varphi, \mathbf{F}^+\varphi\}$ or $\{\mathbf{T}^-\varphi, \mathbf{F}^-\varphi\}$.*

Theorem 1. *A set of signed formulas is unsatisfiable in $\mathcal{M}_4$ (i.e.: does not have an $\mathcal{M}_4$-model) iff it has a closed tableau in $Tab(\mathcal{M}_4)$ (i.e.: a tableau in which every branch is closed).*

---

[3] This system is closely related to the Gentzen-type system for this logic that was presented in [1, 9], and its completeness can be derived from the completeness of that system. It is however more illuminating to prove it directly.

Proof: Obviously, any set T of formulas, for which either of the two closure conditions obtains, is unsatisfiable. It is also straightforward to check that for every expansion rule R, if T has a model in $\mathcal{M}_4$ then so does at least one of the sets which are obtained from T by R. Together these two facts imply the soundness of $Tab(\mathcal{M}_4)$ (i.e.: if T has a closed tableau then it is unsatisfiable). For the converse it suffices to prove that the set of formulas Γ of any fully expanded open branch (in a complete tableau for T) has the following v as a model:

$$v(p) = \begin{cases} f & \mathbf{T}^+p \notin \Gamma, \mathbf{F}^-p \notin \Gamma \\ \bot & \mathbf{F}^+p \in \Gamma, \mathbf{F}^-p \in \Gamma \\ \top & \mathbf{T}^+p \in \Gamma, \mathbf{T}^-p \in \Gamma \\ t & otherwise \end{cases}$$

(note that $v(p) = t$ iff either $\mathbf{T}^+p \in \Gamma, \mathbf{T}^-p \notin \Gamma$ or $\mathbf{F}^+p \notin \Gamma, \mathbf{F}^-p \in \Gamma$). Indeed, the fact that Γ is open (i.e. contains no subset of the form $\{\mathbf{T}^+\varphi, \mathbf{F}^+\varphi\}$ or of the form $\{\mathbf{T}^-\varphi, \mathbf{F}^-\varphi\}$) implies that v is well defined. It also implies that if $Sp \in \Gamma$, where S is one of the four signs and p is atomic, then v is a model of Sp. Using induction on the structure of formulas and the fact that Γ is fully expanded, it is not difficult then to show that v is a model of any formula in Γ.

Corollary 1. *Let φ and ψ be formulas in the language of $\{\neg, \vee, \wedge\}$. Then $\varphi \to \psi$ is a valid first degree entailment of the relevance logic R ([3, 4, 19]) iff $\{\mathbf{T}^+\varphi, \mathbf{F}^+\psi\}$ has a closed tableau in $Tab(\mathcal{M}_4)$.*

Note. Although we use here 4 signs, these signs do *not* correspond to the 4 truth values of the semantics. Indeed, the same signs will be used below (with very similar systems) for 3-valued logics, and even for classical logic.

Note. It is well known from the literature on bilattices (see e.g. [28, 25]) that $\mathcal{M}_4$ can be identified with $\{t, f\} \times \{t, f\}$ (so that $\top$ represents (t, t), $\bot$ represents (f, f), t represents (t, f) and f represents (f, t)). This allows an alternative presentation of the semantics, using *two* valuations in $\{t, f\}$, representing independent information concerning truth and falsity of formulas (see [38]). We shall explain more about this approach in section 3.

2.2 Three-Valued Logics and Classical Logic

We consider next two basic three-valued logics, whose matrices are submatrices of $\mathcal{M}_4$.

Definition 3. *The matrix $\mathcal{M}_3^{\{t\}} = \langle M_3^{\{t\}}, D_3^{\{t\}}, O_3^{\{t\}} \rangle$:*

- $M_3^{\{t\}} = \{t, f, \bot\}$
- $D_3^{\{t\}} = \{t\}$
- *The operations in $O_3^{\{t\}}$ are defined by:*
  1. $\neg t = f$, $\neg f = t$, $\neg \bot = \bot$
  2. $a \vee b = sup_{\leq_t}(a, b)$, $a \wedge b = inf_{\leq_t}(a, b)$,

$$3.\ a \supset b = \begin{cases} b & \text{if } a \in D_3^{\{t\}} \\ t & \text{if } a \notin D_3^{\{t\}} \end{cases}$$

Note. The connective $\supset$ of $\mathcal{M}_3^{\{t\}}$ was originally introduced by Słupecki in [36]. It was independently reintroduced in [32, 39, 35] and [7] (see also [14]). The language of $\mathcal{M}_3^{\{t\}}$ is equivalent ([7]) to that used in the logic LPF of the VDM project ([30]), as well as to the language of Łukasiewicz 3-valued logic L₃ ([31]). It is in fact the language of all 3-valued operations which are *classically closed*. It can be shown that by adding one propositional constant to it (corresponding to the truth value $\perp$) we get a functionally complete set of 3-valued connectives (See [8] for further details and references).

Definition 4. *The matrix* $\mathcal{M}_3^{\{t,\top\}} = \langle M_3^{\{t,\top\}}, D_3^{\{t,\top\}}, O_3^{\{t,\top\}} \rangle$:

- $M_3^{\{t,\top\}} = \{t, f, \top\}$
- $D_3^{\{t,\top\}} = \{t, \top\}$
- *The operations in* $O_3^{\{t,\top\}}$ *are defined by:*
  1. $\neg t = f, \ \neg f = t, \ \neg\top = \top$
  2. $a \vee b = \sup_{\leq_t}(a,b), \ a \wedge b = \inf_{\leq_t}(a,b),$
  3. $a \supset b = \begin{cases} b & \text{if } a \in D_3^{\{t,\top\}} \\ t & \text{if } a \notin D_3^{\{t,\top\}} \end{cases}$

Note that the main difference between $\mathcal{M}_3^{\{t,\top\}}$ and $\mathcal{M}_3^{\{t\}}$ is in the choice of the designated values. As a result, the connective $\supset$ of $\mathcal{M}_3^{\{t,\top\}}$ is not identical to that of $\mathcal{M}_3^{\{t\}}$, despite their similar definitions (the other three connectives of $\mathcal{M}_3^{\{t,\top\}}$ *are* identical to their $\mathcal{M}_3^{\{t\}}$'s counterparts).

Note. The implication connective of $\mathcal{M}_3^{\{t,\top\}}$ was first introduced in [17, 16]. It was independently introduced also in [6]. The language of $\mathcal{M}_3^{\{t,\top\}}$ is equivalent to that used in the standard 3-valued paraconsistent logic J_3 ([18, 6, 34, 20]. In [7] it is called *Pac*), as well as to that used in the semi-relevant system RM_3 ([3, 4, 19]. See also [6, 7]). It is the language of all 3-valued operations which are classically closed and *free* ([8]).

The concepts of $\mathcal{M}_3^{\{t\}}$-model and of $\mathcal{M}_3^{\{t,\top\}}$-model are defined now exactly as in the case of $\mathcal{M}_4$ (but of course only the available truth-values are relevant). Thus practically a valuation v in $\mathcal{M}_3^{\{t\}}$ is an $\mathcal{M}_3^{\{t\}}$-model of $\mathbf{T}^+\varphi$ if $v(\varphi) = t$.

The Tableau System $Tab(\mathcal{M}_3^{\{t\}})$: This is the system obtained from $Tab(\mathcal{M}_4)$ by adding to it the following extra closure condition: a branch is closed also if for some formula φ it contains $\{\mathbf{T}^+\varphi, \mathbf{T}^-\varphi\}$.

The Tableau System $Tab(\mathcal{M}_3^{\{t,\top\}})$: This is the system obtained from $Tab(\mathcal{M}_4)$ by adding to it the following extra closure condition: a branch is closed also if for some formula φ it contains $\{\mathbf{F}^+\varphi, \mathbf{F}^-\varphi\}$.

Theorem 2.   *1. A set of signed formulas is unsatisfiable in $\mathcal{M}_3^{\{t\}}$ iff it has a closed tableau in $Tab(\mathcal{M}_3^{\{t\}})$.*

2. A set of signed formulas is unsatisfiable in $\mathcal{M}_3^{\{t,\top\}}$ iff it has a closed tableau in $Tab(\mathcal{M}_3^{\{t,\top\}})$.

Proof: The proofs of both parts are almost identical to the proof of Theorem 1. The only difference in the case of $Tab(\mathcal{M}_3^{\{t\}})$ is that the extra closure condition rules out the possibility that $v(\varphi) = \top$ for some formula φ, while in the case of $Tab(\mathcal{M}_3^{\{t,\top\}})$ it rules out the possibility that $v(\varphi) = \bot$.

Note.   Because of the strong expressive power of the languages of $\mathcal{M}_3^{\{t\}}$ and $Tab(\mathcal{M}_3^{\{t\}})$ (see [8, 9]), their tableau systems can be used as bases for all other 3-valued logics. For example, $\varphi \to \psi$, where $\to$ is Łukasiewicz 3-valued implication, is equivalent in $\mathcal{M}_3^{\{t\}}$ to $(\varphi \supset \psi) \wedge (\neg\psi \supset \neg\varphi)$. This leads to the following 4 rules for it:

$$(\mathbf{T^+} \to) \frac{\mathbf{T^+}\varphi \to \psi}{\mathbf{T^+}\psi \mid \mathbf{T^-}\varphi \mid \mathbf{F^+}\varphi, \mathbf{F^-}\psi} \qquad\qquad \frac{\mathbf{T^-}\varphi \to \psi}{\mathbf{T^+}\varphi, \mathbf{T^-}\psi} (\mathbf{T^-} \to)$$

$$(\mathbf{F^+} \to) \frac{\mathbf{F^+}\varphi \to \psi}{\mathbf{T^+}\varphi, \mathbf{F^+}\psi \mid \mathbf{T^-}\psi, \mathbf{F^-}\varphi} \qquad\qquad \frac{\mathbf{F^-}\varphi \to \psi}{\mathbf{F^+}\varphi \mid \mathbf{F^-}\psi} (\mathbf{F^-} \to)$$

We end this section with a characterization of classical logic itself:

Theorem 3.   *Let $Tab(\mathcal{M}_2)$ be the system obtained from $Tab(\mathcal{M}_4)$ by adding to it both of the extra closure conditions of $Tab(\mathcal{M}_3^{\{t\}})$ and $Tab(\mathcal{M}_3^{\{t,\top\}})$. Then a set of signed formulas is unsatisfiable in classical logic iff it has a closed tableau in $Tab(\mathcal{M}_2)$.*

Proof: The proof is again almost identical to the proof of Theorem 1. Only this time the two extra closure conditions together rule out the possibility that $v(\varphi) \in \{\top, \bot\}$ for some formula φ.

Note.   It is easy to see that the positive fragments of all the four logics considered in this section are identical (i.e. the fragments in which only the signs $\mathbf{T^+}, \mathbf{F^+}$ and the connectives $\wedge, \vee$, and $\supset$ are used). Hence all these logics are conservative extensions of positive classical logic.

3   Conservative Extensions of Positive Intuitionistic Logic

3.1   Nelson's Logics for Negation

The logics $\mathbf{N^-}$ and $\mathbf{N}$ are conservative extensions of positive intuitionistic logic which were independently introduced by Nelson (see [5]) and Kutschera ([37]). The motivation for their introduction has been the wish to provide an adequate treatment of negative information within the framework of constructive logic. See [38] for further details and references.

The standard semantics of $\mathbf{N}^-$ is based on Kripke frames $\mathcal{I} = \langle I, \leq, v^+, v^- \rangle$ in which v^+ and v^- are valuations from $I \times \mathcal{F}$ into $\{t, f\}$ (where $\mathcal{F}$ is the set of formulas) which satisfy the following two basic conditions:

(H^+) If $a \leq b$ and $v^+(a, \varphi) = t$ then $v^+(b, \varphi) = t$
(H^-) If $a \leq b$ and $v^-(a, \varphi) = t$ then $v^-(b, \varphi) = t$

v^+ and v^- should further satisfy also the following conditions:

$$v^+(a, \varphi \wedge \psi) = t \quad \text{iff} \quad v^+(a, \varphi) = t \text{ and } v^+(a, \psi) = t$$
$$v^-(a, \varphi \wedge \psi) = t \quad \text{iff} \quad v^-(a, \varphi) = t \text{ or } v^-(a, \psi) = t$$
$$v^+(a, \varphi \vee \psi) = t \quad \text{iff} \quad v^+(a, \varphi) = t \text{ or } v^+(a, \psi) = t$$
$$v^-(a, \varphi \vee \psi) = t \quad \text{iff} \quad v^-(a, \varphi) = t \text{ and } v^-(a, \psi) = t$$
$$v^+(a, \varphi \supset \psi) = t \quad \text{iff} \quad \text{for all } b \geq a, \text{ either } v^+(b, \varphi) = f \text{ or } v^+(b, \psi) = t$$
$$v^-(a, \varphi \supset \psi) = t \quad \text{iff} \quad v^+(a, \varphi) = t \text{ and } v^-(a, \psi) = t$$
$$v^+(a, \neg\varphi) = t \quad \text{iff} \quad v^-(a, \varphi) = t$$
$$v^-(a, \neg\varphi) = t \quad \text{iff} \quad v^+(a, \varphi) = t$$

Call a frame $\mathcal{I} = \langle I, \leq, v^+, v^- \rangle$ satisfying the above conditions an $\mathbf{N}^-$-frame. An $\mathbf{N}$-frame is defined similarly, with one extra condition: that $v^+(a, \varphi)$ and $v^-(a, \varphi)$ cannot both be t at the same time.

Note.   It can be shown that it suffices to demand the H(ereditary) conditions (H^+) and (H^-) only for atomic formulas. They are imposed then on the set of all formulas by the other conditions.

The semantics of formulas and of signed formulas is defined now as follows. Let $\mathcal{I} = \langle I, \leq, v^+, v^- \rangle$ be an $\mathbf{N}^-$-frame, and let $a \in I$. Define:

- $(\mathcal{I}, a)$ is a $\mathbf{N}^-$-model of $\mathbf{T}^+\varphi$ if $v^+(a, \varphi) = t$
- $(\mathcal{I}, a)$ is a $\mathbf{N}^-$-model of $\mathbf{T}^-\varphi$ if $v^-(a, \varphi) = t$
- $(\mathcal{I}, a)$ is a $\mathbf{N}^-$-model of $\mathbf{F}^+\varphi$ if $v^+(a, \varphi) = f$
- $(\mathcal{I}, a)$ is a $\mathbf{N}^-$-model of $\mathbf{F}^-\varphi$ if $v^-(a, \varphi) = f$

Call now $(\mathcal{I}, a)$ an $\mathbf{N}^-$-model of an ordinary formula φ iff it is an $\mathbf{N}^-$-model of $\mathbf{T}^+\varphi$ (iff it is not an $\mathbf{N}^-$-model of $\mathbf{F}^+\varphi$). Define $\mathbf{N}$-models of signed formulas and of ordinary formulas in a similar way, using $\mathbf{N}$-frames instead of $\mathbf{N}^-$-frames.

Note.   If we allow only one element in I then what we get is equivalent to the four-valued logic of $\mathcal{M}_4$. As we have already noted above, it is indeed quite common to use two valuations v^+ and v^- from $\mathcal{F}$ to $\{t, f\}$ for an equivalent representation of the semantics of this logic ([38]). The conditions concerning v^+ and v^- are practically identical to those in the case of $\mathbf{N}^-$, with only one exception: instead of the above condition concerning $v^+(a, \varphi \supset \psi) = t$ we have in that logic the simpler condition:

$$v^+(a, \varphi \supset \psi) = t \quad \text{iff} \quad v^+(a, \varphi) = f \text{ or } v^+(a, \psi) = t$$

It is possible then to define the meanings of the signed formulas in this logic in a way which is completely analogous to the way this was done above for $\mathbf{N}^-$.

We present now tableau systems which are sound and complete with respect to $\mathbf{N}^-$ and $\mathbf{N}$.

Definition 5. *The Tableau Systems $Tab(\mathbf{N}^-)$ and $Tab(\mathbf{N})$ are obtained from $Tab(\mathcal{M}_4)$ and $Tab(\mathcal{M}_3^{\{t\}})$ (respectively) by replacing their $(\mathbf{F}^+ \supset)$ rule with the following pair of rules:*

$$(\mathbf{F}^+ \supset)^w \qquad \frac{\mathbf{F}^+\varphi \supset \psi}{\mathbf{F}^+\psi}$$

$$(\mathbf{F}^+ \supset)^I \qquad \frac{\mathcal{S}, \mathbf{F}^+\varphi \supset \psi}{T(\mathcal{S}), \mathbf{T}^+\varphi, \mathbf{F}^+\psi}$$

Here $(\mathbf{F}^+ \supset)^I$ is a variant of the usual special intuitionistic rule for refuting implication: if $\mathcal{S}$ is the set of signed formulas on some branch then $T(\mathcal{S})$ is the set of all the elements in $\mathcal{S}$ whose sign is either $\mathbf{T}^+$ or $\mathbf{T}^-$, and an expansion of a branch by this rule requires the creation of a new tableau for the set $T(\mathcal{S}) \cup \{\mathbf{T}^+\varphi, \mathbf{F}^+\psi\}$.

Theorem 4. *A set of signed formulas is unsatisfiable in $\mathbf{N}^-$ iff it has a closed tableau in $Tab(\mathbf{N}^-)$.*

Proof: The proof is similar to the proofs of the soundness and completeness of the standard tableau system for propositional intuitionistic logic, or of the soundness and completeness of the usual Gentzen-type systems for $\mathbf{N}^-$ and $\mathbf{N}$ (as presented e.g. in [38])[4]. An outline of the proof goes as follows. Call a set Γ of signed formulas *saturated* if it satisfies the following conditions:

1. Γ has no closed tableau in $Tab(\mathbf{N}^-)$.
2. With the exception of $(\mathbf{F}^+ \supset)^I$, Γ respects all the expansion rules of $Tab(\mathbf{N}^-)$ (e.g.: if $\mathbf{T}^+\varphi \wedge \psi \in \Gamma$ then both $\mathbf{T}^+\varphi \in \Gamma$ and $\mathbf{T}^+\psi \in \Gamma$, while if $\mathbf{F}^+\varphi \wedge \psi \in \Gamma$ then either $\mathbf{F}^+\varphi \in \Gamma$ or $\mathbf{F}^+\psi \in \Gamma$).

Obviously, if a set Δ of signed formulas does not have a closed tableau in $Tab(\mathbf{N}^-)$ then it can be extended to a saturated set Δ^*, so that every (ordinary) formula which occurs in Δ^* is a subformula of some formula in Δ. Let I be the set of all saturated sets which have this property. Define a partial order $\leq$ on I by: $\Delta_1 \leq \Delta_2$ if either $\Delta_1 = \Delta_2$ or $T(\Delta_1) \subset T(\Delta_2)$ (where the inclusion should be proper). Define next v^+ and v^- for $\Gamma \in I$ and p atomic by: $v^+(\Gamma, p) = t$ iff $\mathbf{T}^+p \in \Gamma$, while $v^-(\Gamma, p) = t$ iff $\mathbf{T}^-p \in \Gamma$. It is not difficult to show that $\mathcal{I} = \langle I, \leq, v^+, v^- \rangle$ is an $\mathbf{N}^-$-frame, and that for each $\Gamma \in I$, $(\mathcal{I}, \Gamma)$ is a model of all the signed formulas of Γ. In particular: if Δ does not have a closed tableau in $Tab(\mathbf{N}^-)$ then $(\mathcal{I}, \Delta^*)$ is a model of all the signed formulas in Δ.

Theorem 5. *A set of signed formulas is unsatisfiable in $\mathbf{N}$ iff it has a closed tableau in $Tab(\mathbf{N})$.*

Proof: Similar to the proof of Theorem 4.

---

[4] The tableau systems we present here are of course strongly related to these Gentzen-type systems.

3.2 Extensions with Excluded Middle

It is well known that it is impossible to conservatively add to intuitionistic positive logic a negation which is both explosive (i.e.: $\neg\varphi, \varphi \vdash \psi$ for all φ, ψ) and for which LEM (the Law of Excluded Middle: $\neg\varphi \vee \varphi$) is valid. With such an addition we get classical logic. In $\mathbf{N}$ (following the tradition of intuitionistic logic) the choice was on explosiveness. In the paraconsistent logics of da Costa's school ([16, 15]) explosiveness is rejected, while LEM is accepted. Thus da Costa's basic system C_ω is a conservative extension of positive intuitionistic logic which is obtained from any standard Hilbert-type formulation of this logic by adding as axioms $\neg\varphi \vee \varphi$ and $\neg\neg\varphi \supset \varphi$. We present now a Kripke-style semantics for C_ω which is similar to that we have presented above for $\mathbf{N}$ [5].

Definition 6. *A C_ω-frame is a structure $\mathcal{I} = \langle I, \leq, v^+, v^- \rangle$ in which v^+ and v^- are valuations from $I \times \mathcal{F}$ into $\{t, f\}$ such that:*

1. *There are no $a \in I$ and φ for which both $v^+(a, \varphi) = f$ and $v^-(a, \varphi) = f$.*
2. *The basic conditions (H^+) and (H^-) above are satisfied.*
3. *v^+ and v^- satisfy also the following conditions:*

$$
\begin{array}{llll}
v^+(a, \varphi \wedge \psi) = t & \text{iff} & v^+(a, \varphi) = t \text{ and } v^+(a, \psi) = t \\
v^+(a, \varphi \vee \psi) = t & \text{iff} & v^+(a, \varphi) = t \text{ or } v^+(a, \psi) = t \\
v^+(a, \varphi \supset \psi) = t & \text{iff} & \text{for all } b \geq a, \text{ either } v^+(b, \varphi) = f \text{ or } v^+(b, \psi) = t \\
v^+(a, \neg\varphi) = t & \text{iff} & v^-(a, \varphi) = t \\
v^-(a, \neg\varphi) = f & \text{if} & v^+(a, \varphi) = f
\end{array}
$$

Thus the conditions concerning v^+ are identical to those in the case of $\mathbf{N}^-$, and are fully deterministic (given v^-). The values assigned to v^-, in contrast, are in general not determined by the values assigned by v^+ and v^- to its subformulas, and they are only subjected to two *constraints* (this implies, among other things, that it does not suffice to assume conditions (H^+) and (H^-) only for atomic formulas, since this does not enforce them to hold for arbitrary formulas).

The concept of a model of a signed formula, and the associated consequence relation are defined now exactly as in the case of $\mathbf{N}^-$ and $\mathbf{N}$. We present now a corresponding tableau system: [6]

Definition 7. *The tableau system $Tab(C_\omega)$ has the following rules and closure conditions:*

Closure Conditions: Like in the case of $Tab(\mathcal{M}_3^{\{t,\top\}})$, a branch is closed iff for some formula φ it contains either $\{\mathbf{T}^+\varphi, \mathbf{F}^+\varphi\}$, or $\{\mathbf{T}^-\varphi, \mathbf{F}^-\varphi\}$, or $\{\mathbf{F}^+\varphi, \mathbf{F}^-\varphi\}$.

---

[5] This similarity, and the fact that it needs no new ad-hoc constructs (like the function T used in [10]) are the main advantages of our semantics over the older one given in [10]. It is also considerably simpler, requiring no complicated conditions of the sort of condition III.f. from Definition 1 of [10].

[6] This system is closely related, but not identical, to the Gentzen-type system given for C_ω in [33].

Expansion Rules: The rules $(\mathbf{T}^+\neg)$, $(\mathbf{F}^+\neg)$, $(\mathbf{T}^+\wedge)$, $(\mathbf{F}^+\wedge)$, $(\mathbf{T}^+\vee)$, $(\mathbf{F}^+\vee)$, $(\mathbf{T}^+\supset)$, and $(\mathbf{T}^-\neg)$ of $Tab(\mathcal{M}_4)$, as well as $(\mathbf{F}^+\supset)^w$ and $(\mathbf{F}^+\supset)^I$ of $Tab(\mathbf{N}^-)$.

Analytic Cuts:

$$\frac{S\varphi}{\mathbf{T}^+\psi \mid \mathbf{F}^+\psi} \qquad\qquad \frac{S\varphi}{\mathbf{T}^-\psi \mid \mathbf{F}^-\psi}$$

Where $S \in \{\mathbf{T}^+, \mathbf{F}^+, \mathbf{T}^-, \mathbf{F}^-\}$ and ψ is a subformula of φ

Theorem 6. *A set of signed formulas is unsatisfiable in C_ω iff it has a closed tableau in $Tab(C_\omega)$.*

Proof: Again we only give an outline. This time we call a set Γ of signed formulas *saturated* if it satisfies the following conditions:

1. Γ has no closed tableau in $Tab(C_\omega)$.
2. If $S\varphi \in \Gamma$ then for every subformula ψ of φ, either $\mathbf{T}^+\psi \in \Gamma$ or $\mathbf{F}^+\psi \in \Gamma$, and either $\mathbf{T}^-\psi \in \Gamma$ or $\mathbf{F}^-\psi \in \Gamma$.
3. With the exception of $(\mathbf{F}^+\supset)^I$, Γ respects all the expansion rules of $Tab(C_\omega)$.

Because of the presence of the analytic cuts, it is easy to see that if Δ does not have a closed tableau in $Tab(C_\omega)$ then again it can be extended to a saturated set Δ^*, so that every (ordinary) formula which occurs in Δ^* is a subformula of some formula in Δ. Let I be the set of all saturated sets which have this property. Obviously $\Delta^* \in I$. Define $\leq$ on I like in the proof of Theorem 4. Define next v^+ and v^- for $\Gamma \in I$ and $\varphi \in \mathcal{F}$ recursively as follows:

- If φ is atomic, then $v^+(\Gamma, \varphi) = f$ iff $\mathbf{F}^+\varphi \in \Gamma$, and $v^-(\Gamma, \varphi) = f$ iff $\mathbf{F}^-\varphi \in \Gamma$.
- If $\varphi = \psi_1 \wedge \psi_2$ then $v^+(\Gamma, \varphi) = f$ iff either $v^+(\Gamma, \psi_1) = f$ or $v^+(\Gamma, \psi_2) = f$, while $v^-(\Gamma, \varphi) = f$ iff $\mathbf{F}^-\varphi \in \Gamma$.
- If $\varphi = \psi_1 \vee \psi_2$ then $v^+(\Gamma, \varphi) = f$ iff $v^+(\Gamma, \psi_1) = f$ and $v^+(\Gamma, \psi_2) = f$, while $v^-(\Gamma, \varphi) = f$ iff $\mathbf{F}^-\varphi \in \Gamma$.
- If $\varphi = \psi_1 \supset \psi_2$ then $v^+(\Gamma, \varphi) = f$ iff there exists $\Gamma^* \supseteq \Gamma$ in I such that $v^+(\Gamma^*, \psi_1) = t$ and $v^-(\Gamma^*, \psi_2) = f$, while $v^-(\Gamma, \varphi) = f$ iff $\mathbf{F}^-\varphi \in \Gamma$.
- If $\varphi = \neg\psi$ then $v^+(\Gamma, \varphi) = f$ iff $v^-(\Gamma, \psi) = f$, while $v^-(\Gamma, \varphi) = f$ iff either $\mathbf{F}^-\varphi \in \Gamma$ or $v^+(\Gamma, \psi) = f$.

We proceed next to show that $\mathcal{I} = \langle I, \leq, v^+, v^- \rangle$ is a C_ω-frame, and that for each $\Gamma \in I$, $(\mathcal{I}, \Gamma)$ is a model of all the signed formulas of Γ. In particular: if Δ does not have a closed tableau in $Tab(C_\omega)$ then $(\mathcal{I}, \Delta^*)$ is a model of all the signed formulas in Δ.

Note. With the exception of $(\mathbf{F}^+\supset)$ and $(\mathbf{F}^-\supset)$, it is possible to add to $Tab(C_\omega)$ all the other expansion rules of $Tab(\mathcal{M}_4)$, and still get a conservative extension of positive intuitionistic logic. It is possible also to modify the semantics in an appropriate way to get soundness and completeness for the resulting system. On the other hand by adding $(\mathbf{F}^-\supset)$ to $Tab(C_\omega)$ we get classical logic.

Note.   The crucial step in the proof of the last theorem is to show that the resulting $\langle I, \subseteq, v^+, v^- \rangle$ is indeed a frame. It is at this point where the addition of $(\mathbf{F}^- \supset)$ causes the argument to fail.

One final remark. It is possible to conservatively add a propositional constant $\mathbf{f}$ to all the systems we have discussed above, together with the extra closure condition that a branch which contains $\mathbf{T}^+\mathbf{f}$ (and optionally also $\mathbf{F}^-\mathbf{f}$) is closed. Hence we could have assumed that full propositional intuitionistic logic is contained in all these systems. It seems difficult to satisfactorily handle intuitionistic "negation" itself within our framework, but this is not so important anyway, since this negation is best understood in terms of $\supset$ and $\mathbf{f}$.

4   Conclusion and Further Research

We have presented a simple tableau-based framework with four types of signed formulas, corresponding to positive and negative statements about truth and falsity. The framework is essentially a generalization and refinement (using two extra signs) of Fitting's tableau system for Intuitionistic Logic ([21]). We show that it can uniformly cope with various kinds of multi-valued logics (even some with infinitely many truth values). This includes classical logic, three-valued logics, the four-valued logic of logical bilattices, Nelson's logics for constructive negation, and the paraconsistent logic C_ω. This uniformity is obviously due to the close relationships between the logics under consideration, which our framework helps to identify. Accordingly it sheds (so we believe) some proof-theoretical light on these relationships (which are often blurred by semantical approaches). For example: the tableau system for Nelson's logic N is simply obtained from that for the three valued logic with one designated value by constraining the implication rule to discard all formulas with polarity F from the context — exactly the same way one obtains a tableau system for Intuitionistic Logic from that of the two-valued Classical Logic. Therefore the reason for the constructiveness of Nelson's negation is clear when seen from this perspective.

Some problems for future research:

- Is there a natural characterization of the intuitionistic negation itself within the framework presented here?
- In the semantics given to C_ω there is an essential asymmetry between the roles of the positive valuation v^+ and the negative one v^-. What is the deep reason for this difference? What kind of logics will be obtained if a dual treatment is adopted, and in what contexts may such logics be useful?
- For the Belnap logic, there is a second set of connectives that is sometimes considered (the knowledge/information ones). Can these be captured by tableau rules too? Can some version of them be added to other logics that have been considered above?
- In general, what other variations can one impose on the tableau rules, and what will be the corresponding semantics? Is there some natural way of getting the semantic conditions from the tableau rules, beyond a case-by-case treatment?

- What sort of games can be played with the machinery we have developed within our framework?
- Is there a significant difference in terms of proof-length between the systems introduce here and those described in [29] and elsewhere?

Acknowledgment

This research was supported by the ISRAEL SCIENCE FOUNDATION founded by the Israel Academy of Sciences and Humanities.

References

[1] O. Arieli and A. Avron, "Reasoning with logical bilattices," *J. of Logic, Language and Information*, vol. 5, no. 1, pp. 25–63, 1996.

[2] O. Arieli and A. Avron, "The value of four values," *Artificial Intelligence*, vol. 102, no. 1, pp. 97–141, 1998.

[3] A. R. Anderson and N. D. Belnap, *Entailment*, vol. I. Princeton University Press, 1975.

[4] A. R. Anderson and N. D. Belnap, *Entailment*, vol. II. Princeton University Press, 1992.

[5] A. Almukdad and D. Nelson, "Constructible falsity and inexact predicates," *Journal of Symbolic Logic*, vol. 49, pp. 231–333, 1984.

[6] A. Avron, "On an implication connective of RM," *Notre Dame Journal of Formal Logic*, vol. 27, pp. 201–209, 1986.

[7] A. Avron, "Natural 3-valued logics: characterization and proof theory," *J. of Symbolic Logic*, vol. 56, no. 1, pp. 276–294, 1991.

[8] A. Avron, "On the expressive power of three-valued and four-valued languages," *Journal of Logic and Computation*, vol. 9, pp. 977–994, 1999.

[9] A. Avron, "Classical Gentzen-type methods in propositional many-valued logics," in *Beyond Two: Theory and Applications of Multiple-Valued Logic* (M. Fitting and E. Orlowska, eds.), vol. 114 of *Studies in Fuzziness and Soft Computing*, pp. 117–155, Physica Verlag, 2003.

[10] M. Baaz, "Kripke-type semantics for da Costa's paraconsistent logic c_ω," *Notre Dame Journal of Formal Logic*, vol. 27, pp. 523–527, 1986.

[11] N. D. Belnap, "How computers should think," in *Contemporary Aspects of Philosophy* (Gilbert Ryle, ed.), pp. 30–56, Oriel Press, Stocksfield, England, 1977.

[12] N. D. Belnap, "A useful four-valued logic," in *Modern Uses of Multiple-Valued Logic* (G. Epstein and J. M. Dunn, eds.), pp. 7–37, Reidel, Dordrecht, 1977.

[13] M. Baaz, C. G. Fermüller, and G. Salzer, "Automated deduction for many-valued logics," in *Handbook of Automated Reasoning* (A. Robinson and A. Voronkov, eds.), Elsevier Science Publishers, 2000.

[14] D. Busch, "Sequent formalizations of three-valued logic," in *Partiality, Modality, and Nonmonotonicity*, pp. 45–75, Studies in Logic, Language and Information, CSLI Publications, 1996.

[15] W. A. Carnielli and J. Marcos, "A taxonomy of c-systems," in *Paraconsistency — the logical way to the inconsistent* (M. E. Coniglio W. A. Carnielli and I. L. M. D'ottaviano, eds.), vol. 228 of *Lecture notes in pure and applied Mathematics*, pp. 1–94, Marcell Dekker, New York, Basel, 2002.

[16] N. C. A. da Costa, "On the theory of inconsistent formal systems," *Notre Dame Journal of Formal Logic*, vol. 15, pp. 497–510, 1974.

[17] I. L. M. D'ottaviano and N. C. A. da Costa, "Sur un problme de Jaskowski," *C. R.Acad. Sc. Paris*, vol. 270, Sèrie A, pp. 1349–1353, 1970.

[18] I. L. M. D'ottaviano, "The completeness and compactness of a three-valued first-order logic," *Revista Colombiana de Matematicas*, vol. XIX, no. 1-2, pp. 31–42, 1985.

[19] J. M. Dunn, "Relevance logic and entailment," in [26], vol. III, ch. 3, pp. 117–224, 1986.

[20] R. L. Epstein, *The semantic foundation of logic*, vol. I: propositional logics, ch. IX. Kluwer Academic Publisher, 1990.

[21] M. Fitting, *Proof Methods for Modal and Intuitionistic Logics*, Reidel, Dordrecht, 1983.

[22] M. Fitting, "Bilattices in logic programming," in *20th Int. Symp. on Multiple-Valued Logic* (G. Epstein, ed.), pp. 238–246, IEEE Press, 1990.

[23] M. Fitting, "Kleene's logic, generalized," *Journal of Logic and Computation*, vol. 1, pp. 797–810, 1990.

[24] M. Fitting, "Bilattices and the semantics of logic programming," *Journal of Logic Programming*, vol. 11, no. 2, pp. 91–116, 1991.

[25] M. Fitting, "Kleene's three-valued logics and their children," *Fundamenta Informaticae*, vol. 20, pp. 113–131, 1994.

[26] D. M. Gabbay and F. Guenthner, *Handbook of Philosophical Logic*. D. Reidel Publishing company, 1986.

[27] M. L. Ginsberg, "Multiple-valued logics," in *Readings in Non-Monotonic Reasoning* (M. L. Ginsberg, ed.), pp. 251–258, Los-Altos, CA, 1987.

[28] M. L. Ginsberg, "Multivalued logics: a uniform approach to reasoning in AI," *Computer Intelligence*, vol. 4, pp. 256–316, 1988.

[29] R. Hähnle, "Tableaux for multiple-valued logics," in *Handbook of Tableau Methods* (M. D'Agostino, D. M. Gabbay, R. Hähnle, and J. Posegga, eds.), pp. 529–580, Kluwer Publishing Company, 1999.

[30] C. B. Jones, *Systematic Software Development Using VDM*. Prentice-Hall International, U. K., 1986.

[31] J. Lukasiewicz, "On 3-valued logic," in *Polish Logic* (S. McCall, ed.), Oxford University Press, 1967.

[32] A. Monteiro, "Construction des algebres de Lukasiewicz trivalentes dans les algebres de Boole monadiques, i," *Mat. Jap.*, vol. 12, pp. 1–23, 1967.

[33] A. R. Raggio, "Propositional sequence-calculi for inconsistent systems," *Notre Dame Journal of Formal Logic*, vol. 9, pp. 359–366, 1968.

[34] L. I. Rozoner, "On interpretation of inconsistent theories," *Information Sciences*, vol. 47, pp. 243–266, 1989.

[35] P. H. Schmitt, "Computational aspects of three-valued logic," in *Proceedings of the 8th Conference on Automated Deduction*, pp. 190–198, Springer Verlag, LNCS 230, 1986.

[36] J. Słupecki, "Der volle dreiwertige aussagenkalkül," *Com. rend. Soc. Sci. Lett. de Varsovie*, vol. 29, pp. 9–11, 1936.

[37] F. von Kutschera, "Ein verallgemeinerter widerlegungsbegriff für Gentzenkalküle," *Archiv fur Mathematische Logik und Grundlagenforschung*, vol. 12, pp. 104–118, 1969.

[38] H. Wansing, *The Logic of Information Structures*, vol. 681 of *LNAI*. Springer-Verlag, 1993.

[39] R. Wójcicki, *Lectures on Propositional Calculi*. Warsaw: Ossolineum, 1984.

A Labelled Sequent-Calculus
for Observation Logic

Olivier Brunet

Équipe QUI – Laboratoire LEIBNIZ - UMR 5522
46, avenue Félix Viallet - 38031 Grenoble Cedex - France
olivier.brunet@imag.fr

Abstract. We present an overview of *observation logic*, an intuitionistic modal logic designed for reasoning about approximations and multiple contexts, and investigate a sequent-calculus formulation for it. Due to the validity of an axiom (called **T2**) which is a weakening of axiom **T**, one needs a labelled version of the sequent-calculus formalism in order to satisfy some classical properties, such as cut elimination and the subformula property. We thus introduce a sequent-calculus formulation of **OL** relying on labelled terms, and use it to show the decidability of the logic.

1   Introduction

Observation logic (**OL**, [1, 2]) is a formalization of the way information behaves in a *partial-observation* context, when all knowledge comes from possibly partial observations. This logic originated as the axiomatization of a satisfiability predicate defined over a general kind of algebraic structure, called *representation system*. These structures have been designed as an attempt to provide a construction which embodies the notion of approximate representation of a system (since our observations, being partial, can be seen as a partial description of its state), but without having the studied or observed system explicitly represented.

This constitutes a new and general approach to the problem of reasoning about approximations [3, 4, 5, 6] and about multiple contexts and theories [7, 8, 9]. In particular, *observation logic* is a modal intuitionistic logic with a collection of modal operators (denoted K_i) which can all be associated to a partial way of considering information about a system. This can be related to the use of evidence for producing assertions about the system as in Voorbraak's Nonmotonic Observation Logic [10] but our approach does not rely on defeasible observation, so that **OL** is based on intuitionistic rather than on Reiter's default logic [11].

Thus, these modal connectives correspond to an approximation, or similarly to a partial observation method. They behave in an **S4** way, with a few adaptations. The most important one is that axiom **T** : $K_i \varphi \to \varphi$ is not valid, but weaker versions of this axiom are (namely **T2** : $K_i K_j \varphi \to K_j \varphi$ and

M. Cialdea Mayer and F. Pirri (Eds.): TABLEAUX 2003, LNAI 2796, pp. 17–31, 2003.

LT : $K_i (K_i \varphi \rightarrow \varphi)$). If the latter just characterizes the way knowledge behaves internally, the former is a cornerstone of our approach, as it permits to relate knowledge and information between different contexts.

In the present paper, we give an overview of observation logic and the algebraic structures underlying it, and provide a labelled sequent calculus for it. We then show that the calculus is a sound and complete formulation of *observation logic* and that this logic is decidable.

2   Observation Logic

In this section, we provide an overview of *representation systems* and *observation logic* [1, 2]. This logic was designed as an attempt to provide strong theoretical foundations to the study of systems where knowledge originates from observations, and thus where descriptions of the system have to be considered as partial.

The motivation for this originates from the remark that in the practice of science, the notion of partiality is extremely important for descriptions. Should this partiality either come from an impossibility to obtain complete descriptions of the studied object (as in physical sciences, where all knowledge comes from some actual observation process) and thus be unavoidable, or should it be wanted as in many domains of computer science where for efficiency purposes one wants to model relevant data only. This second source of partiality corresponds for instance to the notion of abstraction in articifial intelligence [3]. However, despite its importance, it appears that the notion of partiality of knowledge is not as present as it deserves in knowledge representation theories.

Let us first introduce our basic formalism, namely the *representation systems*, which are a collection of partially-ordered sets, each of them corresponding to a set of partial descriptions obtained with one particular way of observing the system. These posets are related together by a collection of *transformation functions* which express how the different observation methods can be related. Thus, the only elements which can be manipulated are partial descriptions of the system.

2.1   Representation Systems

Intuitively, one may define an approximation process using the following structures: first, the system to be studied and approximated is represented by a poset $\langle \mathcal{P}_S, \leq_S \rangle$ whose elements may for instance be seen as sets of possible states, as in a Kripke's possible worlds approach [12], where the partial order $\leq_S$ is such that if $d_1 \leq_S d_2$, then d_1 is a more precise description of the state of the system than d_2 (in terms of possible worlds, the set of possible states associated to d_1 is included in the set of possible states associated to d_2). The result of the approximation can also be formalized using a poset $\langle \mathcal{P}_A, \leq_A \rangle$. Then, the approximation relation between these two posets can be defined as a Galois surjection [13, 14, 15, 16, 17]:

$$\langle \mathcal{P}_S, \leq_S \rangle \xrightarrow[\alpha]{\gamma} \langle \mathcal{P}_A, \leq_A \rangle$$

This provides a natural way to express approximations, as given an element d of $\mathcal{P}_S$ corresponding to a set of possible states of the system, one associates to it the element $\alpha(d)$ of $\mathcal{P}_A$ which can be seen as an approximate description of the state of the system. In particular, it can be seen as an approximation of d, since from the definition of Galois surjections, one has:

$$\forall d \in \mathcal{P}_S, \ d \leq_S \gamma \circ \alpha(d) \qquad \forall d' \in \mathcal{P}_A, \ d' = \alpha \circ \gamma(d')$$

This construction can be generalized by considering a collection of approximation methods (indexed by elements i of a set $\mathcal{I}$), each defined by a poset $\langle \mathcal{P}_i, \leq_i \rangle$ and a Galois surjection $\langle \alpha_i, \gamma_i \rangle$.

Moreover, it is possible to introduce "transformation functions" relating the different approximate posets $\mathcal{P}_i$ by defining $f_{i|j} = \alpha_i \circ \gamma_j$. With these functions, one can express relationships between the different approximations of a given system, without referring explicitly to the system itself. The definition of Galois surjections can be used to identify properties satisfied by the $f_{i|j}$ functions in our formalism, which we will use as a characterization of our "transformation functions". This leads to the definition of *representation systems*.

Definition 1 (Representation System)
A representation system over a set of indices $\mathcal{I}$ is a pair:

$$\left\langle \{\langle \mathcal{P}_i, \leq_i \rangle\}_{i \in \mathcal{I}}, \{f_{i|j}\}_{i,j \in \mathcal{I}} \right\rangle$$

where for each $i \in \mathcal{I}$, $\langle \mathcal{P}_i, \leq_i \rangle$ is a poset and such that the functions $f_{i|j} : \mathcal{P}_j \to \mathcal{P}_i$ verify:

$$\forall d, \ f_{i|i}(d) = d \qquad\qquad\qquad \text{Identity}$$
$$\forall d \leq_j d', \ f_{i|j}(d) \leq_i f_{i|j}(d') \qquad\qquad \text{Monotony}$$
$$\forall d, \ f_{i|k}(d) \leq f_{i|j} \circ f_{j|k}(d) \qquad\qquad \text{Composition}$$

This definition is sufficient for ensuring that all the representations can be considered as approximations of a single system, since given a representation system $\mathcal{S}$, it is possible to build a poset $\mathcal{P}_S$ and a collection of Galois surjections $\langle \alpha_i, \gamma_i \rangle$ from $\mathcal{P}_S$ to $\mathcal{P}_i$ such that $f_{i|j} = \alpha_i \circ \gamma_j$.

Example: Time on Earth

Consider the Earth, associate a point of view to each time zone, and define each associated poset by taking descriptions of the form $t_1 \nearrow t_2 =$ "the local time is between t_1 and t_2" where t_1 and t_2 stand for integer hours (we do not take minutes into account). These descriptions provide some information about an instant on Earth, and a description d_1 is less precise than d_2 if d_1 overlaps d_2. In this setup, the transformation functions will convert time intervals from one time zone to another. For instance, if one considers the GMT time and the Pacific one (GMT $- 8h$), one has modulo 24 hours:

$$f_{\text{Pac}|\text{GMT}}(t_1 \nearrow t_2) = (t_1 - 8) \nearrow (t_2 - 8)$$

Now, if one considers the time zone of Nepal (GMT+5h45m), the conversion from the GMT time zone to this time zone results in the increase of the size of the interval: a time interval from 1h to 2h in the GMT time zone corresponds to the interval from 6h45 to 7h45 in Nepal, so that the best description one can give is 6 $\nearrow$ 8. More generally, one has:

$$f_{\text{Nep}|\text{GMT}}(t_1 \nearrow t_2) = (t_1+5) \nearrow (t_2+6)$$

This interval increase corresponds to a loss of information and illustrates the *composition* inequality of the previous definition, since one can write :

$$\begin{aligned}
f_{\text{Pac}|\text{GMT}}(t_1 \nearrow t_2) &= (t_1-8) \nearrow (t_2-8) \\
&\leq (t_1-9) \nearrow (t_2-7) \\
&= f_{\text{Pac}|\text{Nep}} \circ f_{\text{Nep}|\text{GMT}}(t_1 \nearrow t_2)
\end{aligned}$$

In this formalism, one manipulates approximations and partial descriptions of a given system, but the system itself is not explicitly present, except through the structure of each poset and the relationship that exist between them. In order to provide a general and flexible way to study the behaviour of knowledge and information in this formalism, we will now introduce a logical formalization of those structures. We obtain a logic, called *observation logic*, which definition comes solely from the use of representation systems as the underlying model.

2.2   Logical Translation

Let us first define our langage $\mathcal{L}_{\mathcal{I},\Psi}$ by the grammar:

$$P = AP \mid \bot \mid P \vee P \mid P \wedge P \mid P \to P \mid K_i P$$

In this definition, a term of the form AP stands for an element ψ in the set of atomic propositions Ψ, and in a term of the form $K_i P$, the index i stands for an element of $\mathcal{I}$. In the following, propositions will be denoted by φ, ψ, ϑ, etc.

To relate $\mathcal{L}_{\mathcal{I},\Psi}$ and a representation system $\mathcal{S} = \langle \{\mathcal{P}_i\}, \{f_{i|j}\} \rangle$ over $\mathcal{I}$, we will define a collection of interpretation functions $[\![\cdot]\!]_i : \mathcal{L}_{\mathcal{I},\Psi} \to \wp^{\downarrow}(\mathcal{P}_i)$ (where $\wp^{\downarrow}(\mathcal{P}_i)$ stands for the set of ideals, i.e. of downward-closed subsets of $\mathcal{P}_i$). Given a proposition φ, its interpretation $[\![\varphi]\!]_i$ corresponds to the set of elements of $\mathcal{P}_i$ which, seen as partial descriptions of the state of the system, provide enough information in order to prove that property φ actually holds. Now, if a element d is in this set, so will be any element $d' \leq_i d$, since d' provides more information than d, which is exactly the definition of an ideal.

The interpretation functions are defined inductively from the structure of terms. For atomic propositions, one has to provide an atomic interpretation $\nu_i : \Psi \to \wp^{\downarrow}(\mathcal{P}_i)$. For classical connectives, the interpretation corresponds to intuitionistic logic, since all propositions are interpreted as ideals of a poset. For modal connectives, the interpretation $[\![K_j \varphi]\!]_i$ relies on the use of the transformation functions $f_{i|j}$, since it is the set of elements of $\mathcal{P}_i$ which, after transformation into $\mathcal{P}_j$ by $f_{j|i}$, lie in the interpretation $[\![\varphi]\!]_j$, so that $[\![K_j \varphi]\!]_i =$

$$[\![\psi]\!]_{\mathcal{M},i} = \nu_i(\psi) \quad \psi \in \Psi$$
$$[\![\varphi \vee \psi]\!]_{\mathcal{M},i} = [\![\varphi]\!]_{\mathcal{M},i} \cup [\![\psi]\!]_{\mathcal{M},i}$$
$$[\![\varphi \wedge \psi]\!]_{\mathcal{M},i} = [\![\varphi]\!]_{\mathcal{M},i} \cap [\![\psi]\!]_{\mathcal{M},i}$$
$$[\![\varphi \rightarrow \psi]\!]_{\mathcal{M},i} = \{ d \mid \forall d' \leq d, \ d' \in [\![\varphi]\!]_{\mathcal{M},i} \Rightarrow d' \in [\![\psi]\!]_{\mathcal{M},i} \}$$
$$[\![\bot]\!]_{\mathcal{M},i} = \emptyset$$
$$[\![K_j \varphi]\!]_{\mathcal{M},i} = \{ d \mid f_{j|i}(d) \in [\![\varphi]\!]_{\mathcal{M},j} \}$$

Fig. 1. Interpretation Function

$\{ d \mid f_{j|i}(d) \in [\![\varphi]\!]_j \}$. A pair $\langle \mathcal{S}, \nu \rangle$ – where $\mathcal{S} = \langle \{\mathcal{P}_i\}, \{f_{i|j}\} \rangle$ is a representation system over $\mathcal{I}$ and $\nu = \{ \nu_i : \Psi \rightarrow \wp^{\downarrow}(\mathcal{P}_i) \}_i$ is an atomic interpretation – will be called a representation model over $\mathcal{I}$, and let $\mathrm{RM}(\mathcal{I}, \Psi)$ be the set of all representation models over $\mathcal{I}$ for atomic propositions Ψ. Given such a representation model $\mathcal{M}$, the definition of $[\![\cdot]\!]_{\mathcal{M},i}$ is summarized in figure 1.

With this interpretation function, it is possible to define a notion of satisfiability, so as to identify which propositions of $\mathcal{L}_{\mathcal{I},\Psi}$ do properly correspond to the behavior of information in our partial description approach.

Definition 2 (Satisfiability)

i. A proposition $\varphi \in \mathcal{L}_{\mathcal{I},\Psi}$ is satisfied by a representation model $\mathcal{M} = \langle \mathcal{S}, \nu \rangle \in \mathrm{RM}(\mathcal{I}, \Psi)$ (which we denote $\mathcal{M} \models_{\mathcal{S},\mathcal{I},\Psi} \varphi$) if and only if $\forall i,\ [\![\varphi]\!]_{\mathcal{M},i} = \mathcal{P}_i$.

ii. A proposition $\varphi \in \mathcal{L}_{\mathcal{I},\Psi}$ is satisfied by representation systems over $\mathcal{I}$ if and only if it is satisfied by all representation models over $\mathcal{I}$:

$$\models_{\mathcal{S},\mathcal{I},\Psi} \varphi \Leftrightarrow \forall \mathcal{M} \in \mathrm{RM}(\mathcal{I}, \Psi),\ \mathcal{M} \models_{\mathcal{S},\mathcal{I},\Psi} \varphi$$

In the following, for readability reasons, we will drop the $\mathcal{I}, \Psi$ subscripts, since we will refer to only one set of indices and one set of atomic propositions.

2.3 Axiomatization

We provide an axiomatization of this satisfaction predicate $\models_{\mathcal{S}}$ for representation systems by defining the logic **OL**, as the intuitionistic logic [18, 19, 20] together with the modal axioms and rules listed in figure 2.

A few comments can be made about **OL**. First, this logic can be seen as a multi-context reasoning logic. As exposed in [8, 21], axiom **K**-modalities is a good candidate for defining formal systems about contexts. Moreover, axiom **T** $(K_i \varphi \rightarrow \varphi)$ is not valid, so that in our logic, facts inside a context need not be true, which emphasizes the fact that we are considering our contexts as approximations [7]. Yet, a weaker axiom (**T2** : $K_i K_j \varphi \rightarrow K_j \varphi$) is valid, which, while not referring to "reality", allows to relate different contexts, and thus to reason with multiple contexts.

$$K_i\,(\varphi \to \psi) \to K_i\,\varphi \to K_i\,\psi \qquad\qquad \mathbf{K}$$
$$K_i\,\varphi \to \neg K_i\,\neg\varphi \qquad\qquad\qquad\quad \mathbf{D}$$
$$K_i\,K_j\,\varphi \to K_j\,\varphi \qquad\qquad\qquad\;\; \mathbf{T2}$$
$$K_i\,(\varphi \leftrightarrow K_i\,\varphi) \qquad\qquad\qquad\quad\; \mathbf{L}$$
$$K_i\,(\varphi \lor \psi) \to K_i\,\varphi \lor K_i\,\psi \qquad\qquad \mathbf{V}$$

$$\frac{\vdash \varphi}{\vdash K_i\,\varphi}\ \text{Nec} \qquad\qquad \frac{\forall i,\ \vdash K_i\,\varphi}{\vdash \varphi}\ \text{Univ}$$

Fig. 2. Modal Axioms and Rules of **OL**

Example: Time on Earth

In the previous example with time zones, a statement of the form "the local time is between t_1 and t_2" (which we denote $t_1 \nearrow t_2$) makes no sense, since even if it is the case in one time zone, a different time zone will correspond to another description of the same instant. And even if an explicit reference to a time zone is provided, the given information cannot be used as is in another time zone:

$$\nvdash K_{\text{GMT}}\,K_{\text{Pac}}\,(t_1 \nearrow t_2) \to K_{\text{GMT}}\,(t_1 \nearrow t_2)$$

But it is still possible to relate different time zones: suppose I stand in the GMT time zone, and from my local time, I can assert that the pacific time is between t_1 and t_2 at the same moment, then it is actually the case:

$$\vdash K_{\text{GMT}}\,K_{\text{Pac}}\,(t_1 \nearrow t_2) \to K_{\text{Pac}}\,(t_1 \nearrow t_2)$$

This illustrates the role of axiom **T2**.

In *observation logic*, valid propositions are exactly those which are valid in every context. The classical Nec-rule tells that valid propositions are valid within every context. But the Univ-rule states the converse: if a proposition is valid in every context, then this proposition is considered as valid "objectively", with no reference to any context. If there is a single context ($\mathcal{I}$ is a singleton $\{\iota\}$), the unique modal operator K_ι has no meaning, since one has: $\forall \varphi,\ \vdash \varphi \leftrightarrow K_\iota\,\varphi$. If $\mathcal{I}$ is finite, this rule is equivalent to the axiom schemata $(\bigwedge_i K_i\,\varphi) \to \varphi$. Finally, we will show later that if $\mathcal{I}$ is infinite, this rule can actually be suppressed, since if an index i does not appear in a formula φ, then proving the validity of $K_i\,\varphi$ is equivalent to proving that of φ itself.

As one might expect, this logic is sound and complete with regards to representation systems, as illustrated by the following proposition:

Proposition 2.1
The logic **OL** *is a sound and complete axiomatization of* $\models_S$.

Proof It is easy to check that $\models_S$ is sound w.r.t. **OL** by checking that all its axioms are valid for representation systems. The completeness proof can be done in a classical way using a canonical model [19, 20]. The specific proof for **OL** can be found in [1, 2]. $\qquad\square$

3 Sequent Calculus

3.1 Words and Orders

We first define a few notations for dealing with labels. Let $\mathcal{I}^*$ denote the monoïd of words over $\mathcal{I}$ modulo idempotency (that is finite words over $\mathcal{I}$ with no subword of the form ii for $i \in \mathcal{I}$). We introduce the following notations: ε is the empty word, $|\Lambda|$ the length of word Λ, $\cdot$ the concatenation operation and given a word $\Lambda = \lambda_1 \ldots \lambda_{|\Lambda|}$, $\Lambda_{a \ldots b}$ is the word $\lambda_a \ldots \lambda_b$ where a and b are integers giving the range of the sub-word of Λ.

Such a labelling word corresponds to a path along contexts, and all terms appearing in sequents will be labelled this way. Intuitively, a term in a sequent of the form $[\varphi]_\Lambda$ will be though of as an equivalent to $K_{\lambda_n} \ldots K_{\lambda_1} \varphi$. The fact that in **OL** $K_i K_i \varphi$ and $K_i \varphi$ are equivalent justifies the fact that words are considered modulo idempotency.

Let us now introduce two partial orders on $\mathcal{I}^*$. The first one ($\leq$) corresponds to the word inclusion relation, while the other one ($\leq_\star$) will be used to capture the behaviour of axiom **T2**.

Definition 3 (Partial Orders on $\mathcal{I}^*$)
Given two words $\Omega = \omega_1 \ldots \omega_n$ and $\Lambda = \lambda_1 \ldots \lambda_m$, $\Omega \leq \Lambda$ if and only if Ω is a sub-word of Λ, that is if and only if there exists an increasing function $\sigma : [1 \ldots n] \to [1 \ldots m]$ such that $\forall i$, $\omega_i = \lambda_{\sigma(i)}$.
Moreover, $\Omega \leq_\star \Lambda$ if and only if $\Omega \leq \Lambda$ and either $\Omega = \Lambda = \varepsilon$ or $\omega_1 = \lambda_1$.

As said above, the partial order $\leq_\star$ has a very close relation to the K_i connectives, since one can show that if $K_\Lambda \varphi$ stands for $K_{\lambda_n} \ldots K_{\lambda_1} \varphi$, then one has:

$$\Omega \leq_\star \Lambda \Leftrightarrow \forall \varphi, \ \vdash K_\Lambda \varphi \to K_\Omega \varphi$$

A last point to be noted is that given a word $\Lambda \in \mathcal{I}^*$, the sets $\{\Omega \mid \Omega \leq \Lambda\}$ and $\{\Omega \mid \Omega \leq_\star \Lambda\}$ are finite. Thus, both partial orders are well-founded.

3.2 Definition of the Calculus

Our sequent calculus is defined by the rules given in figure 3. In this definition, each sequent is of the form $[\gamma_1]_{\Lambda_1} \cdots [\gamma_n]_{\Lambda_n} \Vdash [\varphi]_\Lambda$, so that each proposition appearing in a sequent comes with a label, which we call its *localization*. The use of labels permits to have the *subformula property* verified, since any modal connective can be removed and replaced by an extra index in the localization. This is illustrated in the following example:

$$\frac{\Lambda' \leq_\star \Lambda}{[\varphi]_\Lambda \Vdash [\varphi]_{\Lambda'}} \text{ Axiom} \qquad \frac{\Lambda' \leq \Lambda}{[\bot]_\Lambda \Vdash [\bot]_{\Lambda'}} \bot$$

$$\frac{\Gamma \Vdash [\psi]_\Lambda}{\Gamma, [\varphi]_\Lambda \Vdash [\psi]_\Lambda} \text{ Weak} \qquad \frac{\Gamma, [\varphi]_\Lambda, [\varphi]_\Lambda \Vdash \psi}{\Gamma, [\varphi]_\Lambda \Vdash \psi} \text{ Contract}$$

$$\frac{\Gamma \Vdash [\varphi]_\Lambda \quad \Delta, [\varphi]_{\Lambda'} \Vdash \psi \quad \Lambda' \leq_\star \Lambda}{\Gamma, \Delta \Vdash \psi} \text{ Cut}$$

$$\frac{\forall i, \; \Gamma_i \Vdash [\varphi]_i}{\Gamma \Vdash [\varphi]_\varepsilon} \text{ Univ} \qquad \frac{\Gamma, [\varphi]_{\Lambda'} \Vdash \psi \quad \Lambda' \leq_\star \Lambda}{\Gamma, [\varphi]_\Lambda \Vdash \psi} \text{ Loc}$$

$$\frac{\Gamma, [\varphi]_\Lambda \Vdash \vartheta \quad \Gamma, [\psi]_\Lambda \Vdash \vartheta}{\Gamma, [\varphi \vee \psi]_\Lambda \Vdash \vartheta} \vee L \qquad \frac{\Gamma \Vdash [\varphi]_\Lambda}{\Gamma \Vdash [\varphi \vee \psi]_\Lambda} \vee R_1 \qquad \frac{\Gamma \Vdash [\psi]_\Lambda}{\Gamma \Vdash [\varphi \vee \psi]_\Lambda} \vee R_2$$

$$\frac{\Gamma, [\varphi]_\Lambda \Vdash \vartheta}{\Gamma, [\varphi \wedge \psi]_\Lambda \Vdash \vartheta} \wedge L_1 \qquad \frac{\Gamma, [\psi]_\Lambda \Vdash \vartheta}{\Gamma, [\varphi \wedge \psi]_\Lambda \Vdash \vartheta} \wedge L_2 \qquad \frac{\Gamma \Vdash [\varphi]_\Lambda \quad \Gamma \Vdash [\psi]_\Lambda}{\Gamma \Vdash [\varphi \wedge \psi]_\Lambda} \wedge R$$

$$\frac{\Gamma \Vdash [\varphi]_\Lambda \quad \Gamma, [\psi]_\Lambda \Vdash \vartheta}{\Gamma, [\varphi \rightarrow \psi]_\Lambda \Vdash \vartheta} \rightarrow L \qquad \frac{\Gamma, [\varphi]_\Lambda \Vdash [\psi]_\Lambda}{\Gamma \Vdash [\varphi \rightarrow \psi]_\Lambda} \rightarrow R$$

$$\frac{\Gamma, [\varphi]_{i\cdot\Lambda} \Vdash [\psi]_{\Lambda'} \quad \Lambda' \leq \Lambda}{\Gamma, [K_i\,\varphi]_\Lambda \Vdash [\psi]_{\Lambda'}} \, KL \qquad \frac{\Gamma \Vdash [\varphi]_{i\cdot\Lambda}}{\Gamma \Vdash [K_i\,\varphi]_\Lambda} \, KR$$

Fig. 3.   Sequent-Calculus Rules

$$\frac{\dfrac{\dfrac{\dfrac{2 \leq_\star 2\cdot1}{[\varphi]_{2\cdot1} \Vdash [\varphi]_2}}{[\varphi]_{2\cdot1} \Vdash [\varphi \vee \psi]_2}}{[K_2\,\varphi]_1 \Vdash [K_2\,(\varphi \vee \psi)]_\varepsilon} \quad \dfrac{\dfrac{\dfrac{j \leq_\star 2\cdot1}{[\psi]_{2\cdot1} \Vdash [\psi]_2}}{[\psi]_{2\cdot1} \Vdash [\varphi \vee \psi]_2}}{[K_2\,\psi]_1 \Vdash [K_2\,(\varphi \vee \psi)]_\varepsilon}}{\dfrac{\dfrac{[K_2\,\varphi \vee K_2\,\psi]_1 \Vdash [K_2\,(\varphi \vee \psi)]_\varepsilon}{[K_1\,(K_2\,\varphi \vee K_2\,\psi)]_\varepsilon \Vdash [K_2\,(\varphi \vee \psi)]_\varepsilon}}{\Vdash [K_1\,(K_2\,\varphi \vee K_2\,\psi) \rightarrow K_2\,(\varphi \vee \psi)]_\varepsilon}}$$

First, let us show that our sequent calculus is correct with respects to the observational logic **OL**.

Proposition 3.1

The sequent calculus defined in figure 3 is sound and complete w.r.t. **OL***. More precisely, one has:*

$$\forall \varphi \in \mathcal{L}_{\mathcal{I},\Psi}, \quad \vdash_{\mathbf{OL}} \varphi \Leftrightarrow \emptyset \Vdash [\varphi]_\varepsilon$$

Proof The $\Rightarrow$-implication can be easily proved by checking that all axioms of **OL** can be derived in the sequent-calculus formalism. In particular, for axiom **L**, one has to use the fact that $i \simeq i \cdot i$ for any index $i \in \mathcal{I}$.

To prove the $\Leftarrow$-implication, it suffices to show that all the rules are correct w.r.t. the following translation in terms of representation systems:

$$[\gamma_1]_{\Lambda_1} \cdots [\gamma_n]_{\Lambda n} \Vdash [\varphi]_\Lambda \rightsquigarrow \forall i, \quad \bigwedge_j [\![K_{\Lambda_j} \gamma_j]\!]_i \subseteq [\![K_\Lambda \varphi]\!]_i$$

It follows from this that one has:

$$\forall \varphi, \quad \vdash_{\mathbf{OL}} \varphi \Rightarrow \emptyset \Vdash [\varphi]_\varepsilon \Rightarrow \models_\mathcal{S} \varphi \Rightarrow \vdash_{\mathbf{OL}} \varphi$$

□

Now, in order to use this sequent calculus efficiently for the search of proofs, two rules deserve attention: the Cut-rule which may introduce new formulas, and the Univ-rule, which introduces new indices. In the following section, we will investigate some proofs manipulations, and show that the Cut-rule can always be eliminated, and that the Univ-rule can be used at most once.

4   Proofs Manipulation

4.1   Label Manipulation on Proofs

We will first study the way localizations behave inside proofs of system $\Vdash$. From its rules, one can first remark that for any sequent $\Gamma \Vdash [\varphi]_\Lambda$, all localized proposition $[\gamma]_\Omega$ in Γ is such that $\Lambda \leq \Omega$. This result can be shown by induction on the height of the proof, since this property appears explicitly in rules Axiom and $\bot$, and it is preserved by the application of the other rules (and is also explicitly demanded for rule KL).

Another property can be stated: given a proof Π of a sequent $\Gamma \Vdash [\varphi]_\Lambda$, any sequent $\Delta \Vdash [\psi]_\Omega$ in Π is such that $\Lambda \leq \Omega$. By combining those two properties, one gets the following result:

Proposition 4.1

Given a proof Π of a sequent $\Gamma \Vdash [\varphi]_\Lambda$, any term $[\psi]_{\Lambda'}$ appearing in Π is such that $\Lambda \leq \Lambda'$.

This property suggests an interesting manipulation of the localizations appearing in a proof. Given a proof Π of $\Gamma \Vdash [\varphi]_\Lambda$, proposition 4.1 asserts that any localization Ω in Π can be written as $\Omega_1 \cdot \Omega_2$ with $\Lambda \leq_* \Omega_2$. One would then want to replace this localization by $\Omega_1 \cdot \Lambda$ or even $\Omega_1 \cdot \Lambda'$ for some $\Lambda' \leq_\star \Lambda$. Such manipulation is in fact necessary if one wants to have the cut-elimination property for $\Vdash$, since if one has a proof of Π of $\Gamma \Vdash [\varphi]_\Lambda$, then for $\Lambda' \leq_\star \Lambda$, the following provides a proof of $\Gamma \Vdash [\varphi]_{\Lambda'}$:

$$\frac{\begin{array}{cc} \Pi & \dfrac{\Lambda' \leq_\star \Lambda}{[\varphi]_\Lambda \Vdash [\varphi]_{\Lambda'}} \text{ Axiom} \\ \Gamma \Vdash [\varphi]_\Lambda & \end{array}}{\Gamma \Vdash [\varphi]_{\Lambda'}} \text{ Cut}$$

In that situation, eliminating the cut implies that one has a way to transform Π into a proof of $\Gamma \Vdash [\varphi]_{\Lambda'}$. For this to be done, we introduce an operation on words which does the proper manipulation on localizations.

Definition 4
Given three words $\Lambda' \leq \Lambda \leq \Omega$, we define $\Omega \langle \Lambda \triangleleft \Lambda' \rangle$ as $\Omega_{1...l} \cdot \Lambda'$ where l is the greatest integer such that $\Lambda \leq \Omega_{l+1...|\Omega|}$.

This operation works as follows: given two words Λ and Ω such that $\Lambda \leq \Omega$ (or, stated another way, Λ is included in Ω), one first finds the rightmost way to include Λ into Ω, "cuts" Ω at this position, and appends another word Λ' instead. For instance, $\texttt{adbcdcbad}\langle \texttt{dba} \triangleright \texttt{a} \rangle = \texttt{adbca}$, as illustrated in the following decomposition:

$$\texttt{adbc|dcbad} \quad \leadsto \quad \texttt{adbc|a}$$

Proposition 4.2
Given two words $\Lambda \leq \Lambda'$, one has:

$$\Omega \leq \Omega' \Rightarrow \Omega \langle \Lambda' \triangleright \Lambda \rangle \leq \Omega' \langle \Lambda' \triangleright \Lambda \rangle$$

Moreover, if $\Lambda \leq_\star \Lambda'$, then one has:

$$\Omega \leq_\star \Omega' \Rightarrow \Omega \langle \Lambda' \triangleright \Lambda \rangle \leq_\star \Omega' \langle \Lambda' \triangleright \Lambda \rangle$$
$$\Omega \langle \Lambda' \triangleright \Lambda \rangle \leq_\star \Omega$$
$$\Omega \leq \Omega' \Rightarrow \Omega' \langle \Omega \triangleright \Omega \langle \Lambda' \triangleright \Lambda \rangle \rangle \leq_\star \Omega' \langle \Lambda' \triangleright \Lambda \rangle$$

As one can see, operation $\_\langle \_ \triangleright \_\rangle$ has, considering the previous properties, some connections with the transformations functions $f_{i|j}$ used in the definition of representation systems: the three inequalities for the case $\Lambda \leq_\star \Lambda'$ can be put respectively in correspondance with the monotony, identity and composition properties of transformation functions. The next proposition shows how they can apply to proof manipulation.

Proposition 4.3
Every proof Π of a sequent $\Gamma \Vdash [\varphi]_\Lambda$ can be transformed into a proof Π' of $\Gamma_{\langle \Lambda \triangleright \Lambda' \rangle} \Vdash [\varphi]_{\Lambda'}$ for $\Lambda' \leq_\star \Lambda$. Moreover, the structure of Π' differs only from that of Π by the addition of some applications of the Loc-rule.
The notation $\Gamma_{\langle \Lambda \triangleright \Lambda' \rangle}$ corresponds to replacing each $[\psi]_\Omega$ in Γ by $[\psi]_{\Omega \langle \Lambda \triangleright \Lambda' \rangle}$.

Sketch of Proof This result can be proved by induction on the size of the proof, and relies mainly on the properties of the operation $_-\langle_- \triangleright _-\rangle$. The main rules to be examined are Loc, Cut and $\rightarrow L$.

The validity of Loc comes from the fact that if $\Lambda' \leq_* \Lambda$, then $_-\langle \Lambda \triangleright \Lambda' \rangle$ is $\leq_*$-monotonous. The validity of rules Cut and $\rightarrow L$ is a consequence of the "composition" property: if $\Lambda' \leq_* \Lambda \leq \Omega' \leq \Omega$, then one has $\Omega \langle \Omega' \triangleright \Omega' \langle \Lambda \triangleright \Lambda' \rangle \rangle \leq_* \Omega \langle \Lambda \triangleright \Lambda' \rangle$. □

Corollary 4.3.1
Every proof Π of a sequent $\Gamma \Vdash [\varphi]_\Lambda$ can be transformed into a proof Π' of $\Gamma \Vdash [\varphi]_{\Lambda'}$ for $\Lambda' \leq_* \Lambda$. Moreover, the structure of Π' differs only from that of Π by the addition of some applications of the Loc-rule.

Proof It is a combination of the previous proposition and of multiple applications of the Loc-rule. □

This corollary is the justification of operation $_-\langle_- \triangleright _-\rangle$, since it is the central tool for "lowering" the localization on the right-side proposition of a sequent. This is necessary for achieving cut-elimination, which is our next topic.

4.2 Cut Elimination

The manipulations presented in the previous section are essential for eliminating cut, and can be combined with the classical cut-elimination procedure [22, 23, 24, 25]. The detailed specific proof for **OL** is long and technical, though it contains no special difficulties, and we only provide a sketch of it.

Theorem 4.4 (Cut Elimination)
Given a proof Π of a sequent $\Gamma \Vdash [\varphi]_\Lambda$, it is possible to transform Π into a cut-free proof Π' of the same sequent.

Sketch of Proof The proof of this elimination is mainly an adaptation of that given in [25]. It based on the use of a well-founded order defined on proofs, and on a set of transformation rules which are strictly decreasing w.r.t. this order.

This order is defined lexicographically as $\langle \leq_{cd}, \leq_{cr}, \leq_{ld}, \leq_{ps} \rangle$ where $\leq_{cd}$ compares the *cut depth* of proofs (that is the maximum number of Cut rule instances present in a branch of the proof tree), $\leq_{cr}$ correspond to the *cut rank* of a proof (the size of the biggest active term in a Cut rule instance), $\leq_{ld}$ to its logical depth and $\leq_{ps}$ to its size.

Most rules can be directly adapted from [25], and we present the reduction rule for the case where the cut formula is of the form $K_i \varphi$, and the two premisses are left- and right introductions of the modal connective. Starting from a proof

Π of the form:

$$\dfrac{\dfrac{\Gamma \vdash [\varphi]_{i\cdot\Lambda}}{\Gamma \vdash [K_i\,\varphi]_\Lambda}\ \text{KR} \qquad \dfrac{\Delta, [\varphi]_{i\cdot\Lambda'} \vdash [\vartheta]_{\Lambda''}}{\Delta, [K_i\,\varphi]_{\Lambda'} \vdash [\vartheta]_{\Lambda''}}\ \text{KL}}{\Gamma, \Delta \vdash [\vartheta]_{\Lambda''}}\ \text{Cut}$$

one can transform it into the proof Π':

$$\dfrac{\overset{\pi_1}{\Gamma \vdash [\varphi]_{i\cdot\Lambda}} \qquad \overset{\pi_2}{\Delta, [\varphi]_{i\cdot\Lambda'} \vdash [\vartheta]_{\Lambda''}}}{\Gamma, \Delta \vdash [\vartheta]_{\Lambda''}}\ \text{Cut}$$

The other important transformation, presented in section 4.1, concerns an instance of the Cut rule with the Axiom rule as one is its premisses, and relies on Corollary 4.3.1. The complete proof can be found in [2].   □

4.3   Univ Limitation

Another rule which deserves close examination is the Univ-rule. The main problem with this rule appears in the case of an infinite index set $\mathcal{I}$, since this would lead to infinite proofs. Before tackling this problem, we first show how to have a "normal" form of proofs with regards to the Univ-rule when $\mathcal{I}$ is finite.

First, given a proof Π of a sequent $\Gamma \Vdash [\varphi]_\Lambda$, it is possible to obtain a proof of $\Gamma_{\cdot i} \Vdash [\varphi]_{\Lambda\cdot i}$ where each localization Ω in Π is replaced by $\Omega\cdot i$. All rules except Univ are left unchanged, since they remain valid after adding i on the right of the localizations. For the Univ-rule, one just has to select the ith premise, thus removing the instance of the rule. This way, from a proof Π of $\Gamma \Vdash [\varphi]_\Lambda$, one can get for each $i \in \mathcal{I}$ a proof Π_i of $\Gamma_{\cdot i} \Vdash [\varphi]_{\Lambda\cdot i}$ which, combining together, provide a proof Π' of $\Gamma \Vdash [\varphi]_\Lambda$ with only one occurence of the Univ-rule, at the root.

Proposition 4.5 (Univ Limitation)
If $\mathcal{I}$ is finite, every proof Π of a sequent $\Gamma \Vdash [\varphi]_\Lambda$ can be transformed so as to have at most one instance of the Univ-rule, which instance at the root.

Now, suppose that one has a proof Π of a sequent $\Gamma_{\cdot i} \Vdash [\varphi]_{\Lambda\cdot i}$ where i appears nowhere in sequent $\Gamma \Vdash [\varphi]_\Lambda$. In that case, it appears that index i is not relevant in Π, and one would want to simply erase it from Π. This can actually be done, and one has the following proposition:

Proposition 4.6
Every proof Π of a sequent $\Gamma_{\cdot i} \Vdash [\varphi]_{\Lambda\cdot i}$ such that i is not present in the sequent $\Gamma \Vdash [\varphi]_\Lambda$, can be transformed into a proof of $\Gamma \Vdash [\varphi]_\Lambda$ without changing its structure.

Corollary 4.6.1
Every proof Π of a sequent $\Gamma_{\cdot i} \Vdash [\varphi]_{\Lambda \cdot i}$ such that i is not present in the sequent $\Gamma \Vdash [\varphi]_\Lambda$, can be transformed into a proof of $\Gamma \Vdash [\varphi]_\Lambda$ with no occurence of the Univ-rule.

This corollary comes from the combination of propositions 4.5 and 4.6 since starting from a proof Π of the sequent, one applies prop. 4.5, selects the ith premise of the root and applies prop. 4.6. Since corollary 4.6.1 always applies if $\mathcal{I}$ is infinite, one has finally:

Proposition 4.7 (Univ Elimination)
If $\mathcal{I}$ is infinite, the Univ-rule can be omitted in the search of a proof.

5 Decidability

If one excepts the localization-handling rules, the sequent-calculus system as given in figure 3 is very close to a formulation of intuitionistic logic. Thus suggest to introduce a variant of this system in order to avoid problems related to the Contract-rule. One manipulation consists in changing the rules of the sequent system so as to have in the left-hand side of a sequent at most one copy of each formula. This way, there are no redundancies in sequents, and the Contract-rule becomes useless. This manipulation has been introduced by S. Kleene (system G3a in [26] p.481, one can also refer to system $\mathcal{GK}_i$ in [25] p.36) and is such that for an introduction rule on the left-hand side, the obtained term appears in each premise. This transformation can easily be applied to our system. For instance, rule $\to L$ can be restated as:

$$\frac{\Gamma, [\varphi \to \psi]_\Lambda \Vdash [\varphi]_\Lambda \quad \Gamma, [\varphi \to \psi]_\Lambda, [\psi]_\Lambda \Vdash \vartheta}{\Gamma, [\varphi \to \psi]_\Lambda \Vdash \vartheta} \to L'$$

With this system, if one does not take localizations into account, there are finitely many possible sequents. Now, for the localization-related rules, it can be shown that there are also finitely many possible localizations which may intervene in the proof of a sequent. This comes from the fact that the Univ-rule may be used at most once (prop. 4.5), and the other indices can only be introduced with a KL- or a KR-rule. Thus, finitely many sequents have to be considered, so that one has:

Theorem 5.1 (Termination of Proof Search)
The search of a proof of a sequent with system $\Vdash$ terminates.

Corollary 5.1.1 (Decidability of OL)
OL *is decidable.*

6   Conclusion

In this article, we have presented an overview of the *observation logic*. We have then studied a sequent-calculus formulation of it. Due to the presence of a special modal axiom, namely **T2** : $K_i K_j \varphi \rightarrow K_j \varphi$, some additions to a "classical" formulation are needed in order to have the satisfaction of properties such as cut elimination.

This problem is solved with the use of context path prefixes labelling each terms of a proof, ensuring the subformula property by replacing modal operators by an additional context in term's labels. We thus proposed a sequent-calculus formulation of **OL** and have shown its soundness and correctness. Finally, we have presented the Cut- and Univ- elimination processes, and shown the decidability of this logic.

References

[1] Brunet, O.: A modal logic for observation-based knowledge representation. In: Proceedings of the IMLA'02 workshop (Intuitionnistic Modal Logic and Applications), Copenhaguen, DK (2002)

[2] Brunet, O.: Étude de la connaissance dans le cadre d'observations partielles : La logique de l'observation. PhD thesis, Université Joseph Fourier, Grenoble (France) (2002)

[3] Giunchiglia, F., Villafiorita, A., Walsh, T.: Theories of abstraction. AI Communications **10** (1997) 167–176

[4] Giunchiglia, F., Walsh, T.: A theory of abstraction. Artificial Intelligence **56** (1992) 323–390

[5] Nayak, P. P., Levy, A.: A semantic theory of abstractions. In Mellish, C., ed.: Proceedings of the Fourteenth International Joint Conference on Artificial Intelligence, Morgan Kaufmann (1995) 196–203

[6] McCarthy, J.: Approximate objects and approximate theories (2000)

[7] Nayak, P. P.: Representing multiple theories. In Hayes-Roth, B., Korf, R., eds.: Proceedings of the Twelfth National Conference on Artificial Intelligence, Menlo Park, CA., AAAI Press (1994) 1154 – 1160

[8] McCarthy, J.: Notes on formalizing contexts. In Kehler, T., Rosenschein, S., eds.: Proceedings of the Thirteenth National Conference on Artificial Intelligence, Los Altos, California, Morgan Kaufmann (1993) 555–560

[9] Bellin, G., de Paiva, V., Ritter, E.: Extended curry-howard correspondence for a basic constructive modal logic (2001)

[10] Voorbraak, F.: A nonmonotonic observation logic (1997)

[11] Reiter, R.: A logic for default reasoning. Artificial Intelligence (1980)

[12] Kripke, S. A.: A semantical analysis of modal logic I : normal modal propositional calculi. Zeitschrift für Mathematische Logik und Grundlagen der Mathematik **9** (1963) 67–96

[13] Ore, O.: Galois connections. Transactions of the American Mathematical Society **55** (1944) 493 – 513

[14] Pickert, G.: Bemerkungen über galois-verbindungen. Archv. Math. J. **3** (1952) 285–289

[15] Birkhoff, G.: Lattice Theory. 3rd edn. Colloquim Publications. American Mathematical Society (1967)

[16] Erné, M., Koslowski, J., Melton, A., Strecker, G. E.: A primer on galois connections (1992)

[17] Cousot, P., Cousot, R.: Abstract interpretation and application to logic programs. Journal of Logic Programming **13** (1992) 103–179

[18] Heyting, A.: Intuitionism: An Introduction. North-Holland (1956)

[19] Fitting, M.: Proof Methods for Modal and Intuitionnistic Logics. Volume 169. D. Reidel Publishing (1983)

[20] van Dalen, D.: Intuitionnistic logic. In Gabbay, D., Guenthner, F., eds.: Handbook of Philosophical Logic. Volume III. Reidel (1986) 225–340

[21] Buvač, S., Buvač, V., Mason, I. A.: Metamathematics of contexts. Fundamenta Informaticae **23** (1995) 263–301

[22] Gentzen, G.: Investigations into logical deduction. In Szabo, M. E., ed.: The collected papers of Gerhard Gentzen, North Holland (1969) 68 – 128

[23] Tait, W. W.: Normal derivability in classical logic. In Barwise, J., ed.: The Syntax and Semantics of Infinitary Languages. Springer Verlag (1968)

[24] Girard, J. Y.: Proof Theory and Logical Complexity. Bibliopolis (1987)

[25] Gallier, J.: Constructive logics. part i: A tutorial on proof systems and typed λ-calculi. Theoretical Computer Science **110** (1993) 249 – 339

[26] Kleene, S. C.: Introduction to Metamathematics. Seventh edition edn. North-Holland (1952)

Bounded Łukasiewicz Logics

Agata Ciabattoni[1] and George Metcalfe[2]

[1] Institut für Algebra und Computermathematik
Technische Universität Wien
Wiedner Haupstrasse 8-10/118, A-1040 Wien, Austria
agata@logic.at
[2] Department of Computer Science, King's College London
Strand, London WC2R 2LS, UK
metcalfe@dcs.kcl.ac.uk

Abstract. In this work we investigate *bounded Łukasiewicz logics*, characterised as the intersection of the k-valued Łukasiewicz logics for $k = 2, \ldots, n$ $(n \geq 2)$. These logics formalise a generalisation of Ulam's game with applications in Information Theory. Here we provide an analytic proof calculus $\mathbf{GLB_n}$ for each bounded Łukasiewicz logic, obtained by adding a single rule to $\mathbf{GL}$, a hypersequent calculus for Łukasiewicz infinite-valued logic. We give a first cut-elimination proof for $\mathbf{GL}$ with (suitable forms of) cut rules. We then prove completeness for $\mathbf{GLB_n}$ with cut and show that cut can also be eliminated in this case.

1 Introduction

Łukasiewicz logics were introduced for philosophical reasons by Jan Łukasiewicz in the 1920s [8] and are among the first examples of *many-valued logics*. Currently they are of great importance in several areas of research. Firstly, in *fuzzy logic* [16], where infinite-valued Łukasiewicz logic **Ł**, along with Gödel logic and Product logic, emerges as one of the fundamental "t-norm based"[1] fuzzy logics [7]. From an *algebraic* perspective, Chang's MV-algebras [2] for Łukasiewicz logics are of great interest and form the subject of a recent monograph containing many deep mathematical results [5]. Łukasiewicz logics can also be viewed from a *geometric* perspective via *McNaughton's representation theorem* [9] which establishes that formulae in **Ł** stand to particular geometric functions as formulae in classical logic stand to boolean functions. Finally, various *semantic interpretations* of Łukasiewicz logics have been provided, most importantly via Ulam's game, a variant of the game of Twenty Questions where errors/lies are allowed in the answers [12, 13]. Ulam's game models situations in the processing and sending of information that might be affected by "noise" (see e.g. [14]), and strategies for the game lead naturally to the theory of error-correcting codes.

---

[1] T-norms are widely used to combine vague information in applications for approximate reasoning, knowledge representation and decision making.

M. Cialdea Mayer and F. Pirri (Eds.): TABLEAUX 2003, LNAI 2796, pp. 32–47, 2003.
© Springer-Verlag Berlin Heidelberg 2003

In [11] an analytic proof calculus **GŁ** was defined for **Ł** using *hypersequents*, a natural generalisation of Gentzen sequents introduced by Avron in [1] [2]. Soundness and completeness for **GŁ** were proved *semantically* in [11] via an embedding of **Ł** into Meyer and Slaney's abelian logic **A**, the logic of abelian groups. Hence "cut" rules, which permit the introduction of lemmas or intermediary steps into proofs, were shown to be *admissible* for **GŁ** *without* proving cut-elimination, i.e. without providing an algorithm for obtaining proofs in **GŁ** from proofs in **GŁ** with cut.

Bounded Łukasiewicz logics $\mathbf{LB_n}$ ($n \geq 2$) arise as the *intersection* of k-valued Łukasiewicz logics, for $k = 2 \ldots n$. Informally they capture the notion of having "at most" n truth values, as expressed by the validity of the following sequent-style rule:

$$\frac{\Gamma, \overbrace{A, \ldots, A}^{\text{n times}} \vdash \Delta}{\Gamma, \underbrace{A, \ldots, A}_{\text{n-1 times}} \vdash \Delta} \quad (n\text{-contraction})$$

Note that in the particular cases where $n = 2$ and $n = 3$, $\mathbf{LB_n}$ coincides with classical logic and 3-valued Łukasiewicz logic respectively. Other families of logics satisfying (n-contraction) were investigated in [15, 3], and in [3] also an analytic calculus for $\mathbf{LB_4}$ was defined.

In this work we introduce semantic interpretations and analytic proof calculi for the family of bounded Łukasiewicz logics. We start in Section 2 by introducing finite-valued, infinite-valued and bounded Łukasiewicz logics. We then show in Section 3 that bounded Łukasiewicz logics can be interpreted in terms of a *generalised* version of Ulam's game with applications in Information Theory (see e.g. [4]). In Section 4 we recall the hypersequent calculus **GŁ** presented in [11], and give first proofs of cut-elimination for **GŁ** with two forms of cut rules. Finally in Section 5 we define calculi for the bounded Łukasiewicz logics by adding a single rule to **GŁ** in each case. We then prove completeness syntactically using cut and show that the cut-elimination proofs of Section 4.1 can be extended in the presence of this extra rule.

2 Łukasiewicz Logics

We start by defining the infinite-valued Łukasiewicz logic **Ł**, noting that in this work we identify theoremhood in a logic with derivability in the corresponding Hilbert-style system.

Definition 1 (Łukasiewicz Infinite-Valued Logic, Ł). *A Hilbert-style system for **Ł**, using the connectives $\supset$ and $\perp$, consists of the rule:*

$$(mp) \frac{A \supset B, A}{B} \quad \textit{together with the axioms:}$$

---

[2] Note that a *single* sequent calculus for **Ł** has also been defined in [10]

$L1$ $A \supset (B \supset A)$ $L3$ $((A \supset B) \supset B) \supset ((B \supset A) \supset A)$
$L2$ $(A \supset B) \supset ((B \supset C) \supset (A \supset C))$ $L4$ $((A \supset \bot) \supset (B \supset \bot)) \supset (B \supset A)$

Other connectives are defined as follows: $\neg A =^{def} A \supset \bot$, $A \oplus B =^{def} \neg A \supset B$
and $A \odot B =^{def} \neg(A \supset \neg B)$. *We also adopt the notation below:*

$$A \Leftrightarrow B = (A \supset B) \odot (B \supset A) \qquad n.A = \overbrace{A \oplus \ldots \oplus A}^{n} \qquad A^n = \overbrace{A \odot \ldots \odot A}^{n}$$

Remark 1. An alternative Hilbert-style system for **L** is obtained by adding axioms L1 and L3 to any axiomatization of the multiplicative additive fragment of linear logic[3] (see [15]). In particular, axiom L3 allows the additive connectives $\vee$ and $\wedge$ to be defined over the multiplicative ones ($\odot$ and $\oplus$ or, equivalently, $\supset$) as follows: $A \vee B =^{def} (A \supset B) \supset B$ and $A \wedge B =^{def} \neg(\neg A \vee \neg B)$.

Hilbert-style systems extending **L** were provided for the *finite-valued* Łukasiewicz logics by Grigolia in [6].

Definition 2 (N-valued Łukasiewicz Logic, $\mathbf{L_n}$). *A Hilbert-style system for* $\mathbf{L_n}$ *consists of the same axioms and rules as* **L** *and also:*

$$L_n 5 \ \ n.A \supset (n-1).A \qquad and \qquad L_n 6 \ \ (p.A^{p-1}) \Leftrightarrow n.A^p$$

for every integer $p = 2, \ldots, n-2$ *that does not divide* $n-1$.

Remark 2. Axiom $L_n 5$ corresponds to the (n-contraction) rule (see [15]).

Algebraic structures for the above logics are defined as follows, using the same notation for algebraic operations as the corresponding connectives.

Definition 3 (MV-algebra). *An MV-algebra[4] is an algebra* $A_L = \langle A, \oplus, \neg, \bot \rangle$ *with a binary operation* $\oplus$, *a unary operation* $\neg$ *and a constant* $\bot$, *satisfying the following equations:*

$MV1$ $x \oplus (y \oplus z) = (x \oplus y) \oplus z$ $MV2$ $x \oplus y = y \oplus x$
$MV3$ $x \oplus \bot = x$ $MV4$ $\neg\neg x = x$
$MV5$ $x \oplus \neg\bot = \neg\bot$ $MV6$ $\neg(\neg x \oplus y) \oplus y = \neg(\neg y \oplus x) \oplus x$

We also define: $x \supset y =^{def} \neg x \oplus y$ *and* $\top =^{def} \neg\bot$.

Definition 4 (MV$_\mathbf{n}$-algebra). *An MV_n-algebra is an MV-algebra satisfying the equations:*

$$(E_{n0}) \ \ n.x = (n-1).x \qquad and \qquad (E_{np}) \ \ (p.x^{p-1})^n = n.x^p$$

for every integer $p = 2, \ldots, n-2$ *that does not divide* $n-1$.

---

[3] I.e. linear logic without exponential connectives.
[4] MV stands for many-valued.

Valuations for these algebraic structures are defined in the usual way (see, e.g., [5]). We say that a formula Φ is *valid* in an MV-algebra A (resp. in an MV_n-algebra A) if for all valuations v on A, $v(\Phi) = \top$.

We now introduce some important MV-algebras and MV_n-algebras.

Definition 5 ($[-1,0]_{\mathbf{L}}$, $[-1,0]_{\mathbf{L_n}}$). *Let* $x \oplus y = min(0, x + y + 1)$, $\neg x = -1 - x$ *and* $\bot = -1$, *then* $[-1,0]_L = \langle [-1,0]_{\mathbb{R}}, \oplus, \neg, \bot \rangle$ *is an MV-algebra and* $[-1,0]_{L_n} = \langle [-1, -(n-2)/(n-1), \ldots, -1/(n-1), 0], \oplus, \neg, \bot \rangle$ *is an MV_n-algebra.*

In fact these algebras are *characteristic* for $\mathbf{L}$ and $\mathbf{L_n}$.

Theorem 1 ([2]). *The following are equivalent: (1) ϕ is a theorem of $\mathbf{L}$. (2) ϕ is valid in all MV-algebras. (3) ϕ is valid in $[-1,0]_L$.*

Theorem 2 ([6]). *The following are equivalent: (1) ϕ is a theorem of $\mathbf{L_n}$. (2) ϕ is valid in all MV_n-algebras. (3) ϕ is valid in $[-1,0]_{L_n}$.*

We now introduce *bounded* Łukasiewicz logics.

Definition 6 (N-bounded Łukasiewicz Logic, $\mathbf{LB_n}$). *A Hilbert-style system for n-bounded Łukasiewicz Logic $\mathbf{LB_n}$ consists of the same axioms and rules as $\mathbf{L}$ together with axiom L_n5.*

Remark 3. $\mathbf{LB_2}$ and $\mathbf{LB_3}$ coincide with classical logic and $\mathbf{L_3}$, respectively.

$\mathbf{LB_n}$ is characterised by the following algebraic structures.

Definition 7 (N-bounded MV-algebra). *An n-bounded MV-algebra is an MV-algebra satisfying the equation (E_{n0}).*

Remark 4. In the above definition, equation (E_{n0}) can be equivalently replaced by $\neg x \vee (n - 1).x$.

Theorem 3 ([5]). *The following are equivalent: (1) ϕ is a theorem of $\mathbf{LB_n}$. (2) ϕ is valid in all n-bounded MV-algebras. (3) ϕ is valid in all MV_k-algebras for $k = 2, \ldots, n$.*

Corollary 1. *The following are equivalent: (1) ϕ is a theorem of $\mathbf{LB_n}$. (2) ϕ is a theorem of $\mathbf{L_k}$ for $k = 2, \ldots, n$. (3) ϕ is valid in $[-1,0]_{L_k}$ for $k = 2, \ldots, n$.*

Proof. Follows immediately from Theorems 2 and 3. □

3   Ulam's Game Interpretation of $\mathbf{LB_n}$

In [12, 13] a semantic interpretation of Łukasiewicz finite and infinite-valued logics is defined in terms of *Ulam's game* – a variant of the game of Twenty Questions, where lies, or errors, are allowed in the answers. Here we show that a useful generalisation of Ulam's game is formalised by the family of *bounded* Łukasiewicz logics $\mathbf{LB_n}$.

We first recall the connection between Ulam's game with $n - 2$ lies and n-valued Łukasiewicz logic $\mathbf{L_n}$ presented in [5]. An instance of such a game proceeds as follows. An Answerer A chooses a number x from a finite subset S of natural numbers, called the *search space*. A Questioner Q then asks "questions" in the form of *subsets* of S. A, who is allowed to lie up to $n - 2$ times, responds "yes" meaning that x belongs to the chosen subset, or "no" meaning that it does not. Q's objective is to identify x. It is well-known that Ulam's game models situations in the processing and sending of information that may be altered by some kind of "noise" (e.g. such as transmissions from a satellite). Here Q's aim can be interpreted as discovery of the most efficient way of recovering information in the presence of possible distortions; the *strategies* of Q then lead naturally to the theory of error-correcting codes.

Q's *state of knowledge* regarding x is uniquely determined at each point in time by a "conjunction" of the answers given by A. In general this conjunction fails to obey the rules of classical logic. For example, neither the principle of non-contradiction nor idempotency hold. If A answers "x is 2" and then "x is not 2", this is not inconsistent, it just means that A has one less lie to use. Similarly, repeated assertions that "x is 2" are more informative than one such assertion; indeed $n - 1$ such assertions *guarantee* truth. One way of *describing* Q's state of knowledge is by a function $\tau : S \to \{0, 1/(n-1), \ldots, (n-2)/(n-1), 1\}$ that assigns to each number $y \in S$ the truth-value:

$$\tau(y) = 1 - \frac{\text{answers falsified by } y}{(n - 1)}$$

Intuitively $\tau(y)$ measures, in units of $n - 1$, how far y is from falsifying too many answers. Accordingly the *initial state* of the game is the constant function 1 over S. Moreover, as demonstrated in [12, 13], at each stage of the game both Q's state of knowledge and A's replies can be expressed by formulae in *n-valued Łukasiewicz logic* $\mathbf{L_n}$. Hence Q's ith-state of knowledge is given by the Łukasiewicz conjunction $\odot$ of the formulas expressing the jth-state of knowledge, for $j = 1, \ldots, i - 1$. If we also define, for every state of knowledge τ, a "coarsest" state $\neg\tau$ that is incompatible with τ, in the sense that $\tau \odot \neg\tau = 0$ (with $\neg\tau = 1 - \tau$), then we obtain the following characterisation for $\mathbf{L_n}$.

Proposition 1 ([5]). *A formula Φ is a theorem of $\mathbf{L_n}$ if and only if Φ represents the initial state for every Ulam's game with $n - 2$ lies.*

We now consider a *generalised* version of Ulam's game. In this version A and Q agree to split the search space S into $n - 2$ "parts" $S_1, \ldots, S_{n-2}$, where for each S_i, A is allowed to lie i times. Equivalently, we could permit the initial state to be taken from any of Q's intermediate states of knowledge of an instance of Ulam's game with $n - 2$ lies. Here the formulae representing the initial state for every generalised Ulam's game with $n - 2$ parts coincide with the common tautologies of $\mathbf{L_k}$ for $k = 2, \ldots, n$. More formally:

Proposition 2. *A formula Φ is a theorem of $\mathbf{LB_n}$ if and only if Φ represents the initial state for every Ulam's game with k lies, with $k = 0, \ldots, n - 2$.*

Proof. Follows by Corollary 1 and Proposition 1. □

This generalised version of Ulam's game has important applications in Information Theory, modelling search procedures for information with different probabilities of distortions; for example in transmitting "large" or "short" numbers or using different channels or frequency bands to send information (see e.g. [4]).

4 The Hypersequent Calculus GŁ

In this section we present and prove cut-elimination for the hypersequent calculus **GŁ** defined for **Ł** in [11]. Hypersequents were introduced by Avron in [1] as a natural generalisation of Gentzen sequents, consisting of a *multiset* of sequents and permitting the definition of rules that "exchange information" between different sequents. More precisely:

Definition 8 (Hypersequent). *A hypersequent is a multiset of the form*

$$\Gamma_1 \vdash \Delta_1 \mid \ldots \mid \Gamma_n \vdash \Delta_n$$

where for $i = 1, \ldots, n$, Γ_i and Δ_i are multisets[5] of formulae, and $\Gamma_i \vdash \Delta_i$ is an ordinary sequent, called a component of the hypersequent.

The symbol "|" is intended to denote meta-level *disjunction*. In [11] hypersequents for **Ł** are interpreted using the characteristic model $[-1, 0]_Ł$ as follows:

Definition 9 (Interpretation of Hypersequents for Ł). *We say that a hypersequent $\Gamma_1 \vdash \Delta_1 | \ldots | \Gamma_n \vdash \Delta_n$ is valid in **Ł**, in symbols $\models_Ł^* \Gamma_1 \vdash \Delta_1 | \ldots | \Gamma_n \vdash \Delta_n$, iff for all valuations v for $[-1,0]_Ł$ there exists i such that $\Sigma_{A \in \Gamma_i} v(A) \leq \Sigma_{B \in \Delta_i} v(B)$, where $\Sigma_{A \in \emptyset} v(A) = 0$.*

Remark 5. We emphasize that for *formulae*, this interpretation gives the usual notion of validity for **Ł**, ie we have that a formula A is a theorem of **Ł** iff $\models_Ł^* \vdash A$.

Like ordinary sequent calculi, hypersequent calculi consist of axioms, logical rules and structural rules. However for hypersequent calculi the structural rules are divided into *internal* and *external* rules. The former deal with formulas within components, while the latter manipulate whole components of a hypersequent. Standard external structural rules are *external weakening* (*EW*) and *external contraction* (*EC*):

$$\frac{G}{G|\Gamma \vdash \Delta} \ (EW) \qquad \frac{G|\Gamma \vdash \Delta|\Gamma \vdash \Delta}{G|\Gamma \vdash \Delta} \ (EC)$$

---

[5] Note that by using multisets we avoid the need for exchange rules in our calculi

where G represents a (possibly empty) side hypersequent. These rules do not really increase the expressive power of hypersequent calculi with respect to ordinary sequent calculi since they only apply to one component at a time. A rule that allows *interactions* between different components is the *splitting rule* (S):

$$\frac{G|\Gamma_1,\Gamma_2 \vdash \Delta_1,\Delta_2}{G|\Gamma_1 \vdash \Delta_1|\Gamma_2 \vdash \Delta_2} \ (S)$$

The hypersequent calculus **GL** is defined as follows:

Definition 10 (GL). *GL has the following axioms and rules:*

Axioms
$$(ID) \ A \ \vdash A \qquad\qquad (\Lambda) \vdash \qquad\qquad (\bot) \ \bot \vdash A$$

Internal structural rules
$$\frac{G|\Gamma \vdash \Delta}{G|\Gamma, A \vdash \Delta} \ (WL) \qquad\qquad \frac{G|\Gamma_1 \vdash \Delta_1 \quad G|\Gamma_2 \vdash \Delta_2}{G|\Gamma_1,\Gamma_2 \vdash \Delta_1,\Delta_2} \ (M)$$

External structural rules
$$(EW), \ (EC) \ and \ (S)$$

Logical rules[6]
$$\frac{G|\Gamma, B \vdash A, \Delta|\Gamma \vdash \Delta}{G|\Gamma, A \supset B \vdash \Delta} \ (\supset,l) \qquad\qquad \frac{G|\Gamma \vdash \Delta \quad G|\Gamma, A \vdash B, \Delta}{G|\Gamma \vdash A \supset B, \Delta} \ (\supset,r)$$

Example 1. We give a proof in **GL** of L4, the characteristic axiom for **Ł**:

$$\cfrac{\cfrac{\cfrac{\cfrac{B \vdash B \quad A \vdash A}{B, A \vdash A, B} \ (M)}{B, A \vdash A, B|B \vdash A} \ (EW)}{B, B \supset A \vdash A} \ (\supset,l) \quad \cfrac{\cfrac{\cfrac{B \vdash B \quad A \vdash A}{B, A \vdash A, B} \ (M)}{B, B \supset A, A \vdash A, B} \ (WL)}{B, B \supset A \vdash A, A \supset B} \ (\supset,r)}{B, B \supset A \vdash A, A \supset B} \ (M)}{\cdots}$$

$$\cfrac{\cfrac{\vdash}{(A \supset B) \supset B \vdash} \ (WL) \qquad \cfrac{\cfrac{B, B \supset A \vdash A, A \supset B \vdash A, A \supset B}{B, B \supset A \vdash A, A \supset B|B \supset A \vdash A} \ (EW)}{\cfrac{(A \supset B) \supset B, B \supset A \vdash A}{(A \supset B) \supset B \vdash (B \supset A) \supset A} \ (\supset,r)} \ (\supset,l)}{\cfrac{\vdash}{\vdash ((A \supset B) \supset B) \supset ((B \supset A) \supset A)} \ (\supset,r)}$$

Soundness and completeness of **GL** are proved in [11] by relating **GL** to a hypersequent calculus for abelian logic, and then proving the soundness and completeness for this latter calculus semantically.

Theorem 4 (Soundness and Completeness of GL [11]). *A hypersequent G is derivable in **GL** iff $\models^*_L G$.*

---

[6] In the logical rules, $A \supset B$ is called the *principal formula*.

4.1   Cut-Elimination for GŁ

Cut-elimination is one of the most important procedures in logic. The removal of cuts corresponds to the elimination of "lemmas" from proofs. This renders a proof *analytic*, in the sense that all formuale occurring in the proof are sub-formulae of the formula to be proved.

In [11] it was shown that the following cut rules are *admissible* for GŁ.

$$\frac{G|\Gamma, A \vdash \Delta \quad G|\Pi \vdash A, \Sigma}{G|\Gamma, \Pi \vdash \Delta, \Sigma} \ (cut) \qquad \frac{G|\Gamma, A \vdash A, \Delta}{G|\Gamma \vdash \Delta} \ (gencut)$$

Here we prove a stronger result, namely that in GŁ *cut-elimination* holds for these rules, i.e. that there is an algorithm for transforming a proof of a hyper-sequent G in GŁ + (*gencut*) (resp. GŁ + (*cut*)) into a proof of G in GŁ. First observe that the above cut rules are *interderivable* in GŁ:

$$\frac{\dfrac{\dfrac{\Gamma, A \vdash A, \Delta}{\Gamma \vdash \Delta|\Gamma, A \vdash A, \Delta} \ (EW)}{\Gamma, A \supset A \vdash \Delta} \ (\supset, l) \quad \dfrac{\dfrac{\vdash A \vdash A}{\vdash A \supset A} \ (\supset, r)}{} \ (cut)}{\Gamma \vdash \Delta} \qquad \frac{\Gamma, A \vdash \Delta \quad \Pi \vdash A, \Sigma}{\dfrac{\Gamma, \Pi, A \vdash A, \Delta, \Sigma}{\Gamma, \Pi \vdash \Delta, \Sigma} \ (gencut)} \ (M)$$

We therefore prove cut-elimination for GŁ + (*gencut*) and obtain cut-elimination for GŁ + (*cut*) as a corollary. Our strategy is as follows. We show that the logical rules of GŁ are *invertible* (i.e. that the premises of the rule are derivable if the conclusion is derivable); this allows us to reduce all applications of (*gencut*) to applications of (*gencut*) on *atomic* cut formulae. We then show that applications of (*cut*) on atomic cut formulae can be eliminated (required to deal with the rule (M)), and finally that applications of (*gencut*) on atomic cut formulae can be eliminated.

To aid exposition, we adopt the following conventions, where A is a formula and Σ is a multiset of formulae:

$$\lambda A = \overbrace{\{A, \ldots, A\}}^{\lambda} \text{ and } \lambda\Sigma = \overbrace{\Sigma \cup \ldots \cup \Sigma}^{\lambda} \text{ for } \lambda \geq 0$$

The following lemmas will be useful.

Lemma 1 ([11]). *If $G|\Gamma \vdash \bot, \Delta$ is derivable in GŁ, then $G|\Gamma \vdash A, \Delta$ is derivable in GŁ.*

Lemma 2. *If $G|\Gamma \vdash \Delta, \Sigma$ is derivable in GŁ, then $G|\Gamma \vdash \Delta$ is derivable in GŁ.*

Proof. An easy induction on the height of a derivation of $G|\Gamma \vdash \Delta, \Sigma$. □

In the next two propositions we establish the invertibility of the logical rules.

Proposition 3. *If $G|\Gamma, A \supset B \vdash \Delta$ is derivable in GŁ, then $G|\Gamma \vdash \Delta|\Gamma, B \vdash A, \Delta$ is derivable in GŁ.*

Proof. We actually prove a more general result (required to deal with the rule (EC)), namely that:

If $Q = \Gamma_1, \lambda_1(A \supset B) \vdash \Delta_1 | \ldots | \Gamma_k, \lambda_k(A \supset B) \vdash \Delta_k$ is derivable in **GŁ** for $\lambda_1, \ldots, \lambda_k \geq 0$, then $Q' = \Gamma_1 \vdash \Delta_1 | \Gamma_1, \lambda_1 B \vdash \Delta_1, \lambda_1 A | \ldots | \Gamma_k \vdash \Delta_k | \Gamma_k, \lambda_k B \vdash \Delta_k, \lambda_k A$ is derivable in **GŁ**.

We proceed by induction on h, the height of a proof of Q. We assume $\lambda_i > 0$ for some i, $1 \leq i \leq k$, as otherwise the proof is trivial. If $h = 0$ then $Q = A \supset B \vdash A \supset B$ and $Q' = \vdash A \supset B | B \vdash A \supset B, A$ which is derivable in **GŁ**. For $h > 0$ there are several cases for the last rule r applied.

- r is (WL) and we step to:

$$\Gamma_1, (\lambda_1 - 1)(A \supset B) \vdash \Delta_1 | \ldots | \Gamma_k, \lambda_k(A \supset B) \vdash \Delta_k$$

so by the induction hypothesis:

$$\Gamma_1 \vdash \Delta_1 | \Gamma_1, (\lambda_1 - 1)B \vdash \Delta_1, (\lambda_1 - 1)A | \ldots | \Gamma_k \vdash \Delta_k | \Gamma_k, \lambda_k B \vdash \Delta_k, \lambda_k A$$

is derivable in **GŁ**; hence Q' is derivable by repeatedly applying (M) and (S) $(\lambda_1 - 1)$ times together with (EC) and (EW) as necessary.

- r is (EW), (EC), (S), $(\supset, r)$ or (WL) and the weakened formula is in one of the Γ_i's. These cases involve unproblematic applications of the induction hypothesis followed by applications of the corresponding rule.

- r is (M). We step to:

$$\Gamma_1^1, \mu(A \supset B) \vdash \Delta_1^1 | \ldots | \Gamma_k, \lambda_k(A \supset B) \vdash \Delta_k \text{ and}$$

$$\Gamma_1^2, (\lambda_1 - \mu)(A \supset B) \vdash \Delta_1^2 | \ldots | \Gamma_k, \lambda_k(A \supset B) \vdash \Delta_k$$

hence by the induction hypothesis twice we get that:

$$\Gamma_1^1 \vdash \Delta_1^1 | \Gamma_1^1, \mu B \vdash \Delta_1^1, \mu A | \ldots | \Gamma_k \vdash \Delta_k | \Gamma_k, \lambda_k B \vdash \Delta_k, \lambda_k A \text{ and}$$

$$\Gamma_1^2 \vdash \Delta_1^2 | \Gamma_1^2, (\lambda_1 - \mu)B \vdash \Delta_1^2, (\lambda_1 - \mu)A | \ldots | \Gamma_k \vdash \Delta_k | \Gamma_k, \lambda_k B \vdash \Delta_k, \lambda_k A$$

are derivable in **GŁ**. Hence Q' is derivable in **GŁ** by suitable applications of (M) and (EW).

- r is $(\supset, l)$. There are two subcases:

(a) If the principal formula in $(\supset, l)$ is in Γ_i, for some i, $1 \leq i \leq k$, the claim follows by the induction hypothesis and two applications of $(\supset, l)$.

(b) Otherwise, we step to:

$$\Gamma_1, (\lambda_1 - 1)(A \supset B) \vdash \Delta_1 | \Gamma_1, (\lambda_1 - 1)(A \supset B), B \vdash A, \Delta_1 | \ldots | \Gamma_k, \lambda_k(A \supset B) \vdash \Delta_k$$

By the induction hypothesis we have:

$\Gamma_1 \vdash \Delta_1 | \Gamma_1, (\lambda_1 - 1)B \vdash (\lambda_1 - 1)A, \Delta_1 | \Gamma_1, B \vdash A, \Delta_1 | \Gamma_1, \lambda_1 B \vdash \lambda_1 A, \Delta_1 | \ldots |$
$\Gamma_k \vdash \Delta_k | \Gamma_k, \lambda_k B \vdash \Delta_k, \lambda_k A$

By $(\lambda_1 - 1)$ applications of (M) followed by $(\lambda_1 - 1)$ applications of (S) and (EC) we have:

$\Gamma_1 \vdash \Delta_1 | \Gamma_1, (\lambda_1 - 1)B \vdash (\lambda_1 - 1)A, \Delta_1 | \Gamma_1, \lambda_1 B \vdash A, \lambda_1 \Delta_1 | \Gamma_1, \lambda_1 B \vdash \lambda_1 A, \Delta_1 |$
$\ldots | \Gamma_k \vdash \Delta_k | \Gamma_k, \lambda_k B \vdash \Delta_k, \lambda_k A$

Hence Q' is derivable by repeatedly applying (M) and (S) $(\lambda_1 - 1)$ times together with (EC) and (EW) as necessary. □

Proposition 4. *If $G | \Gamma \vdash A \supset B, \Delta$, then $G | \Gamma \vdash \Delta$ and $G | \Gamma, A \vdash B, \Delta$ are derivable in $\mathbf{GL}$.*

Proof. Similar to that of Proposition 3. □

We now show that applications of (cut) on $atomic^7$ formulae are eliminable.

Proposition 5. *If $G | \Gamma, \lambda q \vdash \Delta$ and $G' | \Pi \vdash \lambda q, \Sigma$ are derivable in $\mathbf{GL}$ for q atomic and $\lambda \geq 0$, then $G | G' | \Gamma, \Pi \vdash \Delta, \Sigma$ is derivable in $\mathbf{GL}$.*

Proof. We prove the more general result that:

If $Q = \Gamma_1, \lambda_1 q \vdash \Delta_1 | \ldots | \Gamma_k, \lambda_k q \vdash \Delta_k$ and $Q'_i = G_i | \Pi_i \vdash \Sigma_i, \mu_i q$ for $i = 1, \ldots, k$ are derivable in $\mathbf{GL}$ for q atomic and $\mu_i \geq \lambda_i \geq 0$ for $i = 1, \ldots, k$, then $Q' = G_1 | \ldots | G_k | \Gamma_1, \Pi_1 \vdash \Delta_1, \Sigma_1, (\mu_1 - \lambda_1)q | \ldots | \Gamma_k, \Pi_k \vdash \Delta_k, \Sigma_k, (\mu_k - \lambda_k)q$ is derivable in $\mathbf{GL}$.

We proceed by induction on h, the height of a proof of Q. We assume that $\mu_i \geq \lambda_i > 0$ for some i, $1 \leq i \leq k$ as otherwise Q is derivable easily using (EW) and (M). For $h = 0$ there are two possibilities. If $Q = q \vdash q$ then $Q' = Q'_1 = G_1 | \Pi_1 \vdash \Sigma_1, \mu_1 q$ which is derivable in $\mathbf{GL}$. If $Q = \bot \vdash A$ then $Q' = G_1 | \Pi_1 \vdash \Sigma_1, A, (\mu_1 - 1)\bot$ which is derivable by Lemma 1. For $h > 0$ we have several cases according to the last rule r applied.

- r is (WL). There are two subcases.

  (a) We step to: $Q'' = \Gamma_1, (\lambda_1 - 1)q \vdash \Delta_1 | \ldots | \Gamma_k, \lambda_k q \vdash \Delta_k$

Since $\mu_1 \geq \lambda_1 \geq 1$ we have by Lemma 2 that $Q''_1 = G_1 | \Pi_1 \vdash \Sigma_1, (\mu_1 - 1)q$ is derivable in $\mathbf{GL}$. Hence by the induction hypothesis applied to $Q'', Q''_1, Q'_2, \ldots, Q'_k$ we get that Q' is derivable in $\mathbf{GL}$.

  (b) We step to: $Q'' = \Gamma_1 - \{A\}, \lambda_1 q \vdash \Delta_1 | \ldots | \Gamma_k, \lambda_k q \vdash \Delta_k$

and the claim follows by the induction hypothesis and (WL).

---

[7] Note that an atomic formula is either a propositional variable or $\bot$.

- r is (EW), (EC), $(\supset, l)$ or $(\supset, r)$. These involve straightforward applications of the induction hypothesis and the corresponding rule.

- r is (M). We step to:

$Q_1 = \Gamma_1^1, \lambda_1^1 q \vdash \Delta_1^1 | \ldots | \Gamma_k, \lambda_k q \vdash \Delta_k$ and

$Q_2 = \Gamma_1^2, \lambda_1^2 q \vdash \Delta_1^2 | \ldots | \Gamma_k, \lambda_k q \vdash \Delta_k$

By the induction hypothesis applied to Q_1 and $Q'_1, \ldots, Q'_k$ we get that:

$Q'' = G|G_1|\Gamma_1^1, \Pi_1 \vdash \Delta_1^1, \Sigma_1, (\mu_1 - \lambda_1^1)q$ where

$G = G_2|\ldots|G_k|\Gamma_2, \Pi_2 \vdash \Delta_2, \Sigma_2, (\mu_2 - \lambda_2)q|\ldots|\Gamma_k, \Pi_k \vdash \Delta_k, \Sigma_k, (\mu_k - \lambda_k)q$

is derivable in **GL**. Now we apply the induction hypothesis *again*, this time to Q_2 and $Q'', Q'_2, \ldots, Q'_k$, giving that:

$G|G_1|\Gamma_1^1, \Gamma_1^2, \Pi_1 \vdash \Delta_1^1, \Delta_1^2, \Sigma_1, ((\mu_1 - \lambda_1^1) - \lambda_1^2)q|G$

is derivable in **GL**. Hence Q' is derivable in **GL** using (EC).

- r is (S). We step to:

$Q_1 = \Gamma_1, \Gamma_2, \lambda_1 q, \lambda_2 q \vdash \Delta_1, \Delta_2 | \ldots | \Gamma_k, \lambda_k q \vdash \Delta_k$

Since $G_i|\Pi_i \vdash \Sigma_i, \mu_i q$ is derivable in **GL** for $i = 1, 2$ we have that:

$Q'_{1,2} = G_1|G_2|\Pi_1, \Pi_2 \vdash \Sigma_1, \Sigma_2, (\mu_1 + \mu_2)q$

is derivable in **GL** by (M) and (EW). So now by the induction hypothesis applied to Q_1 and $Q'_{1,2}, Q'_3, \ldots, Q'_k$ we get that:

$G_1|\ldots|G_k|\Gamma_1, \Gamma_2, \Pi_1, \Pi_2 \vdash \Delta_1, \Delta_2, \Sigma_1, \Sigma_2, ((\mu_1 + \mu_2) - (\lambda_1 + \lambda_2))q|G$

is derivable in **GL** where

$G = \Gamma_3, \Pi_3 \vdash \Delta_3, \Sigma_3, (\mu_3 - \lambda_3)q|\ldots|\Gamma_k, \Pi_k \vdash \Delta_k, \Sigma_k, (\mu_k - \lambda_k)q$

Hence Q' is derivable in **GL** by (S). $\square$

We now show that applications of $(gencut)$ on atomic cut formulae can be eliminated.

Proposition 6. *If $Q = \Gamma_1, \lambda_1 q \vdash \Delta_1, \lambda_1 q | \ldots | \Gamma_k, \lambda_k q \vdash \Delta_k, \lambda_k q$ is derivable in **GL** for q atomic and $\lambda_1, \ldots, \lambda_k \geq 0$, then $Q' = \Gamma_1 \vdash \Delta_1 | \ldots | \Gamma_k \vdash \Delta_k$ is derivable in **GL**.*

Proof. We proceed by induction on h, the height of a proof of Q. We assume that $\lambda_i > 0$ for some i, $1 \leq i \leq k$, as otherwise the claim is trivial. If $h = 0$ then $Q = q \vdash q$ and $Q' = \ \vdash$ which is derivable by (Λ). For $h > 0$ we have several possibilities for the last rule r applied. We outline the only non-trivial case, where r is (M) and we step to:

$Q_1 = \Gamma_1^1, \mu q \vdash \Delta_1^1, \gamma q | \ldots | \Gamma_k, \lambda_k q \vdash \Delta_k, \lambda_k q$ and

$Q_2 = \Gamma_1^2, (\lambda_1 - \mu)q \vdash \Delta_1^2, (\lambda_1 - \gamma)q | \ldots | \Gamma_k, \lambda_k q \vdash \Delta_k, \lambda_k q$

Without loss of generality we assume $\mu \geq \gamma$. By the induction hypothesis twice we get that:

$Q_1' = \Gamma_1^1, (\mu - \gamma)q \vdash \Delta_1^1 | \ldots | \Gamma_k \vdash \Delta_k$ and $Q_2' = \Gamma_1^2 \vdash \Delta_1^2, (\mu - \gamma)q | \ldots | \Gamma_k \vdash \Delta_k$

are derivable in **GŁ**, so Q' is derivable in **GŁ** by Proposition 5 and (EC). $\square$

Finally we show that cut-elimination holds for **GŁ** $+$ $(gencut)$.

Theorem 5. *If $G|\Gamma, A \vdash A, \Delta$ is derivable in **GŁ**, then $G|\Gamma \vdash \Delta$ is derivable in **GŁ**.*

Proof. We proceed by induction on c, the complexity of A. If $c = 1$ then the claim holds by Proposition 6. For $c > 1$ we use Propositions 3 and 4 and apply the appropriate logical rules to $A = B \supset C$ on both sides, giving that:

$Q_1 = G|\Gamma \vdash \Delta|\Gamma, C \vdash \Delta, B$ and $Q_2 = G|\Gamma, B, C \vdash \Delta, B, C|\Gamma, B \vdash \Delta, C$

are derivable in **GŁ**. By the induction hypothesis twice for Q_2 we have that:

$Q_3 = G|\Gamma \vdash \Delta|\Gamma, B \vdash \Delta, C$

is derivable in **GŁ**. By (M) applied to Q_1 and Q_3 we get that:

$Q_4 = G|\Gamma \vdash \Delta|\Gamma, \Gamma, B, C \vdash \Delta, \Delta, B, C$

is derivable in **GŁ**. Hence, applying the induction hypothesis twice to Q_4:

$G|\Gamma \vdash \Delta|\Gamma, \Gamma \vdash \Delta, \Delta$

is derivable in **GŁ**. The claim now follows using (EC) and (S). $\square$

Corollary 2. *Cut-elimination holds for **GŁ** $+$ (cut).*

Proof. By Theorem 5 and the derivability of (cut) in **GŁ** $+$ (gencut). $\square$

5   Hypersequent Calculi for Bounded Łukasiewicz Logics

We now return our attention to bounded Łukasiewicz logics, defining the validity of a hypersequent in **LB$_n$** in terms of its validity in the k-valued Łukasiewicz logics for $k = 2, \ldots, n$.

Definition 11 (Interpretation of Hypersequents for Ł$_k$). *A hypersequent $G = \Gamma_1 \vdash \Delta_1 | \ldots | \Gamma_n \vdash \Delta_n$ is valid in Ł$_k$, in symbols $\models_{Ł_k}^* G$, iff for all valuations v for $[-1, 0]_{Ł_k}$ there exists i such that $\Sigma_{A \in \Gamma_i} v(A) \leq \Sigma_{B \in \Delta_i} v(B)$, where $\Sigma_{A \in \emptyset} v(A) = 0$.*

Definition 12 (Interpretation of Hypersequents for LB$_n$). *A hypersequent G is valid in LB$_n$, in symbols $\models_{LB_n}^* G$, iff $\models_{Ł_k}^* G$ for $k = 2, \ldots, n$.*

Remark 6. It follows from Corollary 1 that A is a theorem of $\mathbf{LB_n}$ iff $\models^*_{LB_n} \vdash A$.

We introduce analytic calculi for $\mathbf{LB_n}$ based on this interpretation.

Definition 13 ($\mathbf{G\mathcal{L}B_n}$). $\mathbf{G\mathcal{L}B_n}$ *has the same axioms and rules as* $\mathbf{G\mathcal{L}}$ *and:*

$$\frac{G| \overbrace{\Pi,\ldots,\Pi}^{n-1},\Gamma,\bot \vdash \overbrace{\Sigma,\ldots,\Sigma}^{n-1},\Delta}{G|\Pi \vdash \Sigma|\Gamma \vdash \Delta} \ (nC)$$

Remark 7. To obtain a true *subformula property* for the calculus we can replace the occurence of $\bot$ in the above rule by an atomic formula q with the added condition that q must occur in $G \cup \Sigma \cup \Delta$.

Example 2. As an example we prove the characteristic axiom of $\mathbf{LB_n}$.

$$\frac{\bot \vdash A \quad \dfrac{\bot \vdash \bot \quad A \vdash A}{\bot, A \vdash A, \bot}\ (M)}{\bot \vdash A, \neg A}\ (\supset, r)$$

$$\frac{A \vdash A \quad \dfrac{\neg A \vdash \neg A \quad \dfrac{(n-3)\neg A, \bot \vdash A, (n-2)\neg A}{\dfrac{(n-2)\neg A, \bot \vdash A, (n-1)\neg A}{(n-2)\neg A \vdash A|\vdash \neg A}\ (nC)}\ (M)}{(n-2)\neg A \vdash A|A \vdash A, \neg A}\ (M)}{(n-2)\neg A \vdash A|A \vdash A, \neg A}\ (\supset, l)$$

$$\frac{\neg A \vdash \neg A \quad \dfrac{\vdots}{(n-2)\neg A \vdash A|(n-1).A, (n-3)\neg A \vdash A}\ (\supset, l)}{\dfrac{(n-2)\neg A \vdash A|(n-1).A, (n-2)\neg A \vdash A, \neg A}{n.A, (n-2)\neg A \vdash A}\ (\supset, l)}\ (M)$$

$$\frac{\dfrac{\vdots}{\dfrac{n.A, \neg A \vdash (n-2).A}{\dfrac{n.A \vdash (n-1).A}{\vdash n.A \supset (n-1).A}\ (\supset, r)}\ (\supset, r)}\ (\supset, r)}{}$$

Theorem 6 (Soundness of $\mathbf{G\mathcal{L}B_n}$). *If a hypersequent G is derivable in* $\mathbf{G\mathcal{L}B_n}$, *then* $\models^*_{LB_n} G$.

Proof. We proceed by induction on the height of a proof of G, checking that the axioms are valid in $\mathbf{G\mathcal{L}B_n}$ and that the rules preserve validity, i.e. that if the premises are valid in $\mathbf{G\mathcal{L}B_n}$ then so is the conclusion. Since all but one of the cases are essentially the same as for the proof of Theorem 4 given in [11], we just check the rule (nC). Suppose that:

(1) $\models^*_{\mathbf{Ł}_k} \overbrace{\Pi, \ldots, \Pi}^{n-1}, \Gamma, \bot \vdash \overbrace{\Sigma, \ldots, \Sigma}^{n-1}, \Delta$ for $k = 2, \ldots, n$

We check that $\models^*_{\mathbf{Ł}_k} \Pi \vdash \Sigma | \Gamma \vdash \Delta$ for $k = 2, \ldots, n$.

Consider a valuation v for $[-1, 0]_{\mathbf{Ł}_k}$, $2 \leq k \leq n$. If $\Sigma_{A \in \Pi} v(A) \leq \Sigma_{B \in \Sigma} v(B)$ then we are done. If $\Sigma_{A \in \Pi} v(A) > \Sigma_{B \in \Sigma} v(B)$ then we get:

$0 > \Sigma_{B \in \Sigma} v(B) - \Sigma_{A \in \Pi} v(A)$ and, since there are just k truth values:

$-1/(n-1) \geq -1/(k-1) \geq \Sigma_{B \in \Sigma} v(B) - \Sigma_{A \in \Pi} v(A)$

But now, since by (1), we have:

$(n-1)(\Sigma_{A \in \Pi} v(A)) + \Sigma_{A \in \Gamma} v(A) + (-1) \leq (n-1)(\Sigma_{B \in \Sigma} v(B)) + \Sigma_{B \in \Delta} v(B)$

we get that $\Sigma_{A \in \Gamma} v(A) \leq \Sigma_{B \in \Delta} v(B)$ and $\models^*_{\mathbf{Ł}_k} \Pi \vdash \Sigma | \Gamma \vdash \Delta$ as required. $\square$

We now turn our attention to proving the *completeness* of $\mathbf{GŁB_n}$. As a first step we show that $\mathbf{GŁB_n} + (gencut)$ is complete.

Proposition 7. *If $\vdash A$ and $\vdash A \supset B$ are derivable in $\mathbf{GŁB_n} + (gencut)$ then $\vdash B$ is derivable in $\mathbf{GŁB_n} + (gencut)$.*

Proof. We have that $\vdash A$ and $A \vdash B$ are derivable in $\mathbf{GŁB_n} + (gencut)$. Hence by (M), $A \vdash A, B$ is derivable in $\mathbf{GŁB_n} + (gencut)$, and by $(gencut)$, $\vdash B$ is derivable in $\mathbf{GŁB_n} + (gencut)$. $\square$

Theorem 7 (Completeness). *If a formula A is a theorem of $\mathbf{ŁB_n}$, then $\vdash A$ is derivable in $\mathbf{GŁB_n} + (gencut)$.*

Proof. We use the completeness of the axiomatisation for $\mathbf{ŁB_n}$ of Definition 6. It is easy to check that $\mathbf{GŁB_n} + (gencut)$ proves all the axioms of $\mathbf{ŁB_n}$, and by Proposition 7 also (mp) is admissible in $\mathbf{GŁB_n} + (gencut)$. $\square$

We now show that $(gencut)$ can be *eliminated* from derivations in $\mathbf{GŁB_n} + (gencut)$, by checking that the rule (nC) does not spoil the cut-elimination procedure for $\mathbf{GŁ}$ outlined in Section 4.1. It is easy to prove that the logical rules remain invertible in $\mathbf{GŁB_n}$ by extending the proofs of Propositions 3 and 4. The next step is to show that applications of (cut) on atomic cut formulae can be eliminated.

Proposition 8. *If $G|\Gamma, \lambda q \vdash \Delta$ and $G'|\Pi \vdash \Sigma, \lambda q$ are derivable in $\mathbf{GŁB_n}$ for q atomic, then $G|G'|\Gamma, \Pi \vdash \Delta, \Sigma$ is derivable in $\mathbf{GŁB_n}$.*

Proof. We follow exactly the proof of Proposition 5 and prove the more general result that:

If $Q = \Gamma_1, \lambda_1 q \vdash \Delta_1 | \ldots | \Gamma_k, \lambda_k q \vdash \Delta_k$ and $Q'_i = G_i | \Pi_i \vdash \Sigma_i, \mu_i q$ for $i = 1, \ldots, k$ are derivable in $\mathbf{GŁB_n}$ for q atomic and $\mu_i \geq \lambda_i \geq 0$ for $i = 1, \ldots, k$, then

$Q' = G_1|\ldots|G_k|\Gamma_1, \Pi_1 \vdash \Delta_1, \Sigma_1, (\mu_1 - \lambda_1)q|\ldots|\Gamma_k, \Pi_k \vdash \Delta_k, \Sigma_k, (\mu_k - \lambda_k)q$ is derivable in $\mathbf{GLB_n}$.

As before we proceed by induction on the height of a proof of Q. Here we just check the case where the last step is an application of the rule (nC). We step to:

$$Q_1 = (n-1)\Gamma_1, \Gamma_2, ((n-1)\lambda_1 + \lambda_2)q, \perp \vdash (n-1)\Delta_1, \Delta_2|\ldots|\Gamma_k, \lambda_k q \vdash \Delta_k$$

Now, since $G_i|\Pi_i \vdash \Sigma_i, \mu_i q$ is derivable in $\mathbf{GLB_n}$ for $i = 1, 2$ we have that:

$$Q'_{1,2} = G_1|G_2|(n-1)\Pi_1, \Pi_2 \vdash (n-1)\Sigma_1, \Sigma_2, ((n-1)\mu_1 + \mu_2)q$$

is derivable in $\mathbf{GLB_n}$ by (M) and (EW). So by the induction hypothesis applied to Q_1 and $Q'_{1,2}, Q'_3, \ldots, Q'_k$ we get that:

$$G_1|\ldots|G_k|(n-1)\Gamma_1, (n-1)\Pi_1, \Gamma_2, \Pi_2, \perp \vdash (n-1)\Delta_1, (n-1)\Sigma_1, \Delta_2, \Sigma_2, (n-1)(\mu_1 - \lambda_1)q, (\mu_2 - \lambda_2)q|\ldots|\Gamma_k, \Pi_k \vdash \Delta_k, \Sigma_k, (\mu_k - \lambda_k)q$$

is derivable in $\mathbf{GLB_n}$. Hence Q' is derivable in $\mathbf{GLB_n}$ by (nC) as required. $\square$

We now check that applications of $(gencut)$ on atomic cut formulae can be eliminated in $\mathbf{GLB_n}$.

Proposition 9. *If* $Q = \Gamma_1, \lambda_1 q \vdash \Delta_1, \lambda_1 q|\ldots|\Gamma_k, \lambda_k q \vdash \Delta_k, \lambda_k q$ *is derivable in* GLB_n *for q atomic and* $\lambda_1, \ldots, \lambda_k \geq 0$, *then* $Q' = \Gamma_1 \vdash \Delta_1|\ldots|\Gamma_k \vdash \Delta_k$ *is derivable in* GLB_n.

Proof. We follow the proof of Proposition 6 and just check the extra case where the last step in the proof of Q is an application of (nC). We step to:

$$(n-1)\Gamma_1, \Gamma_2, ((n-1)\lambda_1 + \lambda_2)q, \perp \vdash (n-1)\Delta_1, \Delta_2, ((n-1)\lambda_1 + \lambda_2)q|\ldots|\Gamma_k, \lambda_k q \vdash \Delta_k, \lambda_k q.$$ By the induction hypothesis we get that:

$$(n-1)\Gamma_1, \Gamma_2, \perp \vdash (n-1)\Delta_1, \Delta_2|\ldots|\Gamma_k \vdash \Delta_k.$$

is derivable in $\mathbf{GLB_n}$. Hence Q is derivable in $\mathbf{GLB_n}$ by (nC). $\square$

We arrive at the following cut-elimination theorem for $\mathbf{GLB_n}$.

Theorem 8. *If* $G|\Gamma, A \vdash \Delta, A$ *is derivable in* $\mathbf{GLB_n}$, *then* $G|\Gamma \vdash \Delta$ *is derivable in* GLB_n.

References

[1] A. Avron. A constructive analysis of RM. *J. of Symbolic Logic*, 52:939–951, 1987.

[2] C. C. Chang. Algebraic analysis of many valued logics. *Trans. Amer. Math. Soc.*, 88:467–490, 1958.

[3] A. Ciabattoni. Bounded contraction in systems with linearity. In *Proceedings of TABLEAUX 99*, volume 1617 of *LNAI*, pages 113–127. Springer, 1999.

[4] F. Cicalese. *Reliable Computation with Unreliable Information*. PhD thesis, University of Salerno, 2001. http://dia.unisa.it/cicalese.dir/thesis.html.

[5] R. L. O. Cignoli, D. Mundici, and I. M. L. D'Ottaviano. *Algebraic Foundations of Many-valued Reasoning*. Kluwer, 2000.

[6] R. Grigolia. Algebraic analysis of Łukasiewicz-Tarski n-valued logical systems. In R. Wòjciki and G. Malinowski, editors, *Selected Papers on Łukasiewicz Sentential Calculi*, pages 81–91. Polish Acad. of Sciences, 1977.

[7] P. Hájek. *Metamathematics of Fuzzy Logic*. Kluwer, 1998.

[8] J. Łukasiewicz. O logice tròjwartościowej. *Ruch Filozoficzny*, 5:169–171, 1920.

[9] R. McNaughton. A theorem about infinite-valued sentential logic. *Journal of Symbolic Logic*, 16(1):1–13, 1951.

[10] G. Metcalfe, N. Olivetti, and D. Gabbay. Analytic sequent calculi for abelian and Łukasiewicz logics. In *Proceedings of TABLEAUX 2002*, volume 2381 of *LNCS*, pages 191–205. Springer, 2002.

[11] G. Metcalfe, N. Olivetti, and D. Gabbay. Sequent and hypersequent calculi for Abelian and Łukasiewicz logics. Submitted. http://arXiv.org/list/cs/0211, 2003.

[12] D. Mundici. The logic of Ulam's game with lies. In C. Bicchieri and M. L. Dalla Chiara, editors, *Knowledge, belief and strategic interaction*, pages 275–284. Cambridge University Press, 1992.

[13] D. Mundici. Ulam's game, Łukasiewicz logic and C*-algebras. *Fundamenta Informaticae*, 18:151–161, 1993.

[14] A. Pelc. Searching with known error probability. *Theoretical Computer Science*, 63:185–202, 1989.

[15] A. Prijatelj. Bounded contraction and Gentzen style formulation of Łukasiewicz logics. *Studia Logica*, 57:437–456, 1996.

[16] L. A. Zadeh. Fuzzy sets. *Information and Control*, 8:338–353, 1965.

Parallel Dialogue Games and Hypersequents for Intermediate Logics

Christian G. Fermüller

Technische Universität Wien, Austria

Abstract. A parallel version of Lorenzen's dialogue theoretic foundation for intuitionistic logic is shown to be adequate for a number of important intermediate logics. The soundness and completeness proofs proceed by relating hypersequent derivations to winning strategies for parallel dialogue games. This also provides a computational interpretation of hypersequents.

1 Introduction

In recent years hypersequent calculi have emerged as a flexible type of proof system for a wide range of logics (see, e.g., [3, 5, 4]). These calculi share many favorable properties with Gentzen's classic sequent calculi **LK** and **LI**. Most importantly, cuts are eliminable and the logical rules are strictly analytic — i.e., they only refer to immediate subformulas of the introduced formula and are context independent. Consequently these calculi are of relevance to automated reasoning. However, the relation between hypersequent derivations and the semantics of the corresponding logics is much more obscure than in the case of classical or intuitionistic sequents. Standard completeness proofs for **LK** and **LI** show how to extract counter models for underivable formulas; it is mainly this feature that allows to call (a particular form of) goal oriented proof search in sequent calculi 'semantic tableaux'. Unfortunately, the hypersequent calculi that have been formulated for intermediate logics like Gödel-Dummett logic $\mathbf{G}_\infty$, the logic **LQ** of weak excluded middle, or finite-valued Gödel logics do not relate directly to a semantic foundation of these logics. To address this concern, we show that hypersequents bear a close relation to an interesting foundational approach that constitutes an alternative to standard Tarski-style semantics: dialogue games.

2 Lorenzen Style Dialogue Games

Logical dialogue games come in many forms and versions, nowadays. Here, we do not use more recent formulations in the style of Blass [2] or Abramsky [1][1],

---

[1] These more modern logical dialogue games differ considerably from the orginal ones of Lorenzen and his school. In particular, parallelism is introduced already at the level of analyzing single connectives. This feature of Blass/Abramsky style games makes them useful for modelling certain features of linear logic and related formalisms, but less well connected to the well motivated foundational intentions of Lorenzen.

M. Cialdea Mayer and F. Pirri (Eds.): TABLEAUX 2003, LNAI 2796, pp. 48–64, 2003.

but rather refer directly to Paul Lorenzen's original idea (dating back to the late 1950s, see e.g., [15]) to identify logical validity of a formula A with the existence of a winning strategy for a *proponent* **P** in an idealized confrontational dialogue, in which **P** tries to uphold A against systematic doubts by an *opponent* **O**. Although the claim that this leads to an alternative characterization — or even: 'justification' — of *intuitionistic logic* was implicit already in Lorenzen's early essays, it took more than twenty years until the first rigorous, complete and error free proof of this central claim was published in [8]. Many variants of Lorenzen's original dialogue games have appeared in the literature since. (Already Lorenzen and his collaborators defined different versions of the game. See, eg., [9, 13] for further references.) Here, we define a version of dialogue games that are: 1) well suited for demonstrating the close relation to analytic Gentzen-type systems; 2) easily shown to be equivalent to other versions of dialogue games for intuitionistic logic, that can be found in the literature; 3) straightforward to consider 'in parallel'.

Notation. An *atomic formula (atom)* is either a propositional variable or $\perp$ *(falsum)*. As usual, *compound formulas* are built up from atoms using the connectives $\supset, \wedge, \vee$; $\neg A$ abbreviates $A \supset \perp$. In addition to formulas, the special signs ?, !?, r? can be stated in a dialogue by the players **P** and **O**, as specified below.

Dialogue games are characterized by two sorts of rules: logical ones and structural ones. The *logical rules* define how to attack a compound formula and how to defend against such an attack. They are summarized in the following table. (If **X** is the proponent **P** then **Y** refers to the opponent **O**, and vice versa.[2])

Logical dialogue rules:

XX:	attack by **Y**	defense by **X**
$A \wedge B$	!? or r? (**Y** chooses)	A or B, accordingly
$A \vee B$	?	A or B (**X** chooses)
$A \supset B$	A	B

We will see below that **O** may also attack *atoms* (including $\perp$) by stating '?'.

A *dialogue* is a sequence of *moves*, which are either attacking or defending statements, in accordance with the logical rules. Each dialogue refers to a finite multiset of formulas that are *initially granted* by **O**, and to an *initial formula* to be defended by **P**.

Moves can be viewed as state transitions. In any state of the dialogue the (multiset of) formulas, that have been either initially granted or stated by **O** so far, are called the *granted formulas* (at this state). The last formula that has been stated by **P** and that either already has been attacked or must be attacked in **O**'s next move is called *active formula*. (Note that the active formula, in general, is *not* the last formula stated by **P**; since **P** may have stated formulas *after* the active formula, that are not attacked by **O**.) With each state of a dialogue we

---

[2] Note that *both* players may launch attacks as well as defending moves during the course of a dialogue. For motivation and detailed exposition of these rules we refer to [9].

thus associate a *dialogue sequent* $\Pi \vdash A$, where Π denotes the granted formulas and A the active formula.

We stipulate that each move carries the information (pointers) necessary to reconstruct which formula is attacked or defended in which way in that move. However, we do not care about the exact way in which this information is coded.

Structural rules (Rahmenregeln in the diction of Lorenzen and his school) regulate the succession of moves. Quite a number of different systems of structural rules have been proposed in the literature (See e.g., [16, 9, 13]; in particular, [13] compares and discusses different systems.). The following rules, together with the winning conditions stated below, amount to a version of dialogues traditionally called *Ei*-dialogues (i.e., Felscher's *E*-dialogues combined with the so-called *ipse dixisti* rule; see, e.g., [13]).

Structural Dialogue Rules:

Start: The first move of the dialogue is carried out by **O** and consists in an attack on the initial formula.

Alternate: Moves strictly alternate between players **O** and **P**.

Atom: Atomic formulas, including $\perp$, may be stated by both players, but can neither be attacked nor defended by **P**.

E: Each (but the first) move of **O** reacts directly to the immediately preceding move by **P**. I.e., if **P** attacks a granted formula then **O**'s next move either defends this formula or attacks the formula used by **P** to launch this attack. If, on the other hand, **P**'s last move was a defending one then **O** has to attack immediately the formula stated by **P** in that defense move.

Winning Conditions (for P):

W: The game ends with **P** winning if **O** has attacked a formula that has already been granted (either initially or in a later move) by **O**.

W$\perp$: The game ends with **P** winning if **O** has granted $\perp$.

A *dialogue tree* τ *for* $\Pi \vdash C$ is a rooted directed tree with nodes labelled by dialogue sequents and edges corresponding to moves, such that each branch of τ is a dialogue with initially granted formulas Π and initial formula C. We thus identify the nodes of a dialogue tree with states of a dialogue. We distinguish **P**-nodes and **O**-nodes, according to whether it is **P**'s or **O**'s turn to move at the corresponding state.

A finite dialogue tree is a *winning strategy* (for **P**) if the following conditions hold:

1. Every **P**-node has at most one successor node.
2. All leaf nodes are **P**-nodes in which the winning conditions for **P** are fulfilled.
3. Every **O**-node has a successor node for each move by **O** that is a permissible continuation of the dialogue (according to the rules) at this stage.

Winning strategies for a player in a non-cooperative two-person game are more commonly described as *functions* assigning a move for that player to each state of the game, taking into account all possible moves of the opponent. Observe that our tree form of a winning strategy just describes the corresponding function in a manner that makes the step-wise evolution of permissible dialogues more explicit.

As already mentioned, a dialogue game may be viewed as a state transition system, where moves in a dialogue correspond to transitions between **P**-nodes and **O**-nodes. A dialogue then is a possible trace in the system; and a winning strategy can be obtained by a systematic 'unraveling' of all possible traces.

To illustrate this point, consider the implicational fragment of the language; i.e., the set of formulas not containing $\wedge$ or $\vee$. Henceforth we use the following notation: For every compound formula F of form $C \supset D$, F_p denotes C and F_c denotes D. If F is atomic then F_p is empty (and F_c remains undefined). F_{pp} is C_p if $F = C \supset D$. The figure, below, represents all permissible moves in a dialogue for this fragment. By labelling a transition with $\Pi \overset{+}{\leftarrow} F$ we denote that F is added to the multiset Π of granted formulas. $A \leftarrow C$ means that C, as a result of the corresponding move, is the new active formula.

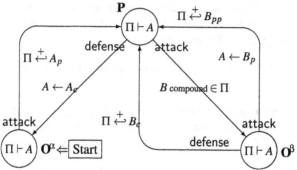

The encircled labels denote the dialogue sequent at the corresponding state. The edges from the **P**-node to the two **O**-nodes correspond to the principal choice of player **P**: either to defend the active formula or to attack a compound formula B from the granted formulas. (The fact that A_c is undefined if A is atomic means that in this case the transition from node **P** to node $\mathbf{O}^\alpha$ is not possible. This corresponds to the stipulation of rule *Atom*, that atomic formulas cannot be defended by **P**. However, remember that the dialogue is already in a winning state for **P** if the active formula A is among the granted formulas Π.)

On the other hand, according to the structural rule E, player **O** has no choice but to attack the last formula of **P** if **P**'s last move was a defense (i.e., if the dialogue is in state $\mathbf{O}^\alpha$). In state $\mathbf{O}^\beta$, however, **O** may either defend the attacked formula or attack the formula used by **P** in launching the last attack ('counterattack').

The winning conditions have to be checked at state **P** only. If $\bot \in \Pi$ or $A \in \Pi$ then the game ends in that state with **P** winning.

Adding $\wedge$ and $\vee$ to the language amounts to adding further possible transitions (between the nodes **P** and $\mathbf{O}^\alpha$, and **P** and $\mathbf{O}^\beta$, respectively).

Basic Adequateness of Dialogues

Proving the adequateness of dialogue games for intuitionistic logic consists in showing that winning strategies can be transformed into (analytic) proofs of Gentzen's well known sequent calculus **LI** for intuitionistic logic, and vice versa.[3]

To this aim, we use the following variant **LI'** of **LI**:

Axioms: $\bot, \Pi \longrightarrow C$ and $A, \Pi \longrightarrow A$

Logical Rules:[4]

$$\frac{A, A \vee B, \Pi \longrightarrow C \quad B, A \vee B, \Pi \longrightarrow C}{A \vee B, \Pi \longrightarrow C} \ (\vee, l) \qquad \frac{\Pi \longrightarrow A_i}{\Pi \longrightarrow A_1 \vee A_2} \ (\vee_i, r)$$

$$\frac{A_i, A_1 \wedge A_2, \Pi \longrightarrow C}{A_1 \wedge A_2, \Pi \longrightarrow C} \ (\wedge_i, l) \qquad \frac{\Pi \longrightarrow A \quad \Pi \longrightarrow B}{\Pi \longrightarrow A \wedge B} \ (\wedge, r)$$

$$\frac{A \supset B, \Pi \longrightarrow A \quad B, A \supset B, \Pi \longrightarrow C}{A \supset B, \Pi \longrightarrow C} \ (\supset, l) \qquad \frac{A, \Pi \longrightarrow B}{\Pi \longrightarrow A \supset B} \ (\supset, r)$$

Structural Rules: These are the usual weakening, contraction and cut rules.

It is straightforward to check that **LI'** is sound and complete for intuitionistic logic. As a corollary, the following holds:

Proposition 1. $A, \Gamma \longrightarrow A \supset B$ *is provable in* **LI'** *only if* $\Gamma \longrightarrow A \supset B$ *is provable.*

Theorem 1. *Every winning strategy* τ *for* $\Gamma \vdash C$ *(i.e., for dialogues with initial formula* C, *where player* **O** *initially grants the formulas in* Γ*) can be transformed into an* **LI'**-*proof of* $\Gamma \longrightarrow C$.

Proof. We prove by induction on the depth d of τ that for every **P**-node of τ there is an **LI'**-proof of the sequent corresponding to the dialogue sequent at this node. That this implies the theorem is obvious for the cases where C is either atomic, or a disjunction, or a conjunction; because, in those cases, the dialogue sequent at the **P**-node(s) immediately succeeding the root node is (are) identical to $\Gamma \vdash C$. In the case where $C = A \supset B$, the **P**-node succeeding the root carries $A, \Gamma \vdash A \supset B$ as dialogue sequent; and thus the theorem follows from Proposition 1.

The base case, $d = 1$, follows from the fact that the **P**-node (or, in case of C being a conjunction, the two **P**-nodes) succeeding the root is a (are) leaf node(s). This implies that one of the winning conditions — $C \in \Gamma$ or $\bot \in \Gamma$ — must hold. Consequently, the corresponding sequent $\Gamma \longrightarrow C$ is an axiom.

For $d > 1$ we have to distinguish cases according to the form of the active formula that is defended or the (compound) formula that is attacked by **P**.

---

[3] Quite a few proofs of the adequateness of dialogue games for characterizing intuitionistic logic can be found in the literature. Since we will build directly on such a proof — also in going beyond intuitionistic logic — we have to present our own version of it, which draws on ideas from [13, 14] and [8] but differs in a number of essential details.

[4] Since $\bot$ is in the language, we do not have to consider empty right hand sides of sequents.

To keep the proof concise, we will only elaborate it for the implicational fragment of the language; it is straightforward to augment the proof to cover also conjunctions and disjunctions.

1. **P** **defends** $A \supset B$: Let $A, \Pi \vdash A \supset B$ be the dialogue sequent at the current **P**-node. **P** moves from the **P**-node to the $\mathbf{O}^\alpha$-node by stating B. **O** has to reply with a move attacking B. We distinguish two cases:

   (a) If B is an atom then the attack consists in stating '?'. Thus we return to a **P**-node with dialogue sequent $A, \Pi \vdash B$. By the induction hypothesis there is an $\mathbf{LI'}$-proof of $A, \Pi \longrightarrow B$, which can be extended to a proof of $A, \Pi \longrightarrow A \supset B$ by applying rule $(\supset, r)$ and weakening.

   (b) If B is of form $B_p \supset B_c$ then **O** has to attack B by adding B_p to the granted formulas Π. Thus we return to a **P**-node with dialogue sequent $A, B_p, \Pi \vdash B$. By the induction hypothesis there is an $\mathbf{LI'}$-proof of $A, B_p, \Pi \longrightarrow B$. By Proposition 1 we obtain an $\mathbf{LI'}$-proof of $A, \Pi \longrightarrow B$. The required proof of $A, \Pi \longrightarrow A \supset B$ is obtained by applying rule $(\supset, r)$ and weakening.

2. **P** **attacks** $D \supset E$: Let $D \supset E, \Pi \vdash A$ be the dialogue sequent at the current **P**-node. **P**'s attack consists in stating D. (The move refers to the edge from node **P** to node $\mathbf{O}^\beta$ in the state transition diagram, above.) The strategy then branches since **O** may either defend the implication or attack D.

   (a) If **O** chooses to attack D then D_p is added to the granted formulas if $D = D_p \supset D_c$. If D is atomic the multiset of granted formulas remains unchanged. In any case, D is the new active formula at the succeeding **P**-node. The corresponding dialogue sequent is (1) $D_p, D \supset E, \Pi \vdash D$, where D_p is empty if D is atomic.

   (b) If, on the other hand, **O** chooses to defend $D \supset E$ then it has to grant E. The active formula at the succeeding **P**-node remains A. The corresponding dialogue sequent is (2) $E, D \supset E, \Pi \vdash A$.

   By the induction hypothesis there are $\mathbf{LI'}$-proofs of the sequents corresponding to (1) and (2). By Proposition 1 we may remove D_p from the left hand side of the sequent corresponding to (1). Therefore we obtain a proof of $D \supset E, \Pi \longrightarrow A$ by combining the two proofs with an application of rule $(\supset, l)$. $\square$

Remark 1. For proving the soundness of dialogue games (by this we mean that winning strategies only exist for intuitionistically valid sequents) it would in fact not have been necessary to refer to formal derivations. It rather suffices to check that intuitionistic validity transfers from the leaves of a dialogue tree upwards to the root. However for the following completeness proof the special format of the intuitionistic proofs is essential.

The 'weakening friendly' formulation of the axioms and rules of $\mathbf{LI'}$ allows to eliminate applications of the weakening rule. (Weakenings in $\mathbf{LI'}$-proofs can be moved upwards to the axioms, where they are obviously redundant.) Also the contraction rule becomes redundant if we disregard multiple occurrences of the same formula in the left hand side of a sequent. Most importantly, **LI′** is complete also without cut. Let as refer to a proof that does not contain any applications of structural rules as *strongly analytic*. The following proposition then sums up the just made observations.

Proposition 2. *There is a strongly analytic proof in* **LI′** *for* $\Gamma \longrightarrow C$ *if and only if* $\Gamma' \longrightarrow C$ *is provable in* **LI′**, *where* Γ' *equals* Γ *if taken as set (i.e., if multiple memberships of the same element are discarded).*

Theorem 2. *Every strongly analytic* **LI′***-proof* π *of* $\Gamma \longrightarrow C$ *can be transformed into a winning strategy for* $\Gamma \vdash C$.

Proof. We proceed by induction on the depth of π. Again, we show the theorem only for the implicational fragment of the language.

If $\Gamma \longrightarrow C$ is an axiom the winning strategy (consisting of two nodes) is obvious. There are two cases to consider for the induction step.

1. π **Ends with an Application of** $(\supset, r)$: The end sequent is of form $\Gamma \longrightarrow A \supset B$.
   By the induction hypothesis there is a winning strategy τ for $A, \Gamma \vdash B$. τ can be extended to a winning strategy for $\Gamma \vdash A \supset B$ as follows. We define a new root node; i.e., an $\mathbf{O}^\alpha$-node with dialogue sequent $\Gamma \vdash A \supset B$. To this root we attach an edge that leads to a new **P**-node. The corresponding move of $\mathbf{O}$ consists in granting A as an attack on $A \supset B$. Therefore the dialogue sequent at the new **P**-node is $A, \Gamma \vdash A \supset B$. We now only have to add an edge from this node to the root node of τ. This edge corresponds to **P** stating B in defense of $A \supset B$.

2. π **Ends with** $(\supset, l)$: The end sequent is of form $A \supset B, \Gamma \longrightarrow C$.
   By the induction hypothesis there is a winning strategy τ_1 for $A \supset B, \Gamma \vdash A$, and a winning strategy τ_2 for $B, A \supset B, \Gamma \vdash C$. Let τ_1^- be the tree, rooted in a **P**-node with dialogue sequent $A \supset B, C_p, A_p, \Gamma \vdash A$, that is obtained from τ_1 by removing its root and adding C_p to the granted formulas. We appeal to the general fact that a winning strategy for $\Pi \vdash F$ is also a winning strategy for $C, \Pi \vdash F$. Similarly let τ_2^- be the tree obtained form τ_2 that is rooted in a **P**-node with dialogue sequent $B, C_p, A \supset B, \Gamma \vdash C$. The construction of the winning strategy for $A \supset B, \Gamma \vdash C$ is illustrated in the following picture that refers to the state transition diagram, presented above.

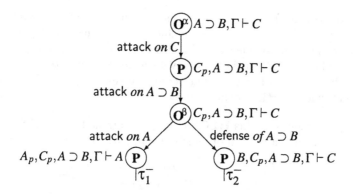

From now on we use the term **I**-*dialogues* to denote the dialogues that have been described in this section.

3   Hypersequent Calculi for Intermediate Logics

Intermediate logics (when identified with the set of its valid formulas) include intuitionistic logic and are included in classical logic. To introduce communicating parallel dialogues that are adequate for some well known intermediate logics we have to switch from sequent to *hypersequent* calculi.

Hypersequent calculi arise by generalizing standard sequent calculi to refer to whole contexts of sequents instead of single sequents. In our context, a hypersequent is defined as a finite, non-empty multiset of **LI′**-sequents, called *components*; written in form

$$\Gamma_1 \longrightarrow C_1 \mid \dots \mid \Gamma_n \longrightarrow C_n.$$

The symbol "|" is intended to denote disjunction at the meta-level.

Like ordinary sequent calculi, hypersequent calculi consist in axioms as well as logical and structural rules. The latter are divided into *internal* and *external* *rules*. The internal structural rules deal with formulas within components, while the external ones manipulate whole components of a hypersequent. The standard **external structural rules** are external weakening and external contraction:

$$\frac{\mathcal{H}}{\Pi \longrightarrow C \mid \mathcal{H}} \; (EW) \qquad \frac{\Pi \longrightarrow C \mid \Pi \longrightarrow C \mid \mathcal{H}}{\Pi \longrightarrow C \mid \mathcal{H}} \; (EC)$$

We can disregard (EW) by taking as axioms all hypersequent that contain an **LI′**-axiom as component.

The logical rules of the hypersequent **HLI′** for intuitionistic logic, are essentially the same as in **LI′**. The only difference is the presence of a side hypersequent $\mathcal{H}$, representing a (possibly empty) hypersequent. For instance, the hypersequent version of the **LI′**-rule $(\supset, l)$ is

$$\frac{A \supset B, \Pi \longrightarrow A \mid \mathcal{H} \quad B, A \supset B, \Pi \longrightarrow C \mid \mathcal{H}}{A \supset B, \Pi \longrightarrow C \mid \mathcal{H}} \; (\supset, l)$$

The hypersequent framework allows one to define analytic calculi for several important intermediate logics. These include Gödel-Dummett logic $\mathbf{G}_\infty$ (also called $\mathbf{LC}$) [6, 11], finite-valued Gödel logics $\mathbf{G}_n$ [11], and the logic $\mathbf{LQ}$ of weak excluded middle [12], also called Jankov logic in reference to [12]. Adequate calculi are obtained by adding just one structural rule, respectively, to the basic hypersequent calculus $\mathbf{HLI}'$, defined above.

- The hypersequent calculus $\mathbf{HLC}'$ for $\mathbf{G}_\infty$ is obtained from $\mathbf{HLI}'$ by adding the following rule, a version of which has already been defined in [3]:

$$\frac{\Pi_1, \Pi_2 \longrightarrow C_1 \mid \mathcal{H} \qquad \Pi_1, \Pi_2 \longrightarrow C_2 \mid \mathcal{H}}{\Pi_1 \longrightarrow C_1 \mid \Pi_2 \longrightarrow C_2 \mid \mathcal{H}} \ (com)$$

- The hypersequent calculi $\mathbf{HG}'_{k+1}$ for $\mathbf{G}_{k+1}$, for all $k \geq 1$, are obtained by adding to $\mathbf{HLI}'$ the following rules, respectively, which (essentially) were defined in [4]:

$$\frac{\mathcal{H} \mid \Gamma_1, \Gamma_2 \longrightarrow A_1 \quad \mathcal{H} \mid \Gamma_2, \Gamma_3 \longrightarrow A_2 \quad \ldots \quad \mathcal{H} \mid \Gamma_k, \Gamma_{k+1} \longrightarrow A_k}{\mathcal{H} \mid \Gamma_1 \longrightarrow A_1 \mid \ldots \mid \Gamma_k \longrightarrow A_k \mid \Gamma_{k+1} \longrightarrow \bot} \ (G_{k+1})$$

Note that $\mathbf{G}_2$ is nothing but classical logic $\mathbf{Cl}$.

- The hypersequent calculus $\mathbf{HLQ}'$ — a variant of which was defined in [5] — is obtained from $\mathbf{HLI}'$ by adding the following rule:

$$\frac{\mathcal{H} \mid \Gamma, \Pi \longrightarrow \bot}{\mathcal{H} \mid \Gamma \longrightarrow \bot \mid \Pi \longrightarrow \bot} \ (lq)$$

Theorem 3. $\mathbf{HLC}'$, $\mathbf{HG}'_n$, and $\mathbf{HLQ}'$ are sound and complete for the logics $\mathbf{G}_\infty$, $\mathbf{G}_n$, and $\mathbf{LQ}$, respectively.

Proof. Follows essentially from the soundness and cut-free completeness of the original calculi proved in [3], [4], and [5], respectively. □

4   Parallel Dialogue Games

To extend the close correspondence between strongly analytic sequent proofs and winning strategies for Lorenzen style dialogues to the hypersequent level we ask the following: what happens to the winning powers of $\mathbf{P}$ if we consider games where dialogues may proceed in parallel? Of course, this question can only be answered once we have defined more precisely what we mean by 'parallel dialogue games'. Many options are open for exploration. Here, we investigate parallel versions of $\mathbf{I}$-dialogue games, that share the following features:

1. The logical and structural rules of $\mathbf{I}$-games remain unchanged. Indeed, ordinary $\mathbf{I}$-game dialogues appear as sub-case of the more general parallel framework.
2. The proponent $\mathbf{P}$ may initiate additional $\mathbf{I}$-dialogues by 'cloning' the dialogue sequent of one of the parallel $\mathbf{I}$-dialogues, in which it is her turn to move.

3. To win a set of parallel dialogues the proponent **P** has to win at least one of the component dialogues.

These items reflect basic decisions concerning 'parallelization'. In particular, it should be clear that we want to separate the level of individual dialogue moves strictly from the initiation of new dialogues and the interaction between dialogues. Moreover, we like to consider **P** as the (sole) 'scheduler' of parallel dialogues. (These features should be contrasted with alternative concepts of dialogue games, like the ones in [1, 2].)

Before exploring rules for the synchronization of parallel dialogues, we will investigate parallel **I**-dialogues as specified by conditions 1-3, alone. We will see that this results in a game that does not change the winning powers of **P** over the (single) **I**-dialogue game.

Notation. A *parallel **I**-dialogue (P-**I**-dialogue)* is a sequence of nodes connected by moves. Each node ν is labelled by a *global state* $\Sigma(\nu)$. A global state is a non-empty finite set $\{\Pi_1 \vdash_{\iota 1} C_1, \ldots, \Pi_n \vdash_{\iota n} C_n\}$ of *indexed **I**-dialogue sequents*. Each index ιk uniquely names one of the n elements, called *component dialogue sequents* or simply *components*, of the global state. In each of the components it is either **P**'s or **O**'s turn to move. We will speak of a **P**-component or an **O**-component, accordingly. We distinguish *internal* and *external* moves.

Internal Moves combine single **I**-dialogue moves for some (possibly also none or all) of the components of the current global state. An internal move from global state $\{\Pi_1 \vdash_{\iota 1} C_1, \ldots, \Pi_n \vdash_{\iota n} C_n\}$ to global state $\{\Pi'_1 \vdash_{\iota 1} C'_1, \ldots, \Pi'_n \vdash_{\iota n} C'_n\}$ consists in a set of indexed **I**-dialogue moves $\{\iota i_1 : \mathsf{move}_1, \ldots, \iota i_m : \mathsf{move}_m\}$, such that the indices ιi_j, $1 \leq j \leq m$, are pairwise distinct elements of $\{\iota 1, \ldots, \iota n\}$. $\Pi'_k \vdash_{\iota k} C'_k$ denotes the component corresponding to the result of move_k applied to the component indexed by ιk if $k \in \{i_1, \ldots, i_m\}$; otherwise $\Pi_k = \Pi'_k$ and $C_k = C'_k$.

External Moves, in contrast to internal moves, may add or remove components of a global state, but do not change the local status (**P** or **O**) of existing components.

For now, we define only one external move, called fork.

> fork is a move by **P** and consists in duplicating one of the **P**-components of the current global state and assigning a new unique index to the added component.

Clearly, fork corresponds to item 2 in the above list of basic features of our parallel dialogue games. We call the new index generated by fork a *child* of the original index of the duplicated component.

The central condition in the definition of a P-**I**-dialogue is the following:

– each sequence of **I**-dialogue moves, that arises by picking at most one element $\iota i : \mathsf{move}_i$ from each of the consecutive internal moves, such that for all $1 \leq i < n$ either $\iota[i+1] = \iota i$ or $\iota[i+1]$ is child of ιi, forms an **I**-dialogue.

The *initial global state* $\Sigma(\nu)$ — that is the state labelling the root node ν of a P-I-dialogue — consists only of $\mathbf{O}$-components. We speak of a P-I-dialogue *for* $\Sigma(\nu)$.

Example 1. We exhibit a P-I-dialogue for $\neg a \lor a$, for some atom a. Remember that $\neg a$ abbreviates $a \supset \bot$:

$$
\begin{array}{ll}
\boxed{\mathbf{O}_1^\alpha} & \{\vdash_1 \neg a \lor a\} \\
\big\downarrow \{1:\ ?[\text{attack} \lor]\} & \\
\boxed{\mathbf{P}_1} & \{\vdash_1 \neg a \lor a\} \\
\big\downarrow \text{fork}:1 & \\
\boxed{\mathbf{P}_1 \| \mathbf{P}_2} & \{\vdash_1 \neg a \lor a,\ \vdash_2 \neg a \lor a\} \\
\big\downarrow \{1:\ \neg a\,[\text{defense} \lor \mathsf{l}],\ 2:\ a\,[\text{defense} \lor \mathsf{r}]\} & \\
\boxed{\mathbf{O}_1^\alpha \| \mathbf{O}_2^\alpha} & \{\vdash_1 \neg a,\ \vdash_2 a\} \\
\big\downarrow \{1:\ a\,[\text{attack} \supset],\ 2:\ a\,[\text{attack}\ atom]\} & \\
\boxed{\mathbf{P}_1 \| \mathbf{P}_2} & \{a \vdash_1 \neg a,\ \vdash_2 a\} \\
\big\downarrow \{1:\ \bot\,[\text{defense} \supset]\} & \\
\boxed{\mathbf{O}_1^\alpha \| \mathbf{P}_2} & \{a \vdash_1 \bot,\ \vdash_2 a\} \\
\big\downarrow \{1:\ ?[\text{attack}\ atom]\} & \\
\boxed{\mathbf{P}_1 \| \mathbf{P}_2} & \{a \vdash_1 \bot,\ \vdash_2 a\}
\end{array}
$$

Alternative P-I-dialogues for $\neg a \lor a$ are possible; but it is easy to check that none of these dialogues lead to a state where player $\mathbf{P}$ is winning. However, we will see below that a special synchronization rule, which is adequate for classical logic, allows to extend this dialogue to a winning strategy for $\neg a \lor a$.

The parallel version of the dialogue game may be viewed as a finite collection of state transition systems that are coordinated by referring to a global, discrete flow of time. At each time step some (possibly also none or all) of the component dialogues advance by one move. In a fork-move the component dialogues remain in their individual current states but a new dialogue, that copies the state of one of the old ones, is created.

Observe that our definition of a P-I-dialogue game allows for considerable flexibility in 'implementing' the involved parallelism. We may, for example, require that *all* component dialogues have to advance at each time step; or, alternatively, that at most k dialogues may advance simultaneously (even if there are more than k components.) The latter option might, e.g., be understood as modeling a dialogue game where $\mathbf{P}$ and $\mathbf{O}$, are not single persons, but rather consist of teams of k players each, and where each component dialogue is conducted by a different pair of opposite players. If, instead, we stick with a single proponent and a single opponent (i.e., $k = 1$) it seems natural to 'sequentialize' by dove-tailing the components of parallel moves. This motivates the following definition:

- A P-I-dialogue is called *sequentialized* if every internal move is a singleton set.

In the proof of Theorem 1 it was essential that full cycles of moves in a winning strategy — from a **P**-state to an **O**-state and back again to a **P**-state with an immediately responding move of **O** — correspond to a single inference step in **LI'**. However, even in sequentialized P-I-dialogues such cycles may be interrupted by internal moves referring to other components or by external moves. We therefore define a P-I-dialogue to be *normal* if the following condition holds. Every internal move that contains a **P**-move indexed with ιk

- is immediately followed by another internal move with a ιk-indexed element (**O**-reply),
- or, else, is the last move in the component dialogue referred to by ιk.

Remark 2. In combination with structural rule E (see Section 2), the conditions for normality can be understood as the stipulation that the proponent of a parallel dialogue game is the sole 'scheduler'. In other words — although **P** has no control over choices of **O** as long as they are immediate replies to her own previous move — **P** always determines at which dialogue component the game is to be continued.

Theorem 4. *Every finite P-I-dialogue δ for Σ can be translated into a sequentialized normal P-I-dialogue for Σ ending in the same global state as δ.*

Proof. Sequentialization is easily achieved by replacing every internal move $\{\iota 1 : \mathsf{move}_1, \ldots, \iota n : \mathsf{move}_n\}$ by a sequence $\{\iota 1 : \mathsf{move}_1\}, \ldots, \{\iota n : \mathsf{move}_n\}$ of internal moves. (Observe that, by the definition of an internal move, the indices ιi are pairwise different and therefore refer to different components of a global state.)

To obtain a normal dialogue, assume that δ is already sequentialized. Unless δ is already normal, it contains a subsequence of at least three moves $\{\iota 1 : \mathsf{move}_1\}, \{\iota 2 : \mathsf{move}_2\} \ldots, \{\iota n : \mathsf{move}_n\}$, where $\iota 1 = \iota n$, but $\iota i \neq \iota 1$ for all $2 \leq i < n$, and where move_n is an I-dialogue move by **O**, that directly reacts to move_1 by **P**. Clearly, reordering the sequence of moves into $\{\iota 1 : \mathsf{move}_1\}, \{\iota n : \mathsf{move}_n\}, \{\iota 2 : \mathsf{move}_2\}, \ldots, \{\iota [n-1] : \mathsf{move}_{n-1}\}$ results in the same final global state. Note that — disregarding proper notation — the moves $\{\iota 2 : \mathsf{move}_2\}$ to $\{\iota [n - 1] : \mathsf{move}_{n-1}\}$ may actually also be external moves without affecting the result. The claim follows by repeating this rearrangement of moves as often as possible.   □

Note [Important]. For the rest of the paper we will consider all parallel dialogues to be sequentialized and normal. Sequentialization implies that, just like for I-dialogues, we can speak of **P**-moves and **O**-moves of P-I-dialogues. (fork also is a **P**-move.) Since the set parentheses are redundant in denoting moves of sequentialized dialogues, we will omit them from now on.

A P-**I**-dialogue tree τ for Σ is a rooted, directed tree with global states as nodes and edges labelled by (internal or external) moves such that each branch of τ is a P-**I**-dialogue for Σ.

A finite P-**I**-dialogue tree is called a *winning strategy* if the following condition is satisfied for every node ν:

(p) either ν has a single successor node, the edge to which is labelled by a **P**-move,

(o) or for each **O**-move that is a permissible continuation of the dialogue at global state $\Sigma(\nu)$ there is an edge leaving ν that is labelled by this move,

(w) or ν is a leaf node and at least one of the components of $\Sigma(\nu)$ fulfills the winning conditions (for **P**).

Nodes satisfying (p) are called **P**-nodes; and nodes satisfying (o) are called **O**-nodes. Observe that, by normality, **P**-moves and **O**-moves strictly alternate in each branch, except for the initial segment (consisting of more than one consecutive **O**-nodes, in general) and external moves (which, in general, result in consecutive **P**-nodes).

Theorem 5. *Every winning strategy τ for sequentialized normal P-**I**-dialogues with initial global state $\{\Gamma \vdash C\}$ can be transformed into an* **HLI**′*-proof of $\Gamma \longrightarrow C$.*

Proof. We show by induction on the depth of τ that for every **P**-node of τ labelled with global state Σ, there is an **HLI**′-proof of the corresponding hypersequent $[\Sigma]$. Since the branches of τ are normal and sequential dialogues, edges of τ that correspond to *internal moves* are translated into corresponding inference steps using logical rules of **HLI**′, exactly as in the proof of Theorem 1. Moreover, if the winning condition is fulfilled for one of the component dialogues, then the global state clearly corresponds to an axiom of **HLI**′.

It remains to show that also fork translates into external contraction (EC): Suppose $\textcircled{$\nu$} \longrightarrow \textcircled{ν'}$ is an edge of τ corresponding to a last fork-move of a branch of τ. Then the global state at $\Sigma(\nu')$ is like $\Sigma(\nu)$ except for an additional dialogue sequent $\Gamma \vdash_{\iota i} A$, where the index ιi is not yet used at ν, but where, for some ιj, $\Gamma \vdash_{\iota j} A$ is an element of $\Sigma(\nu)$. Clearly, the required **HLI**′-proof of $[\Sigma(\nu)]$ is obtained from $\pi_{\nu'}$ by adding an appropriate application of (EC) as the last inference. $\qquad\square$

Again, we call cut-free proofs without applications of (internal or external) weakening or internal contraction *strongly analytic*.

Theorem 6. *Every strongly analytic* **HLI**′*-proof π of the hypersequent $\Gamma \longrightarrow C$ can be transformed into a winning strategy τ for P-**I**-dialogues for $\{\Gamma \longrightarrow C\}$.*

Proof. Since π is strongly analytic, there are no applications of internal structural rules. The logical rules of **HLI**′ translate into full **P–O–P**-cycles of internal moves, exactly as in the proof of Theorem 2. It remains to show that external weakenings correspond to fork-moves for P-$\mathbf{I}$-dialogues. It suffices to consider a sub-proof of π ending with the inference

$$\frac{\Pi \longrightarrow D \mid \Pi \longrightarrow D \mid \mathcal{H}}{\Pi \longrightarrow D \mid \mathcal{H}} \ (EC)$$

By induction hypothesis there exists a winning strategy τ' for $\{\Pi \vdash_{\iota 1} D\} \cup \{\Pi \vdash_{\iota 2} D\} \cup \langle \mathcal{H} \rangle$, which has to be extended to one for $\{\Pi \vdash_{\iota 1} D\} \cup \langle \mathcal{H} \rangle$, where $\langle \mathcal{H} \rangle$ denotes the set of dialogue sequents corresponding to the components of $\mathcal{H}$. This is easily achieved by inserting a new edge corresponding to an appropriate instance of the fork-move immediately after the initial (internal $\mathbf{O}$-)move of τ'.

$\square$

5   Synchronizing Dialogues

Synchronization between $\mathbf{I}$-dialogues is formalized as *merging* of two or more dialogues into one according to the following general principle: $\mathbf{P}$ selects (for merging) some $\mathbf{P}$-components from the global state. The picked components are then merged into a new dialogue in some straightforward way. For some synchronization rules, there are different possible ways to merge the components picked by $\mathbf{P}$. In those cases $\mathbf{O}$ may choose one of them.

In [10] the following (two-part) synchronization rule for Gödel-Dummett logic $\mathbf{G}_\infty$ was already discussed briefly:

---

lc– $\mathbf{P}$-*part*: $\mathbf{P}$ picks two (indices of) $\mathbf{P}$-components $\Pi_1 \vdash_{\iota 1} C_1$ and $\Pi_2 \vdash_{\iota 2} C_2$ from the current global state and thus indicates that $\Pi_1 \cup \Pi_2$ will be the granted formulas of the resulting merged dialogue sequent.

lc– $\mathbf{O}$-*part*: In response to this external $\mathbf{P}$-move, $\mathbf{O}$ chooses either C_1 or C_2 as the active formula of the merged component, which is indexed by $\iota 1$ or $\iota 2$, correspondingly.

Not only infinite valued Gödel logic can be characterized by an appropriate parallel dialogue game, but also each of the n-valued Gödel logics $\mathbf{G}_n$. Here is the appropriate synchronization rule, parameterized by n, where $n \geq 2$:

$\mathbf{g}_n$ – $\mathbf{P}$-*part*: $\mathbf{P}$ picks $n-1$ $\mathbf{P}$-components $\Pi_1 \vdash_{\iota 1} C_1, \dots \Pi_{n-1} \vdash_{\iota[n-1]} C_{n-1}$, and a $\mathbf{P}$-component of form $\Pi_n \vdash_{\iota n} \bot$ from the current global state for merging.

$\mathbf{g}_n$ – $\mathbf{O}$-*part*: $\mathbf{O}$ chooses one of the components $\Pi_1 \cup \Pi_2 \vdash_{\iota 1} C_1$, $\Pi_2 \cup \Pi_3 \vdash_{\iota 2} C_2, \dots$ or $\Pi_{n-1} \cup \Pi_n \vdash_{\iota[n-1]} C_{n-1}$ as the merged component, that replaces the components picked by $\mathbf{P}$.

Note that for the case of two truth values, i.e., for classical logic, no proper choice is left for $\mathbf{O}$; hence $\mathbf{g}_2$ can be stated simpler as follows:

---

cl ($= \mathbf{g}_2$): If the global state contains a $\mathbf{P}$-component of form $\Pi \vdash_\iota \bot$ then $\mathbf{P}$ may remove this component and add Π to the granted formulas of another $\mathbf{P}$-component of the global state.

In other words, if **P** detects that in one of the components she faces the task to defend *falsum*, then she may cancel the corresponding **I**-dialogue while transferring its currently granted formulas to another **P**-component of her choice.

Example 2. Rule cl allows **P** to continue the parallel dialogue for $\neg a \ \vee a$ in Example 1 as follows:

$$\boxed{\mathbf{P}_1 \| \mathbf{P}_2}\ \{a \vdash_1 \bot, \vdash_2 a\}$$

$$\downarrow \text{cl} : 1,2$$

$$\left(\ \mathbf{P}_1\ \right)\ \{a \vdash_1 a\}$$

P wins!

Replacing cl by the following subtle variant, allows to characterize Jankov logic **LQ**, which allows **P** to win every dialogue for a formula of form $\neg A \ \vee \neg\neg A$:

> lq: If the global state contains a **P**-component of form $\Pi \vdash_\iota \bot$ then **P** may remove this component and add Π to the granted formulas of another **P**-component of the global state, which also has $\bot$ as active formula.

We summarize the above synchronization rules and provide names to the resulting systems of parallel dialogue games in the following table.

Parallel dialogue games extending *P*-I-games:

(All dialogue sequents exhibited in the table are **P**-components)

System	Rule	Synchronization (external merging move)
P-**G**	lc	**P** wants to merge $\Pi_1 \vdash_{\iota 1} C_1$ and $\Pi_2 \vdash_{\iota 2} C_2$
		O chooses either $\Pi_1 \cup \Pi_2 \vdash_{\iota 1} C_1$ or $\Pi_1 \cup \Pi_2 \vdash_{\iota 2} C_2$
P-**G$_n$**	g$_n$	**P** wants to merge $\Pi_1 \vdash_{\iota 1} C_1$, and $\dots \Pi_{n-1} \vdash_{\iota[n-1]} C_{n-1}$, and $\Pi_n \vdash_{\iota n}$
		O chooses either $\Pi_1 \cup \Pi_2 \vdash_{\iota 1} C_1$, $\Pi_2 \cup \Pi_3 \vdash_{\iota 2} C_2$, $\dots$
		$\qquad\qquad\qquad\qquad\qquad$ or $\Pi_{n-1} \cup \Pi_n \vdash_{\iota[n-1]} C_{n-1}$
P-**Cl**	cl= g$_2$	**P** merges $\Pi \vdash_{\iota 1} \bot$ and $\Gamma \vdash_{\iota 2} C$ into $\Pi \cup \Gamma \vdash_{\iota 2} C$
P-**LQ**	lq	**P** merges $\Pi \vdash_{\iota 1} \bot$ and $\Gamma \vdash_{\iota 2} \bot$ into $\Pi \cup \Gamma \vdash_{\iota 2} \bot$

Let us call parallel dialogues that are defined exactly as *P*-I-dialogues, except for including one of the rules lc, g$_n$, or lq, *P*-**G**-, *P*-**G$_n$**-, and *P*-**LQ**-dialogues, respectively.

Theorem 7. *Every winning strategy τ for sequentialized normal P-**G**- (P-**G$_n$**-, or P-**LQ**-)dialogues with initial global state $\{\Gamma \vdash_1 A\}$ can be transformed into an* **HLC$'$**- *(**HG$'_n$**-, or **HLQ$'$**-)proof π of the corresponding hypersequent $\Gamma \longrightarrow A$, and vice versa.*

Proof. Given the proofs of Theorems 5 and 6, it remains to show that the synchronization rules lc, g$_n$, and lq correspond to the hypersequent rules (com), (g$_n$), and (lq), respectively. We present the case for lc/(com); the other cases are similar or simpler.

($\Rightarrow$). Suppose $v_0 \longrightarrow v$ is an edge of τ which corresponds to an instance of lc. The relevant part of τ looks as follows. We use F_p to denote A if F is of form $A \supset B$; otherwise F_p is empty:

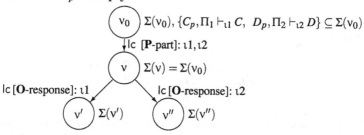

where $\Sigma(v') = \Sigma(v) - \{C_p, \Pi_1 \vdash_{\iota 1} C, \ D_p, \Pi_2 \vdash_{\iota 2} D\} \cup \{C_p, \Pi_1, D_p, \Pi_2 \vdash_{\iota 1} C\}$ and $\Sigma(v'') = \Sigma(v) - \{C_p, \Pi_1 \vdash_{\iota 1} C, \ D_p, \Pi_2 \vdash_{\iota 2} D\} \cup \{C_p, \Pi_1, D_p, \Pi_2 \vdash_{\iota 2} D\}$. By induction hypothesis there exist $\mathbf{HLC'}$-proofs $\pi_{v'}$ and $\pi_{v''}$ of the corresponding hypersequents $[\Sigma(v')]$ and $[\Sigma(v'')]$, respectively. Clearly, $\pi_{v'}$ and $\pi_{v''}$ can be joined by an application of (com) to obtain the required proof of $[\Sigma(v_0)]$.

($\Leftarrow$) Suppose π contains a subproof that ends in an application of the communication rule. (To make the proof more transparent we disregard side hypersequents.)

$$\frac{\quad \vdots \quad \qquad \vdots \quad}{\dfrac{\Pi_1, \Pi_2 \longrightarrow C \qquad \Pi_1, \Pi_2 \longrightarrow D}{\Pi_1 \longrightarrow C \mid \Pi_2 \longrightarrow D}} \text{ (com)}$$

By induction hypothesis there exist winning strategies τ_1 and τ_2 for $\{\Pi_1, \Pi_2 \vdash_{\iota 1} C\}$ and $\{\Pi_1, \Pi_2 \vdash_{\iota 2} D\}$, respectively, that are of following form:

$$\boxed{\mathbf{O}^\alpha}\ \{\Pi_1, \Pi_2 \vdash_{\iota 1} C\}$$
$$\downarrow 1: \text{attack } on\ C$$
$$\mu_1 : \boxed{\mathbf{P}}\ \{C_p, \Pi_1, \Pi_2 \vdash_{\iota 1} C\}$$
$$\tau_1'$$

$$\boxed{\mathbf{O}^\alpha}\ \{\Pi_1, \Pi_2 \vdash_{\iota 2} D\}$$
$$\downarrow 2: \text{attack } on\ D$$
$$\mu_2 : \boxed{\mathbf{P}}\ \{D_p, \Pi_1, \Pi_2 \vdash_{\iota 2} D\}$$
$$\tau_2'$$

A winning strategy for $\{\Pi_1 \vdash_{\iota 1} C, \ \Pi_2 \vdash_{\iota 2} D\}$ is obtained by attaching to

$$\boxed{\mathbf{O}_1^\alpha \| \mathbf{O}_2^\alpha}\ \{\Pi_1 \vdash_{\iota 1} C, \ \Pi_2 \vdash_{\iota 2} D\}$$
$$1: \text{attack on } C \downarrow$$
$$\boxed{\mathbf{P}_1 \| \mathbf{O}_2^\alpha}\ \{C_p, \Pi_1 \vdash_{\iota 1} C, \ \Pi_2 \vdash_{\iota 2} D\}$$
$$2: \text{attack on } D \downarrow$$
$$\mu : \boxed{\mathbf{P}_1 \| \mathbf{P}_2}\ \{C_p, \Pi_1 \vdash_{\iota 1} C, \ D_p, \Pi_2 \vdash_{\iota 2} D\}$$

the nodes for the two parts of external move lc illustrated in case ($\Rightarrow$), above, where node μ is identified with node v_0. Finally, the sub-strategies τ_1' and τ_2' are attached by identifying node v' with node μ_1 and node v'' with node μ_2. $\qquad \square$

Acknowledgement

I thank the referees and Agata Ciabattoni for useful comments.

References

[1] S. Abramsky, R. Jagadeesan: Games and Full Completeness for Multiplicative Linear Logic. *J. Symbolic Logic*, 59(2) (1994), 543–574.

[2] A. Blass: A Game Semantics for Linear Logic. *Annals of Pure and Applied Logic*, 56(1992), 183–220.

[3] A. Avron. Hypersequents, logical consequence and intermediate logics for concurrency. *Annals of Mathematics and Artificial Intelligence*, 4(1991), 225-248.

[4] A. Ciabattoni, M. Ferrari. Hypersequent calculi for some intermediate logics with bounded Kripke models. *Journal of Logic and Computation*, 2(11), pp. 283-294, 2001.

[5] A. Ciabattoni, D. M. Gabbay, N. Olivetti. Cut-free proof systems for logics of weak excluded middle. *Soft Computing*, 2(4), pp 147-156, 1998.

[6] M. Dummett. A propositional calculus with denumerable matrix. *J. Symbolic Logic*, 24(1959), 97–106.

[7] J. M. Dunn, R. K. Meyer. Algebraic completeness results for Dummett's *LC* and its extensions. *Z. Math. Logik Grundlagen Math.*, 17 (1971), 225–230.

[8] W. Felscher: Dialogues, Strategies, and Intuitionistic Provability. *Annals of Pure and Applied Logic*, 28(1985), 217–254.

[9] W. Felscher: Dialogues as Foundation for Intuitionistic Logic. In: D. Gabbay and F. Günther (eds.), *Handbook of Philosophical Logic, III*, Reidel, 1986, 341–372.

[10] C. G. Fermüller, A. Ciabattoni. From Intuitionistic Logic to Gödel-Dummett Logic via Parallel Dialogue Games. 33rd Intl. Symp. on Multiple-Valued Logic, Tokyo May 2003, IEEE Press, 188–193.

[11] K. Gödel: Zum intuitionistischen Aussagenkalkül. *Anz. Akad. Wiss. Wien*, 69(1932), 65-66.

[12] V. Jankov. The calculus of the weak "law of excluded middle". *Mathematics of the USSR* 8, pp. 648–658, 1968.

[13] E. C. W. Krabbe: Formal Systems of Dialogue Rules. *Synthese*, 63(1985), 295–328.

[14] E. C. W. Krabbe: Dialogue Sequents and Quick Proofs of Completeness. In: J.Ph. Hoepelman (ed.), *Representation and Reasoning*. Max Niemeyer Verlag, 1988, 135–140.

[15] P. Lorenzen: Logik und Agon. In: *Atti Congr. Internat. di Filosofia*, Vol. 4 (Sansoni, Firenze, 1960), 187–194.

[16] S. Rahman: *Über Dialoge, Protologische Kategorien und andere Seltenheiten*. Europäische Hochschulschriften, Peter Lang, 1993.

Simplification Rules
for Constrained Formula Tableaux

Martin Giese

Chalmers University of Technology
Department of Computing Science
S-41296 Gothenburg, Sweden
giese@ira.uka.de

Abstract. Several variants of a first-order simplification rule for non-normal form tableaux using syntactic constraints are presented. These can be used as a framework for porting methods like unit resolution or hyper tableaux to non-normal form free variable tableaux.

1 Introduction

Non-clausal form analytic tableaux have a number of advantages over the proof procedures for clausal form implemented in most successful automated theorem provers. For instance, when the logic is enhanced by modal operators, clausal form cannot be used without previously translating the problems into first-order. Another case is the integration of automated and interactive theorem proving [1, 8], where normal forms would be counter-intuitive. Unfortunately, standard non-normal form tableaux tend to be rather inefficient, as many of the refinements available to clausal procedures are hard to adapt. Typical cases in point are unit resolution, especially for propositional provers like the Davis-Putnam-Logemann-Loveland (DPLL) procedure [7], the β^c rules of the KE calculus [6], the application of 'result substitutions' in Stålmarcks Procedure [17], and hyper tableaux [2]. The common feature of these techniques is that they involve inferences between several formulae derived from the formula to be proved, either by using one formula to simplify another one, or—for hyper tableau—making tableau expansions depend on the presence of certain literals on a branch.

In [14], Massacci presents a simplification rule for propositional and modal tableau calculi. This rule is of the form

$$\frac{\psi}{\phi} \quad \underset{\rightsquigarrow}{simp} \quad \frac{\psi[\phi]}{\phi}$$

where $\psi[\phi]$ is the formula that results from first replacing all occurrences of ϕ in ψ by *true*, and if $\phi = \neg\phi'$, all occurrences of ϕ' by *false*, and then applying a set of boolean simplifications of the form

$$\neg true \rightarrow false, \quad \neg false \rightarrow true, \quad true \wedge \phi \rightarrow \phi, \quad false \wedge \phi \rightarrow false, \quad \text{etc.}$$

M. Cialdea Mayer and F. Pirri (Eds.): TABLEAUX 2003, LNAI 2796, pp. 65–80, 2003.

to eliminate all occurrences of truth constants. Massacci shows that proof procedures using this rule can subsume a number of other theorem proving techniques for propositional logic, e.g. the unit rule of DPLL [7], the β^c rules of KE [6], the regularity restriction, and hyper-tableaux [2]. This is done mainly by specifying the strategy of when and where to apply the *simp* rule.

While DPLL and hyper-tableaux are originally formulated for problems in clause normal form (CNF), the simplification rule is applicable to arbitrary predicate logic formulae, making it a good framework to generalize CNF techniques to the non-normal form case. Massacci gives variants of the simplification rule for various modal logics. In [13], he also gives a variant of the rule for first-order free-variable tableaux. Unfortunately, this rule does *not* in general subsume first-order versions of unit-resolution, hyper-tableaux etc., because it places strong restrictions on the instantiation of free variables.

This paper presents variants of the simplification rule for first order logic which overcome this limitation. The rules were first introduced in [9], but the proofs of most of the theorems were only sketched there.

2   Simplification with Global Instantiation

Consider a free-variable tableau branch with the formulae $p(X) \lor q(X)$ and $\neg p(a)$, where X is a free variable. If X were instantiated with a, the disjunction could be simplified to $q(a)$. Our task is to find a version of this ground simplification that works with free variables. The step from a ground version to a free variable version of a rule or a proof is usually referred to as *lifting*.

One possibility for lifting the simplification rule consists in applying a substitution to the whole proof, that unifies certain subformulae, so that a simplification becomes possible.

Such a rule would be formulated using the most general unifier (mgu) of the simplifying formula and some subformula of the simplified formula. A little care must be taken to prevent the instantiation of bound variables by such a unifier. We call (an occurrence of) a subformula ξ of ϕ *simplifiable*, if no variable occurring free in ξ is bound by ϕ. So $p(x)$ is simplifiable in $\exists y.(q(y) \land p(x))$, but not in $\exists x.(p(y) \land p(x))$. It is also simplifiable in $(\exists x.q(x)) \land p(x)$, because the quantifier does not bind the x occurring free in $p(x)$.

Using this notion, a simplification rule with global instantiation can be given:[1]

$$\begin{array}{cc} \psi & \xrightarrow{\;simp\;} \quad \mu(\psi)[\mu(\phi)] \\ \phi & \qquad\quad\;\; \mu(\phi) \end{array}$$

where μ is a mgu of ϕ and some simplifiable subformula of ψ, and μ is applied on all open branches.

---

[1] We use a non-standard notation for tableau rules: the formulae on the left are required to be on the branch and are *replaced* by the ones on the right. This notation has the advantage of making clear which formulae need to be retained after the rule application.

While this approach is sound, it relies on the application of a global instantiation for the free variables. The problem with such a rule is that it introduces a new backtracking point, because the applied unifier might not lead to a proof. Not only does this make the rule unsuitable for a backtracking free proof procedure. It is also problematic for a backtracking prover as efficiency will suffer if more backtracking points than necessary are introduced.

3   Lifting with Constrained Formulae

A universal technique for avoiding the global application of substitutions is to decorate the formulae on tableau branches with unification constraints. A unification constraint C is a conjunction of syntactic equalities between terms or formulae, written as

$$s_1 \equiv t_1 \,\&\, \ldots \,\&\, s_k \equiv t_k \quad .$$

We use the symbol $\equiv$ for syntactic equality in the constraint language to avoid confusion with the meta-level $=$. Let

$$\text{Sat}(C) = \{\sigma \mid \text{for all } i,\ \sigma(s_i) \text{ equals } \sigma(t_i) \text{ syntactically}\}$$

be the set of ground substitutions satisfying a constraint. A constraint is called satisfiable, if $\text{Sat}(C)$ is not empty, which means that there is a simultaneous unifier for the pairs $\{s_i, t_i\}$. A constraint C subsumes a constraint D, iff $\text{Sat}(D) \subseteq \text{Sat}(C)$. Two constraints C and D are equivalent, iff $\text{Sat}(C) = \text{Sat}(D)$.

A *constrained formula* is an ordered pair $\phi \ll C$ of a formula ϕ and a constraint C. The intuition is to consider the formula ϕ as present, only if the free variables are instantiated in a way that satisfies the constraint. The empty constraint, which is satisfied by all ground substitutions, is usually omitted. Instead of globally applying a mgu of two formulae ϕ and ψ to the proof, when a rule application requires some instantiation of free variables, we can annotate the formulae resulting from the rule application with a constraint $\phi \equiv \psi$, which is a local operation that does not lead to a backtracking choicepoint. For instance, simplification of $p(X) \vee q(X)$ with $\neg p(a)$ requires instantiation of X with a leading to *false* $\vee\, q(a) \ll X \equiv a$, which is rewritten to $q(a) \ll X \equiv a$.

Obviously, if formulae $\phi_i \ll C_i$ which already carry constraints are involved in a rule application, the conjunction $C_0 \,\&\, C_1 \ldots$ has to be passed on to the resulting formulae. This is referred to as constraint *propagation*. Constraints are propagated through rule applications, until a branch is closed. Closure between two literals $L \ll C$ and $\neg L' \ll C'$ is only allowed if the constraint $C \,\&\, C' \,\&\, L \equiv L'$ is satisfiable.

Using unification constraints, the simplification rule takes the form

$$\begin{array}{ccc} \psi \ll C & \overset{simp^{c0}}{\rightsquigarrow} & \psi \ll C \\ \phi \ll D & & \phi \ll D \\ & & \mu(\psi)[\mu(\phi)] \ll (C \,\&\, D \,\&\, \phi \equiv \xi) \end{array}$$

where ξ is a simplifiable subformula of ψ, μ is a mgu of ξ and ϕ, and $C \& D \& \phi \equiv \xi$ is satisfiable.

The $simp^{c0}$ rule keeps an unsimplified copy of $\psi \ll C$ on the branch. This will change in later versions. An immediate consequence of keeping both original formulae is that completeness follows trivially from the completeness of the calculus without simplification. We only need to ascertain soundness.

Theorem 1. *The tableau calculus with constrained formulae using the $simp^{c0}$ rule is sound, i.e. if a proof exists for a finite set of formulae, then that set is unsatisfiable.*

Proof. Let σ be a closing substitution for the tableau. This means that σ assigns a ground term to each free variable that occurs on the tableau, so that under σ, every branch contains a complementary pair of literals with constraints satisfied by σ. Consider the ground proof-tree obtained by replacing each formula ϕ on the tableau by $\sigma(\phi)$. In particular, this implies omitting any formulae with constraints that are not satisfied by σ. Tableau expansions for formulae with unsatisfied constraints are left out. For a β-expansion this means that only one of the branches needs to be kept, it doesn't matter which.

There is then a complementary pair on each branch of the resulting ground proof. Furthermore, as constraints can only be strengthened by rule applications, all proof steps needed to derive the complementary pair are still present in the reduced proof. Simplification steps are transformed into instances of the ground simplification rule

$$
\begin{array}{ccc}
\psi & & \psi \\
\phi & \rightsquigarrow & \phi \\
& & \psi[\phi]
\end{array}
$$

It remains to show that this ground version of the rule is sound. For this, it is sufficient to show that in any model where ψ and ϕ hold, $\psi[\phi]$ also holds, which immediately follows from the definition of $\psi[\phi]$. $\square$

It is a little misleading to call $simp^{c0}$ a simplification rule, because the original formula $\psi \ll C$ has to be retained for completeness. Indeed, one cannot simply delete the original formula, because there is no guarantee that the closing instantiation of the proof will be such that the simplification is possible.

There is however an important special case: if the 'new' part of the constraint $D \& \phi \equiv \xi$ subsumes the 'old' part C, the original formula $\psi \ll C$ may be discarded, because this means that the simplification step is valid in all ground instances of ψ allowed by the constraint C. Let $simp^{c1}$ be the rule obtained with this modification.

Theorem 2. *The $simp^{c1}$ rule is sound. It is also complete, in the sense that a branch that can be closed under some σ after applying a sequence R of expansion steps, can still be closed under σ by a modified sequence R' after an* *application of the $simp^{c1}$ rule. Moreover, there is such an R' that is at most as long as the original R.[2]*

Proof. Soundness may be shown as for $simp^{c0}$, see Theorem 1.

Completeness would be difficult to show by a Hintikka-style argument, because of the destructive nature of $simp^{c1}$. Apart from that, such a proof would not yield the statement about the proof sizes. We shall construct R' from R by a proof transformation, in which rule applications on (descendants of) a discarded original formula $\psi \ll C$ can either be applied to (descendants of) the simplified or simplifying formula, or be discarded altogether.

In case the $simp^{c1}$ application does not discard the original formula, we can simply take $R' = R$. Assume that the original formula ψ *is* discarded. In that case the new constraint C & D & $\phi \equiv \xi$ is equivalent to C. We also assume that the closing substitution σ satisfies C, because otherwise, the simplification step could not be useful to close the branch, and R' could be constructed by simply leaving it away. In particular, we thus have $\sigma \in Sat(D)$ and $\sigma(\phi) = \sigma(\xi)$. We now 'factor' the replacement of ψ by $\mu(\psi)[\mu(\phi)]$ into a sequence of simpler replacements and show for each of these how R is transformed.

Firstly, as σ satisfies C, $\sigma(\mu(\psi)) = \sigma(\psi)$. This implies that ψ can be replaced by $\mu(\psi)$ in the original branch, and the derivation R still closes it under σ.

After this, the calculation $\mu(\psi)[\mu(\phi)]$ from $\mu(\psi)$ consists in replacing occurrences of a sub-formula of $\mu(\psi)$ by *true* or *false*, and performing a number of boolean simplifications in the result.

With the formula $\phi \ll D$ on the branch, let us replace an occurrence of $\mu(\phi)$ in $\mu(\psi)$ by *true*. Let R' mimic all the proof steps in R except those which concern the sub-formula which has been replaced by *true*. If the replaced occurrence in ψ has positive polarity, i.e. it is in the scope of an even number of negations, then R produces the formula $\mu(\phi) \ll C$, while R' produces *true* $\ll C$. The proof steps of R on $\mu(\phi) \ll C$ are transformed to R' by applying them on the formula $\phi \ll D$, which is also on the branch. The applicability of tableau rules depends only on the top level junctors and quantifiers, so all rules that are applied on $\mu(\phi)$ can also be applied on ϕ. The constraint D on ϕ is also no problem, as the closing substitution σ satisfies D. If the occurrence is of negative polarity, R' produces the formula *false* $\ll C$, which immediately closes the branch under σ. The dual case, where $\neg\phi \ll D$ is on the branch, and a sub-formula is replaced by *false* is exactly symmetric.

It remains to show how that boolean simplification steps don't affect completeness. We will look only at some representative cases. Assume that an occurrence of $A \wedge true$ is replaced by A in ψ. Again, we let R' mimic the original proof steps until a rule must be applied on the simplified sub-formula. Depending on polarity, we now have only A instead of $A \wedge true$, resp. $\neg A$ instead of $\neg(A \wedge true)$. In the first case, an α rule application in R only leads to an additional literal *true*, which is useless for the proof, so all later proof steps in R can be applied

---

[2] Note that the $simp^{c1}$ application is not counted in R'. So the overall proof size may increase by 1.

in R'. In the second case, the β rule application in R leads to one branch with $\neg A$ and one with $\neg true$. We can use the proof steps of the former to finish R'.

We now consider the case where an occurrence of $A \wedge false$ is replaced by $false$ in ψ. Again depending on polarity, the proof steps of R now produce $false$ instead of $A \wedge false$, resp. $true$ instead of $\neg(A \wedge false)$. In the first case, the $false$ literal can be used to close the derivation R' immediately. In the second case, the β split in R produces one branch with $\neg A$ and one with $\neg false$. As the latter formula cannot be used to close the branch, we can take the rule applications of R on that branch to complete R'.

Other boolean simplification steps for quantifiers and negation can be handled similarly. □

The $simp^c$ rules enjoy an interesting relative termination property, which it shares with the α and β rules, namely that only a finite number of simplification steps can be performed without intervening γ-expansions, under certain side conditions. Call two constrained formulae $\phi \ll C$, $\psi \ll D$ variants, if C and D are equivalent and for all $\sigma \in \text{Sat}(C)$, $\sigma(\phi) = \sigma(\psi)$. E.g. $p(X) \ll X \equiv a$ and $p(a) \ll X \equiv a$ are variants.

Theorem 3. *Starting from a given tableau branch, only a finite number of α, β, and $simp^{c0}$ rule applications without intervening applications of the γ rule are possible, if $simp^{c0}$ is never applied twice to the same pair of constrained formulae, and any formula which is a variant of a formula already present on a branch is discarded. The same is true for the $simp^{c1}$ rule.*

Proof. A formula ϕ can only be simplified by setting one of its subterms to $true$ or $false$, and the resulting simplified formulae are all smaller than ϕ. So the number of distinct formulae that can be generated is finite. On the other hand, all constraints that could be generated are conjunctive combinations of existing constraints and syntactic equations between subformulae of formulae on a branch, so there can be only finitely many non-equivalent constraints. Accordingly, the number of non-variant constrained formulae must be finite. For $simp^{c1}$, formulae are occasionally discarded from a branch. This implies that even less rule applications are possible, so the same argument holds. □

As a practical consequence of this finiteness property, there is no need to interleave γ and $simp^c$ applications in a proof procedure to guarantee fairness. It is possible to apply all possible simplifications before considering an application of the γ rule.

4   Dis-unification Constraints

Although the $simp^{c1}$ rule permits the original formula $\psi \ll C$ to be deleted from the branch in some cases, it will usually have to be kept. This can lead to redundancies as exemplified by the following tableau branch for the set of formulae $\{p(a), q(a), \neg p(X) \lor \neg q(X) \lor r(X)\}$:

$$1 : p(a)$$
$$2 : q(a)$$
$$3 : \neg p(X) \lor \neg q(X) \lor r(X)$$
$$4 = simp(3,1) : \neg q(a) \lor r(a) \ll X \equiv a$$
$$5 = simp(3,2) : \neg p(a) \lor r(a) \ll X \equiv a$$

After generation of 4, formula 5 is redundant, because if X is actually instantiated with a as the constraint of 5 demands, formula 3 could have been discarded after generating 4. $q(a)$ only needs to be used to simplify 4, leading to $r(a) \ll X \equiv a$. In the presence of a large formula and many simplifying literals, a large number of such redundant formulae may be generated.

One way of overcoming this problem is to record instantiations under which a formula could have been discarded in the constraint. To do this, we have to require the constraint language to be closed under negation (denoted '!') as well as conjunction. The resulting constraint satisfiability problems are known as *dis-unification* problems, see e.g. [5], so I will talk of dis-unification or DU constraints.

A little care has to be taken with the semantics of DU constraints: Some DU constraints that are not satisfiable in the current signature might become satisfiable when the signature is extended. E.g., $!X \equiv a$, is not satisfiable in a signature consisting only of the constant symbol a, but it is satisfiable in any extended signature. In our context, satisfiability should be considered with respect to a possibly extended signature, because new skolem symbols might be introduced at a later point. In practice, it turns out that the satisfiability and subsumption checks actually get simpler with this definition. The same effect for term ordering constraints was noted in [15].

Using DU constraints, the simplification rule can be reformulated as follows:

$$\frac{\psi \ll C}{\phi \ll D} \quad \overset{simp^{c2}}{\leadsto} \quad \frac{\psi \ll C \,\&\, !(D \,\&\, \phi \equiv \xi)}{\phi \ll D}$$
$$\mu(\psi)[\mu(\phi)] \ll (C \,\&\, D \,\&\, \phi \equiv \xi)$$

where ξ is a simplifiable subformula of ψ, μ is a mgu of ξ and ϕ, and $C \,\& D \,\& \phi \equiv \xi$ is satisfiable.

This rule differs from $simp^{c0}$ in that the constraint of the original formulae ψ is changed by adding $!(D \,\& \phi \equiv \xi)$. What this means is that the formula is no longer available for simplification steps requiring an instantiation under which *this* simplification would have been possible.

We now allow formulae with unsatisfiable constraints to be discarded, as they cannot contribute to tableau closure anyway. One easily checks, that this makes it possible to discard ψ at least in all those cases, where $simp^{c1}$ allows it.

The example above now becomes

$$
\begin{array}{ll}
1 : p(a) & 1 : p(a) \\
2 : q(a) & 2 : q(a) \\
3 : \neg p(X) \vee \neg q(X) \vee r(X) \quad\leadsto\quad & 3 : \neg p(X) \vee \neg q(X) \vee r(X) \ll\, !\, X \equiv a \\
& 4 : \neg q(a) \vee r(a) \ll X \equiv a
\end{array}
$$

The constraint $!\, X \equiv a$ now prevents the simplification of 3 with 2. But we can perform a second simplification step by simplifying 4 with 2, which changes the constraint of 4 to $X \equiv a\ \&\, !\, X \equiv a$, which is unsatisfiable, so 4 can be discarded after adding the literal $r(a) \ll X \equiv a$.

The $simp^{c2}$ rule enjoys similar properties as $simp^{c1}$, as the following theorem shows.

Theorem 4. *The $simp^{c2}$ rule is sound. It is also complete in the sense of Theorem 2.*

Proof. Soundness follows from Theorem 1, as strictly less rule applications are possible than with $simp^{c0}$. Completeness is shown using the same technique as for Theorem 2. We take the addition of the DU constraint into account by considering three cases, depending on which of the constraints involved in the $simp^{c2}$ application are satisfied by the closing substitution σ. If σ does not satisfy C, any proof steps on $\psi \ll C$ can be left out anyway, as they do not contribute to the closure of the subtableau. If $\sigma \in \mathrm{Sat}(C)$, but $\sigma \notin \mathrm{Sat}(D\ \&\ \phi \equiv \xi)$, we can perform all extensions as in R, because σ satisfies the changed constraint $C\ \&\, !(D\ \&\ \phi \equiv \xi)$. Finally, if $\sigma \in \mathrm{Sat}(C)$ and $\sigma \in \mathrm{Sat}(D\ \&\ \phi \equiv \xi)$, we perform the proof transformation as in the proof of Theorem 2, considering the original formula to be deleted, because its constraint and the constraints of any formulae derived from it is not satisfied by σ. $\qquad\square$

The principal drawback of the $simp^{c2}$ version of our simplification rule is the high complexity of dis-unification. As a compromise, it is possible to keep unification (U) and dis-unification (DU) parts of constraints separate and to weaken the DU part of constraints if convenient. The unification part has to be left alone, as it is relevant for soundness. The DU part only serves to reduce the necessary proof search, so it may be thrown away without losing correctness.

One possible approach consists in restricting oneself to *conjunctive dis-unification constraints* [11], which are constraints of the form $C_0\ \&\, !\, C_1\ \&\, !\, C_2 \ldots$, where the C_i are conjunctive unification constraints as in Sec. 3. Here, C_0 is the U part and $!\, C_1\ \&\, !\, C_2 \ldots$ the DU part of the constraint. The DU part of the constraint of a formula is discarded before it is used to simplify another one, in order to maintain this form for all constraints. Satisfiability and subsumption (for possibly extended signatures) are fairly easy to check for these constraints. In fact, it is shown in [11], that the conjunctive DU-constraint is satisfiable in a possibly extended signature, exactly if C_0 is satisfiable and C_0 is not subsumed by any of the C_i.

5 Using Universal Variables

In practice, the simplification rules as outlined above tend to require a lot of instances of γ-formulae. E.g., given the formulae $\{p(a), p(b), p(c), \forall x.\neg p(x) \vee q(x)\}$, one can produce after one γ expansion the literals $q(a) \ll X \equiv a$, $q(b) \ll X \equiv b$, and $q(c) \ll X \equiv c$. But these literals have mutually contradictory constraints, so any further rule application or closure can involve at most one of these literals. One needs three instances of the γ formula to produce the compatible literals $q(a) \ll X_1 \equiv a$, $q(b) \ll X_2 \equiv b$, and $q(c) \ll X_3 \equiv c$. But with three instances, not only these three useful literals are deducible, but a total of nine q-literals coming from the simplification of each instance $\neg p(X_i) \vee q(X_i)$ with each of the three p-literals. As all of these will subsequently be used to simplify any q-subformula on the branch, this can quickly lead to a huge (though finite) number of rule applications.

One way to reduce the number of distinct instances of γ formulae is to use universal variables, see e.g. [4]. A free variable x is called *universal* with respect to a formula ϕ on a tableau branch, if $\forall x.\phi$ is a logical consequence of the formulae on a branch. All other free variables are called *rigid*. This property is of course undecidable. In practice, one uses simple sufficient criteria to detect universality of free variables, the most common one being to flag all free variables introduced in a γ extension as universal, and to preserve universality through all non-splitting rule applications. After a β rule application, those free variables which occur on more than one of the subformulae become rigid. The benefit of universal variables is that they may be instantiated independently for all formulae and may also be renamed as needed, whereas rigid variables have to be instantiated identically on all branches.

I shall write $[\bar{X}]\phi \ll C$ for a constrained formula with universal variables $\bar{X}$. Using universal variables, the following derivation is possible:

$$
\begin{array}{ll}
\begin{array}{l}
p(a) \\
p(b) \\
p(c) \\
\forall x.\neg p(x) \vee q(x) \\
[X]\neg p(X) \vee q(X)
\end{array}
&
\overset{3\times simp}{\rightsquigarrow}
\qquad
\begin{array}{l}
p(a) \\
p(b) \\
p(c) \\
\forall x.\neg p(x) \vee q(x) \\
[X]\neg p(X) \vee q(X) \ll\ !\,X \equiv a\ \&\ !\,X \equiv b\ \&\ !\,X \equiv c \\
[X]q(a) \ll X \equiv a \\
[X]q(b) \ll X \equiv b \\
[X]q(c) \ll X \equiv c
\end{array}
\end{array}
$$

The resulting literals are no longer incompatible, because X may be instantiated differently for each of them. It is of course possible to eliminate the universal variable and constraint altogether in these literals, but that is a technical optimization which is not strictly necessary.

Formally, in a simplification, all free variables in the result that were universal in one of the original formulae may be flagged as universal in the result [11]:

$$[\bar{X}]\psi \ll C \qquad simp^{c2u} \qquad [\bar{X}]\psi \ll C \,\&\, !(D \,\&\, \phi \equiv \xi)$$
$$[\bar{Y}]\phi \ll D \qquad \rightsquigarrow \qquad [\bar{Y}]\phi \ll D$$
$$[\bar{X} \cup \bar{Y}]\mu(\psi)[\mu(\phi)] \ll (C \,\&\, D \,\&\, \phi \equiv \xi)$$

where ξ is a simplifiable[3] subformula of ψ, μ is a mgu of ξ and ϕ, and $C \& D \& \phi \equiv \xi$ is satisfiable.

This rule is sound and complete for the free-variable tableau calculus with universal variables. Completeness can be shown by a combination of the technique used in [8], Sect. 7.4, for showing completeness of tableaux with universal variables, and the proof transformation technique of Theorems 2 and 4. By contrast, the termination property does not hold anymore, if universal variables are used. To apply the $simp^{c2u}$ rule, it is necessary in general to rename universal variables in the original formulae to make them disjoint. But this renaming destroys termination. Consider for instance the formulae $p(a)$ and $[X]\neg p(X) \vee p(f(X))$. With simplification and renaming of universal variables, it is possible to consecutively deduce

$$[X_1]p(f(a)) \ll X_1 \equiv a$$
$$[X_1, X_2]p(f(f(a))) \ll X_1 \equiv a \,\&\, X_2 \equiv f(a)$$
$$[X_1, X_2, X_3]p(f(f(f(a)))) \ll X_1 \equiv a \,\&\, X_2 \equiv f(a) \,\&\, X_3 \equiv f(f(a))$$
$$\text{etc.}$$

This means, that in general simplification and γ instantiation need to be interleaved to retain fairness. As the $simp^{c2u}$ rule *without* renaming obviously enjoys the finiteness property, one might alternatively interleave renaming and γ instantiation, but that would amount to ignoring universality for most of the time.

It is interesting to note that there are many problems, Schubert's 'Steamroller' [18] being a particularly prominent example, in which simplification with universal variables actually *does* terminate. This is true, in particular, when some *simplification strategy*, like the hyper strategy discussed in the next section is used, which does not apply arbitrary simplification steps. To handle such cases efficiently, it is advisable to equip a proof procedure with some sort of cycle detection that only interleaves simplifier applications with γ rules, if they threaten to lead to infinite simplification sequences. One possibility is to set a limit to the size of inferred formulae, which can be incrementally increased as the tableau is expanded. This would always allow rule applications which really simplify a formula in the sense of making it smaller.

---

[3] The 'simplifiable subformula' condition could be relaxed to permit, e.g. the simplification of $\exists y.p(y)$ with $[X].p(X)$, but this becomes rather technical, so we won't do it in this paper.

6 Simplification Strategies

Although we have identified cases in which we can discard the original formula in a simplification step, we should not forget that this is not possible in general. With the $simp^{c2}$ and $simp^{c2u}$ rules, we can at least strengthen the constraint of the original formula, but this does not change the fact that our so-called simplification rule actually makes branches larger in most cases. The reason of using the name simplification is the analogy to the ground and propositional simplification rules which our first-order version subsumes.

In order for the simplification rules to be useful in a prover, one needs a *simplification strategy*, that is a strategy that prescribes when to apply which kinds of simplification steps.

We claimed in the introduction that our simplification rules are capable of simulating first-order versions of various refinements, including hyper tableaux, and regularity. We have yet to show that this has been achieved. In this section, we shall describe a simplification strategy that implements a non-clausal analogue of hyper tableaux. The details of this strategy and corresponding proofs can be found in [11]. In that work, there is also a description of how the simplification rules may be used to introduce a first order, non-clausal version of regularity.

Hyper tableaux are defined for problems stated in clause normal form (CNF), see [12, 2]. For clause tableaux, it is customary not to include the clauses in the tableau itself. Instead, one only uses the literals which result from expanding the tableau with a clause. Hyper tableaux permit an expansion with a clause only if all new branches which receive negative[4] literals of the clause are immediately closed. All inner leaves are thus positive literals.

Alternatively, one can take the view of interpreting the clauses as tableau expansion rules themselves. In this view, a clause is 'fired' if there is a positive literal on a branch for every negative literal of the clause. The tableau is then extended by one new branch for each of the positive literals of the clause. One usually writes clauses as implications to support this view.

In the first-order case, one has to apply a substitution to unify the negative literals of the clause with corresponding positive literals on the branch. The way variables are handled differs between the various presentations of hyper tableaux. While [2] uses universal variables in branch literals where possible, that version of hyper tableaux does not use rigid variables. Instead, it uses 'purifying substitutions' which generate ground instances of clauses if necessary. This happens whenever a variable is shared between two positive literals of a clause without occurring in any of the negative literals. A version described in [12] uses rigid variables in such situations, using copies of clauses to avoid destructive instantiation. In [20], a variant with rigid variables and constraints is proposed, but constraints are attached to branches instead of formulae as is done in our calculi.

---

[4] We consider *positive* hyper tableaux here. It is possible to exchange the roles of positive and negative literals, which leads to negative hyper tableaux.

We can define a version of first-order hyper tableaux using constrained formulae. As usual, we use rigid variables when necessary, and constraints to capture necessary instantiations. Here is an example of this approach:

Clause Set:
$$p(a, b)$$
$$p(x, z) \rightarrow p(x, y) \vee p(y, z)$$
$$p(x, f(x)) \rightarrow q(x)$$

After putting the literal $p(a, b)$ on the branch using the first clause, we expand the tableau using the second clause, where $p(x, z)$ is instantiated with $p(a, b)$. As the two branches share the new variable Y, this has to be rigid. Subsequent expansion of the left branch with the third clause is possible only if Y is instantiated to $f(a)$. This restriction is captured in the constraint of the generated literal.

These rigid variable, constrained formula hyper tableaux can be emulated using our simplification rule with universal variables and a suitable simplification strategy. From now on, we shall consider normal analytic tableaux again. The set of clauses is given as a set of universally quantified disjunctions of literals. The following simplification strategy then emulates hyper tableaux:

Use simplification only to simplify any *leftmost* negative literals inside disjunctions with positive literals occurring on the branch. Use β-expansion only for disjunctions which contain no negative literals.

With this strategy, the emulation of a hyper tableau expansion will require exactly one intermediate simplification step for every negative literal in the clause/disjunction in question.

There is obviously not much merit in using this emulation of hyper tableaux in an actual implementation, if problems are given as clause sets. It would be simpler and more efficient to implement a rigid variable constrained formula hyper tableau calculus directly, instead of implementing non-clausal tableaux and simplification, and then restricting it to clauses. The interesting point about the emulation is that it suggests a way of generalizing hyper tableaux to non-clausal problems. We show how this works for negation normal form (NNF).

The idea is to look at *disjunctive paths* (d-paths) through formulae instead of clauses. The set of d-paths of a formula ϕ, denoted $dp(\phi)$ is defined by induction over the structure of ϕ as follows.

- If ϕ is a literal or a universally quantified formula, then $dp(\phi) := \{\langle \phi \rangle\}$.
- If $\phi = \alpha_1 \wedge \alpha_2$ is a conjunction, then $dp(\phi) := dp(\alpha_1) \cup dp(\alpha_2)$.
- If $\phi = \beta_1 \vee \beta_2$ is a disjunction, then $dp(\phi) := \{pq \mid p \in dp(\beta_1), q \in dp(\beta_2)\}$.

For instance, for the formula $\phi = (p \wedge \neg p) \vee (q \wedge \neg q)$, this definition gives:

$$dp(p \wedge \neg p) = \{\langle p \rangle, \langle \neg p \rangle\}$$
$$dp(q \wedge \neg q) = \{\langle q \rangle, \langle \neg q \rangle\}$$
$$dp(\phi) = \{\langle p, q \rangle, \langle p, \neg q \rangle, \langle \neg p, q \rangle, \langle \neg p, \neg q \rangle\}$$

As d-paths correspond closely to clauses, it is not surprising that the correct generalization of our simplification strategy may be formulated like this:

Use simplification only to simplify any *leftmost* negative literal of some d-path of a formula on the branch. Use β-expansion only for disjunctions which have at least one d-path that does not contain a negative literal.

For our formula ϕ, β-expansion will thus be applied because of the d-path $\langle p, q \rangle$. Let us call this strategy the *NNF hyper tableau strategy*.

Theorem 5. *The constrained variable tableau calculus with universal variables and the simp^{c2u} rule is complete if restricted according to the NNF hyper tableau strategy.*

Proof. See [11]. $\qquad\qquad\qquad\qquad\qquad\qquad\qquad\qquad\qquad\qquad\qquad\quad$ □

Sometimes, a simplification step permits discarding the original formula ψ. In such cases, a prover using the NNF hyper tableaux strategy has an advantage over usual clausal hyper tableaux, even if the problem is given in clausal form: it can simplify the clause set while proof search is under way. Essentially, unit resolution between a universal branch literal and a clause is performed. For instance, given a literal $[X]p(X)$ and a universal disjunction $[Y]\neg pY \vee rY$, the latter can be destructively simplified to $[Y]rY$ for that branch. This can not be done in normal hyper tableaux, as these do not keep separate clause sets per branch. Note that these separate clause sets do not imply higher memory consumption, because the representation of clauses can easily be shared between branches in an implementation.

The NNF hyper tableau strategy was implemented in the prototypical non-backtracking tableau prover PrInS [10, 11]. We are not going to list statistics here, as the power of hyper tableaux has previously been asserted, e.g. in [12]. We shall only state two results on problems found in the TPTP problem library. [19].

With the given strategy, PrInS is able to solve the *Steamroller* problem in the full first order formalization PUZ031+1 in less than 150 ms. This used to be considered a hard problem for a long time, although today, no state-of-the-art theorem prover has difficulties with it. In particular, hyper tableaux are a good way of quickly finding a proof. The interesting aspect of PrInS solving PUZ031+1 is that it does not use CNF transformation. To our knowledge, PrInS is the first *non clausal* theorem prover to have solved the Steamroller problem.

The problem known as *Andrews Challenge* is an example for the advantage of not needing a clause normal form. The full first order formalization SYN036+2 of that problem had a rating of 0.33 up till version 2.4.0 of the TPTP library, meaning that one third of the provers considered state-of-the-art were *not* able to solve it. The reason for this is that the clause normal form for this problem, if computed in the standard way, consists of 128 clauses of length 8. The full first order version in SYN036+2 is built from the equivalence junctor and quantifiers only, and is very small. The NNF PrInS works on is of course significantly larger, because $p \leftrightarrow q$ has to be translated to $(p \vee \neg q) \wedge (\neg p \vee q)$. But the NNF still helps in keeping large parts of the formula nested below the top level operators which are handled first. With the NNF hyper tableaux strategy, PrInS solves SYN036+2 in less than 200 ms. The prover performs 488 α, β and γ expansions and 322 simplification steps. By contrast, a simple version of PrInS without simplification needs 3938 rule applications and about 11 seconds for SYN036+2.

7   Related Work

The idea of using formulae on a branch to simplify other formulae independently been developed by Peltier [16]. The problem of dealing with the instantiation of rigid variables is solved differently however. While we use ordinary first order formulae and attach a syntactic constraint to them, Peltier intertwines constraints and formulae. The possibility of attaching different constraints to different parts of a larger formula might be an advantage of Peltier's approach, but we have not investigated this. Keeping formulae and constraints apart, as we do certainly makes the calculus easier to understand, and easier to reason about.

Recent work by Baumgartner and Tinelli [3] attempts to lift the unit propagation of the Davis Putnam procedure to first order logic. Their *model evolution* calculus does not use rigid variables however, and accordingly does not need constraints.

8   Conclusion

Several possibilities for a first-order version of the simplification rule of Massacci [13, 14] were presented. Instead of globally applying unifying substitutions, syntactic constraints are used. Besides soundness and completeness, a finiteness property is discussed, which is important for the design of fair proof procedures. Experimental results are quoted, which show that an efficient proof procedure can be implemented using non-clausal tableaux with a simplification rule. We refer the reader to [11] for a more precise discussion of some of the issues we could only mention briefly here.

Future work includes the refinement of cyclicity tests and development of more goal-oriented simplification strategies than the described hyper-tableaux variant.

Acknowledgements

I am grateful to Reiner Hähnle, Bernhard Beckert, Wolfgang Ahrendt, Magnus Björk and the anonymous referees for their many useful comments.

References

[1] Wolfgang Ahrendt, Bernhard Beckert, Reiner Hähnle, Wolfram Menzel, Wolfgang Reif, Gerhard Schellhorn, and Peter H. Schmitt. Integration of automated and interactive theorem proving. In W. Bibel and P. Schmitt, editors, *Automated Deduction: A Basis for Applications*, volume II, chapter 4, pages 97–116. Kluwer, 1998.

[2] Peter Baumgartner. Hyper Tableaux — The Next Generation. In Harrie de Swart, editor, *Proc. International Conference on Automated Reasoning with Analytic Tableaux and Related Methods, Oosterwijk, The Netherlands*, number 1397 in LNCS, pages 60–76. Springer-Verlag, 1998.

[3] Peter Baumgartner and Cesare Tinelli. The Model Evolution Calculus. Fachberichte Informatik 1–2003, Universität Koblenz-Landau, Universität Koblenz-Landau, Institut für Informatik, Rheinau 1, D-56075 Koblenz, 2003.

[4] Bernhard Beckert and Reiner Hähnle. Analytic tableaux. In W. Bibel and P. Schmitt, editors, *Automated Deduction: A Basis for Applications*, volume I, chapter 1, pages 11–41. Kluwer, 1998.

[5] Hubert Comon. Disunification: a survey. In Jean-Louis Lassez and Gordon Plotkin, editors, *Computational Logic: Essays in Honor of Alan Robinson*, chapter 9, pages 322–359. MIT Press, Cambridge, MA, USA, 1991.

[6] Marcello D'Agostino and Marco Mondadori. The taming of the cut. *Journal of Logic and Computation*, 4(3):285–319, 1994.

[7] M. Davis, G. Logemann, and D. Loveland. A machine program for theorem-proving. *Communications of the ACM*, 5:394–397, 1962.

[8] Martin Giese. Integriertes automatisches und interaktives Beweisen: Die Kalkülebene. Diploma Thesis, Fakultät für Informatik, Universität Karlsruhe, June 1998.

[9] Martin Giese. A first-order simplification rule with constraints. In Peter Baumgartner and Hantao Zhang, editors, *3rd Int. Workshop on First-Order Theorem Proving (FTP), St. Andrews, Scotland, TR 5/2000 Univ. of Koblenz*, pages 113–121, 2000.

[10] Martin Giese. Incremental closure of free variable tableaux. In Rajeev Goré, Alexander Leitsch, and Tobias Nipkow, editors, *Proc. Intl. Joint Conf. on Automated Reasoning, Siena, Italy*, volume 2083 of *LNCS*, pages 545–560. Springer-Verlag, 2001.

[11] Martin Giese. *Proof Search without Backtracking for Free Variable Tableaux*. PhD thesis, Fakultät für Informatik, Universität Karlsruhe, July 2002.

[12] Michael Kühn. Rigid hypertableaux. In *Proc. of KI '97: Advances in Artificial Intelligence*, volume 1303 of *LNAI*, pages 87–98. Springer-Verlag, 1997.

[13] Fabio Massacci. Simplification with renaming: A general proof technique for tableau and sequent-based provers. Technical Report 424, Computer Laboratory, Univ. of Cambridge (UK), 1997.

[14] Fabio Massacci. Simplification: A general constraint propagation technique for propositional and modal tableaux. In Harrie de Swart, editor, *Proc. International Conference on Automated Reasoning with Analytic Tableaux and Related Methods, Oosterwijk, The Netherlands*, volume 1397 of *LNCS*, pages 217–232. Springer-Verlag, 1998.

[15] Robert Nieuwenhuis and Albert Rubio. Theorem proving with ordering and equality constrained clauses. *Journal of Symbolic Computation*, 19(4):321–352, 1995.

[16] Nicolas Peltier. Pruning the search space and extracting more models in tableaux. *Logic Journal of the IGPL*, 7(2):217–251, 1999. Available online at http://www3.oup.co.uk/igpl/Volume_07/Issue_02/.

[17] Mary Sheeran and Gunnar Stålmarck. A tutorial on Stålmarck's proof procedure for propositional logic. In Ganesh Gopalakrishnan and Phillip J. Windley, editors, *Proc. Formal Methods in Computer-Aided Design, Second International Conference, FMCAD'98, Palo Alto/CA, USA*, volume 1522 of *LNCS*, pages 82–99, 1998.

[18] Mark E. Stickel. Schubert's steamroller problem: Formulations and solutions. *Journal of Automated Reasoning*, 2:89–101, 1986.

[19] Christian B. Suttner and Geoff Sutcliffe. The TPTP problem library — v2.1.0. Technical Report JCU-CS-97/8, Department of Computer Science, James Cook University, 15 December 1997.

[20] Jan van Eijck. Constrained hyper tableaux. In Laurent Fribourg, editor, *Computer Science Logic, 15th International Workshop, CSL 2001. 10th Annual Conference of the EACSL, Paris, France, September 10-13, 2001, Proceedings*, volume 2142 of *LNCS*, pages 232–246. Springer-Verlag, 2001.

Tableau Calculi for Preference-Based Conditional Logics

Laura Giordano[1], Valentina Gliozzi[2], Nicola Olivetti[2], and Camilla Schwind[3]

[1] Diparimento di Informatica, Università del Piemonte Orientale, Alessandria, Italy
[2] Diparimento di Informatica, Università di Torino, Torino, Italy
[3] MAP, CNRS, Marseille, France

Abstract. In this paper we develop labelled and uniform tableau methods for some fundamental system of propositional conditional logics. We consider the well-known system **CE** (that can be seen as a generalization of preferential nonmonotonic logic), and some related systems. Our tableau proof procedures are based on a possible-worlds structures endowed with a family of preference relations. The tableau procedure gives the first practical decision procedure for **CE**.

1 Introduction

Conditional logics have a long history. They have been studied first by Lewis [21, 22] to formalize a kind of hypothetical reasoning (if A were the case then B) that cannot be captured by classical logic with its material implication. More recently, they have been rediscovered in computer science and artificial intelligence for their potential application in a number of areas (see [5]), such as knowledge representation, non-monotonic reasoning, deductive databases, and natural language semantics. In knowledge representation, conditional logics have been used to reason about prototypical properties [13], to model database update [16], belief revision [3, 14], causal inference in action planning [24] and diagnosis [13]. Moreover conditional logics can provide an axiomatic foundation of non-monotonic reasoning [18], as it turns out that all forms of inferences studied in the framework of non-monotonic logics are particular cases of conditional axioms [6]. The conditional logic **CE** closely corresponds to preferential logic **P** as defined in [18]: the latter coincides with the first-degree fragment of **CE** rational non-monotonic logic corresponds to **CE+CV**, and so on.

In spite of their significance, very few proof systems have been proposed for conditional logics: we just mention [20, 17, 4, 2, 12, 8]. One possible reason of the underdevelopment of proof-methods for conditional logics is the lack of a universally accepted semantics for them. This is in sharp contrast, for instance, with modal or temporal logics which have a consolidated semantics based on a standard kind of Kripke structures.

Similarly to modal logics, the semantics of conditional logics can be defined in terms of possible world structures. The intuition is that a conditional $A \Rightarrow B$ is true in a world w just in case B is true in the A-worlds that are *most*

M. Cialdea Mayer and F. Pirri (Eds.): TABLEAUX 2003, LNAI 2796, pp. 81–101, 2003.

similar/most preferred/closest to w. We consider here conditional logics defined by a *preference-based* semantics. The idea is that every world w has associated a preference relation $\leq_w$ (in general a preorder relation) on the class of worlds. A conditional is true at w if B is true in all minimal A-worlds with respect to the relation $\leq_w$. This semantics characterizes the basic system **CE** we consider in this paper.

The preference semantics is related to the most popular semantics for conditional logics, namely the *sphere semantics* [21] and the *selection function semantics* [22]. The relation is investigated by Grahne [16] who has shown their equivalence for some systems. The preference-based semantics is also taken as the "official" semantics of conditional logics by Friedman and Halpern [10] with, however, one important difference from our setting. Our semantics, like preferential semantics of non-monotonic logics [18] embodies the *limit assumption* (corresponding to the *smoothness condition* in preferential semantics): every non-empty set of worlds has a minimal element with respect to each preorder relation $\leq_x$. This property is not assumed in [10].

The selection function semantics is the most general one and it is suitable for all systems. Proof systems have been developed for the minimal normal conditional logic CK and some extensions of it, based on the selection function semantics [23]. For other stronger logics, such as **CE** ãnd its main extensions) the selection function is not adequate to the purpose of developing a proof system, as there seem to be no way of expressing the specific semantic conditions on the selection function by analytic rules. On the other hand the sphere semantics seems unwarrantly complex to develop proof systems and it does not work for the basic system **CE**.

In this work we propose a labelled tableau caclulus for **CE** and some of its extension. As far as we know this is the first tableau calculus for this logic. *Explicit* or labelled proof systems have been provided for a wide range of modal and substructural logics and go back at least to Fitting's tableaux for modal logics [9]. A systematic development of labelled proof systems has been proposed in [25] and [11]. However the development of this kind of proof systems for conditional logics, with the exception of [2] and [1] is still unexplored.

Outline of the paper: in section 2 we introduce some background on conditional logic **CE**. In section 3, we present a tableau procedure for **CE**. In section 4, we prove its soundness, completeness, and we show how to make it terminating. In section 5, we present some extensions of it. Finally, in section 6 we discuss some related approaches.

2   Background

We consider a propositional language $\mathcal{L}$ over a set of propositional variables ATM. Formulas of $\mathcal{L}$ are built from propositional variables by means of the connectives $\neg, \wedge, \bot, \Rightarrow$; the last one $\Rightarrow$ is the conditional operator.

Definition 1 (Semantic of CE). *A model M has the form $(W, \{\leq_x\}_{x \in W}, I)$, where W is a non-empty set (of worlds), and I is a function $W \rightarrow Pow(ATM)$,* and $\{\leq_x\}_{x \in W}$ is a family of relation on W for each element $x \in W$. For $S \subseteq W$ we define

$$Min_x(S) = \{a \in S \mid \forall b \in S(b \leq_x a \ \rightarrow \ a \leq_x b)\}.$$

We say that $Min_x(S)$ is the set of $\leq_x$-minimal elements of S.

We assume the following facts, for every $x \in W$: (1) $\leq_x$ is a reflexive and transitive relation on W, (2) for every non-empty $S \subseteq W$, $Min_x(S) \neq \emptyset$. We define the truth conditions of formulas wrt. worlds in a model M, by the relation $M, x \models \phi$, as follows:

1. $M, x \models p$, for p atomic, if $p \in I(x)$,
2. $M, x \not\models \bot$,
3. $M, x \models \neg\phi$ if $M, x \not\models \phi$,
4. $M, x \models \phi \wedge \psi$ if $M, x \models \phi$ and $M, x \models \psi$.
5. $M, x \models \phi \Rightarrow \psi$ if for all $y \in Min_x(\phi)$, $M, y \models \psi$, where $Min_x(\phi)$ stands for $Min_x(\{y \in W \mid M, y \models \phi\})$.

We say that ϕ is valid in M if $M, x \models \phi$ for every $x \in W$. We say that ϕ is **CE**-valid if it is valid in every model.

We can also define the *strict relation* $y <_x z$ iff $y \leq_x z \wedge \neg(z \leq_x y)$. Observe that $Min_x(S) = \{a \in S \mid \neg\exists b \in S\ b <_x a\}$. The set of valid formulas according to the previous semantics is axiomatized by considering the following axioms and rules.

Definition 2 (Axiom System CE). *The system* **CE** *is defined by*

(TAUT) All classical tautologies and the Modus Ponens rule.
  (ID) $\phi \Rightarrow \phi$
  (CA) $(\phi \Rightarrow \chi) \wedge (\psi \Rightarrow \chi) \rightarrow (\phi \vee \psi \Rightarrow \chi)$
 (CSO) $(\phi \Rightarrow \psi) \wedge (\psi \Rightarrow \phi) \rightarrow (\phi \Rightarrow \chi) \leftrightarrow (\psi \Rightarrow \chi)$
  (A0) $(\neg\phi \Rightarrow \bot) \rightarrow \phi$
  (A1) $(\neg\phi \Rightarrow \bot) \rightarrow \neg(\neg\phi \Rightarrow \bot) \Rightarrow \bot$
  (A2) $\neg(\phi \Rightarrow \bot) \rightarrow (\phi \Rightarrow \bot) \Rightarrow \bot$
(RCEA) if $\vdash \phi \leftrightarrow \psi$ then $\vdash (\phi \Rightarrow \chi) \leftrightarrow (\psi \Rightarrow \chi)$
 (RCK) if $\vdash (\phi_1 \wedge \ldots \wedge \phi_n) \rightarrow \chi$ then $\vdash (\psi \Rightarrow \phi_1 \wedge \ldots \wedge \psi \Rightarrow \phi_n) \rightarrow (\psi \Rightarrow \chi)$

An alternative axiomatization of **CE** is given by replacing (CSO) with the following axiom: (AC) $(\phi \Rightarrow \psi) \wedge (\phi \Rightarrow \chi) \rightarrow (\phi \wedge \chi \Rightarrow \psi)$. Observe that another well known axiom (RT) $(\phi \wedge \chi \Rightarrow \psi) \wedge (\phi \Rightarrow \chi) \rightarrow (\phi \Rightarrow \psi)$ is derivable in **CE**.

All **CE** axioms (except (A0),(A1) and (A2)) correspond to well-known properties of nonmonotonic systems: (AC) is called *cumulativity*, (RT) is called *non-monotonic cut* [18]. More precisely, the first-degree fragment of this logic corresponds to preferential logic **P**. To understand axioms (A0), (A1) and (A2), define the "internal" modality operator $\Box A$ as $\neg A \Rightarrow \bot$, then (A0), (A1) and (A2) are nothing more than the usual S5-axioms and encode the assumption that each relation $\leq_x$ ranges on the same set of worlds[1]

---

[1] These axioms corresponds to *Uniformity Property* in [10, 16].

Theorem 1 (Soundness and Completeness, [10, 16]). *A formula ϕ is* **CE**-*valid iff it is derivable in* **CE**.

In section 5, we shall present some extensions of **CE**.

3   Tableau Calculus for CE

In order to develop a tableau calculus for **CE**, we extend the language by a kind of *hybrid* formulas. Given a model $M = (W, \{\leq_x\}_{x \in W}, I)$, we introduce pseudo-formulas of the form $\Box_x \phi$, for every formula ϕ and world $x \in W$, whose meaning is defined by stipulating:

$$M, y \models \Box_x \phi \text{ iff for every } z \in W \text{ if } z <_x y \text{ then } M, z \not\models \phi.$$

Observe that for any formula ϕ, we have:

$$y \in Min_x(\phi) \text{ iff } M, y \models \phi \wedge \Box_x \neg \phi.^2$$

The use of modal formulas to interpret the semantics of conditionals is not new. Boutilier [3] introduces a bi-modal logic to define some conditional logics strongly related to **CE** and **CV**. However, there are two important differences: first in his semantics there is only one modality, rather than a family of modal operators indexed on worlds. For this reason his logic is unable to represent nested conditionals properly ($\beta \Rightarrow \gamma$ implies $\alpha \Rightarrow (\beta \Rightarrow \gamma)$). As a second difference he does not accept the limit assumption, and thus has to change the truth definition of conditionals as Lewis [21] and Halpern and Friedman [10].

We use modal-pseudo formulas to give tableau rules for conditional logic **CE**. The tableau formulas are of the following kinds: (a) $x : \phi$, where ϕ is a formula or a pseudo-formula; (b) $x <_y z$.

A branch is a set of tableau formulas. Given a branch B, we denote by W_B the set of labels occurring in B. Figure 1 contains the tableau rules.

A branch is closed if it contains both $x : \phi$ and $x : \neg\phi$, or it contains $x : \bot$. A tableau is closed if every branch is closed. A non-closed branch is called open. We say that a formula ϕ is T-provable if the tableau for $x : \neg\phi$ is closed.

The rules do not need an explanation being a direct encoding of the semantics, with perhaps the exception of ($F\Box$) rule. This rule takes into account the minimality requisite imposed by the limit assumption: if $y \models \neg\Box_x \alpha$ then we conclude that there is a *minimal smaller* $z <_x y$ such that $z \models \neg\alpha$. This rule does the same job as the corresponding rule (due to Fitting [9]) for modal system G, the extension of K4 by Löb axiom $\Box(\Box\alpha \rightarrow \alpha) \rightarrow \Box\alpha$.

Example 1. We show that (CSO) $(\phi \Rightarrow \psi) \wedge (\psi \Rightarrow \phi) \rightarrow (\phi \Rightarrow \chi) \equiv (\psi \Rightarrow \chi)$ is valid. Figure 2 shows one half of the proof tree being symmetrical.

---

[2] We could go further and try to define the translation of a conditional as something like: $M, x \models \phi \Rightarrow \psi$ iff $M, x \models [\mathbf{U}]((\phi \wedge \Box_x \neg\phi) \rightarrow \psi)$, where $[\mathbf{U}]$ represents the universal modality. We are grateful to one of the referees for this suggestion. However, given the presence of the modality $\Box_x$, containing a world index, the above one cannot work as a pure syntactic translation.

$$(T\wedge) \quad \frac{x:\phi\wedge\psi}{\begin{array}{c}x:\phi\\x:\psi\end{array}} \qquad\qquad (F\wedge) \quad \frac{x:\neg(\phi\wedge\psi)}{x:\neg\phi \mid x:\neg\psi}$$

$$(T\vee) \quad \frac{x:\phi\vee\psi}{x:\phi \mid x:\psi} \qquad\qquad (F\vee) \quad \frac{x:\neg(\phi\vee\psi)}{\begin{array}{c}x:\neg\phi\\x:\neg\psi\end{array}}$$

$$(T\to) \quad \frac{x:\phi\to\psi}{x:\neg\phi \mid x:\psi} \qquad\qquad (F\to) \quad \frac{x:\neg(\phi\to\psi)}{\begin{array}{c}x:\phi\\x:\neg\psi\end{array}}$$

$$(NEG) \quad \frac{x:\neg\neg\psi}{x:\psi}$$

$$(T\Rightarrow)(*) \quad \frac{x:\phi\Rightarrow\psi}{y:\neg\phi \mid y:\neg\Box_x\neg\phi \mid y:\psi} \qquad (F\Rightarrow)(**) \quad \frac{x:\neg(\phi\Rightarrow\psi)}{\begin{array}{c}y:\phi\\y:\Box_x\neg\phi\\y:\neg\psi\end{array}}$$

$$(T\Box)(*) \quad \frac{\begin{array}{c}z:\Box_x\phi\\y<_x z\end{array}}{y:\phi} \qquad (F\Box)(**) \quad \frac{z:\neg\Box_x\phi}{\begin{array}{c}y<_x z\\y:\neg\phi\\y:\Box_x\phi\end{array}}$$

$$(Trans) \quad \frac{\begin{array}{c}y<_x z\\z<_x u\end{array}}{y<_x u}$$

(*) y is a label occurring in the branch.
(**) y is a new label not occurring in the branch.

Fig. 1. Tableau rules for **CE**

4   Soundness, Completeness, and Termination for CE

In order to prove soundness and completeness, we need to define the notion of satisfiability of a branch. Let $M = (W, \{\leq_x\}_{x\in W}, I)$ be a model. Given a branch B, we say that $f : W_B \to W$ is a **CE**-mapping if for every $y <_x z \in B$, $f(y) <_{f(x)} f(z)$ holds in M.

Definition 3. *Given a branch B of a tableau, a model M, and a (**CE**)-mapping f from W_B to W, we say that B is satisfiable under f in M if the following holds:*

1. if $x : \phi \in B$ then $M, f(x) \models \phi$,
2. if $x : \neg\phi \in B$ then $M, f(x) \not\models \phi$,

where ϕ is a conditional formula or a pseudo-formula.

B is *satisfiable* if it is satisfiable in some model M under some mapping f. A tableau is satisfiable if one of its branches is satisfiable.

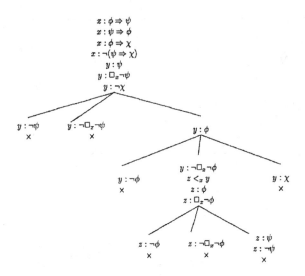

Fig. 2. Proof tree illustrating the proof of CSO

In order to show that the tableaux rules only prove correct formulas, we first show that they preserve satisfiability. The proof of Proposition 1 is in the Appendix.

Proposition 1. *Let T be a satisfiable tableau and let T' be obtained from T by applying one of the rules given for* **CE** *in the table above. Then T' is also satisfiable.*

Theorem 2 (Soundness). *If ϕ is provable then it is valid.*

Proof. Suppose that ϕ is not valid, then $\neg\phi$ is satisfiable in a model M, thus the tableau beginning with $x : \neg\phi$ is satisfiable. By the previous lemma any expansion of the tableau will contain a satisfiable branch B; B cannot be closed, for otherwise we would have $z : \psi, z : \neg\psi \in B$ (for some formula ψ) or $z : \bot \in B$, whence $M, f(z) \models \psi$ and $M, f(z) \models \neg\psi$, or $M, f(z) \models \bot$, and we get a contradiction.

4.1 Termination

The tableau calculus given in figure 1 is non-terminating due to repeated applications of the rules $(F \Rightarrow)$ and $(F\Box)$, which may generate infinite labels. For instance, the tableau construction for the formula $x : \top \Rightarrow \neg(\top \Rightarrow p)$ can produce an infinite branch (containing $x : \neg(\top \Rightarrow p)$, $y : \top$, $y : \Box_x \neg p$, $y : \neg p$, $y : \neg(\top \Rightarrow p)$, and so on). In the following we will show that, under the simple assumption that there are no redundant applications of the tableau rules, we can define a systematic procedure to build a tableau which terminates by introducing suitable restrictions on the applications of the $(F \Rightarrow)$ and $(F\Box)$ rules.

We first define redundant applications of a rule as follows.

Redundant application of a rule. Let B be a tableau branch, let R be a tableau rule applied to B, which produces the extensions of B, $B_1, \ldots, B_k$ (with $k \leq 3$), then the application of R is *redundant* if for some i, $B_i = B$ (as a set of labelled formulas). It is easy to see that a branch B can be extended to a closed branch just in case it can be extended to a closed branch without any redundant application of the rules.

In the following we assume that the tableau construction does not contain redundant applications of the rules. In particular, we assume the following restriction on the $(T \Rightarrow)$ rule:

(RIC): Apply $(T \Rightarrow)$ exactly once to each formula with each parameter from the branch.

The reason is the following: if we apply it twice on a branch with the same parameter, the second application is redundant. Additionally, we assume that a branch does not contain repetitions of labelled formulas: labelled formulas which are already on the branch are not added again when applying new rules.

In the completeness proof of the calculus, we shall prove that there cannot be infinite descending chains of labels related by the same order relation $<_x$ (such as $y_1 <_x y_0, \ldots, y_{i+1} <_x y_i, \ldots$). However, we cannot exclude that there is an infinite branching of the form $x_0 <_x z$, $x_1 <_x z$, $x_2 <_x z \ldots$ Similarly, we cannot exclude that there are infinite descending chains of the form $y_1 <_{x_1} y_0$, $\ldots, y_{i+1} <_{x_i} y_i, \ldots$, when $x_1, \ldots x_i, \ldots$ are not forced to be the same label. In general, the tableau construction could produce infinite sequences of labels by repeatedly generating new labels, with the $(F \square)$ and $(F \Rightarrow)$ rules, and then applying the $(T \Rightarrow)$ rule to the new labels. In order to avoid the generation of infinite branches, we introduce a systematic procedure to build a tableau and we put suitable restrictions on the applications of the $(F \Rightarrow)$ and $(F \square)$ rules. We show that the systematic procedure terminates, whence it does not lead to generate infinitely many labels. Moreover, we show that the completeness of the calculus is not lost if we adopt the systematic procedure with the mentioned restrictions.

Consider the following *systematic procedure* for constructing the tableau for a given formula $x : \alpha$. The procedure executes repeatedly two steps: **(step a)** applies the propositional rules and the $(T \Rightarrow)$, $(T \square)$ and $(Trans)$ rules as far as possible; **(step b)** applies the rules $(F \square)$ and $(F \Rightarrow)$ to the new formulas generated in the previous step. In other words, following the terminology of [15], in (step a) we apply the *static rules*, whereas in (step b) we apply the *dynamic rules*.

The fact that (step a) terminates after a finite number of rule applications is obvious, as no new world is generated in that step. After (step a) is terminated the branch is downward saturated except for the formulas of the form $w : \neg(\phi \Rightarrow \psi)$ and $w : \neg \square_x \phi$. Moreover, for each label z currently on the branch, the only way to add new formulas with that label is through the application of the $(T \Rightarrow)$ rule. In fact, if a new labelled conditional $y : \gamma \Rightarrow \delta$ appears on the branch, the $(T \Rightarrow)$ rule can be applied on that conditional with all the previous labels as parameters (including z). In (step b) we apply the rule $(F \Rightarrow)$ to all the formulas $y : \neg(\gamma \Rightarrow \delta)$ on the branch and rule $(F\Box)$ to all the formulas $y : \neg\Box_x\gamma$ on the branch, where y must be a label generated in the previous (step a).

To avoid that (step a) and (step b) in the systematic procedure repeat forever, by continuing to generate new worlds, we put the following restrictions on the application of the rules $(F \Rightarrow)$ and $(F\Box)$, which are essentially loop checking conditions.

(Restriction 1) the rule $(F \Rightarrow)$ can be applied to the formula $y : \neg(\gamma \Rightarrow \delta)$ on a branch only if there is no label z on the branch such that:

(1a) the branch contains the formula $z : \neg(\gamma \Rightarrow \delta)$ and the rule $(F \Rightarrow)$ has already been applied to that formula;

(1b) the positive conditional formulas labelled by y are a subset of the positive conditional formulas labelled by z.

The idea behind (Restriction 1) is that applying the rule $(F \Rightarrow)$ to $y : \neg(\gamma \Rightarrow \delta)$ cannot add anything more on the branch than what has been obtained by applying it to $z : \neg(\gamma \Rightarrow \delta)$, if all the positive conditional formulas that hold at y also hold at z. In such a case we can avoid to generate a new label y' from y.

(Restriction 2) the rule $(F\Box)$ can be applied to the formula $y : \neg\Box_x\gamma$ on a branch only if we cannot find two labels z and w on the branch such that:

(2a) the branch contains a formula $z : \neg\Box_w\gamma$ and the rule $(F\Box)$ has already been applied to that formula;

(2b) the positive conditional formulas labelled by x on the branch are a subset of the positive conditional formulas labelled by w;

(2c) for each (positive) modal pseudo-formula $y_1 : \Box_x\phi$ occurring on the branch, such that either $y_1 = y$ holds or $y <_x y_1$ occurs on the branch, there is a (positive) modal pseudo-formula $z_1 : \Box_w\phi$ on the branch, such that either $z_1 = z$ holds or $z <_w z_1$ occurs on the branch.

The idea behind (Restriction 2) is that applying the rule $(F\Box)$ to $y : \neg\Box_x\gamma$ cannot add anything more on the branch than what has been obtained by applying it to $z : \neg\Box_w\gamma$, under the assumptions that: all the positive conditional formulas that hold at x also hold at w; and all the positive modal pseudo-formulas on the branch that would be applicable to the world y' generated from y, can also be applied to the world z' generated from z.

The systematic procedure with the two restrictions preserves completeness.

Lemma 1. *If a branch B closes, it still closes under (Restriction 1)*

Lemma 2. *If a branch B closes, it still closes under (Restriction 2)*

The proofs of the lemmas are in the Appendix.

Theorem 3. *The systematic procedure with (Restriction 1) and (Restriction 2) terminates.*

Proof. Let us suppose that a systematic attempt at proving a formula goes forever. Then, there must be an infinite branch S containing infinitely many different labels because it does not contain repetitions and only a finite number of formulas can appear in the tableau (by the subformula property).

The branch must either contain infinitely many applications of the rule $(F \Rightarrow)$ to a formula $\neg(\phi \Rightarrow \psi)$ or infinitely many applications of the rule $(F\Box)$ to a formula $\neg\Box_x\phi$ (or both).

For the first case, it is not possible to apply rule $(F \Rightarrow)$ infinitely many times to the same formula $\neg(\phi \Rightarrow \psi)$ without violating (Restriction 1). To see this, each time the rule is applied to $y : \neg(\phi \Rightarrow \psi)$, by (Restriction 1) for all possible $z : \neg(\phi \Rightarrow \psi)$ on the branch there is some positive conditional formulas labelled by y that is not labelled by z. This is not possible, since by the subformula property there is only a finite number of positive conditionals.

For the second case, it is not possible to apply the rule $(F\Box)$ infinitely many times to the same formula $\neg\Box_x\phi$ without violating (Restriction 2). In fact, each time the rule $(F\Box)$ is applied to $y : \neg\Box_x\phi$, by (Restriction 2) for all possible $z : \neg\Box_w\phi$ on the branch: either there is some positive conditional formula labelled by x that is not labelled by w (but this cannot occur an infinite number of times for the reasons above) or there is a (positive) modal pseudo-formula $y_1 : \Box_x\alpha$ occurring on the branch such that either $y_1 = y$ holds $y <_x y_1$ and there is no corresponding $z_1 : \Box_w\alpha$ on the branch, such that either $z_1 = z$ holds or $z <_w z_1$ occurs on the branch. This second condition says, in essence, that a formula $\Box_x\alpha$ must hold at y while it does not hold at the worlds z such that the $(F\Box)$ rule has already been applied to $z : \neg\Box_w\alpha$. As the number of formulas α which may occur in the tableau is finite (note that α cannot contain modalities), also this condition cannot be true for an infinite number of times.

4.2 Completeness

To prove completeness we restrict our attention to tableaux which can be generated starting from an input formula ψ.

We first show that no tableau may contain infinite descending chains of labels related by the same relation $<_x$, provided it does not contain an infinite number of labeled conditional formulas with the same label. This is of course true if tableaux start with a finite number of formulas. It is interesting to notice that this holds independently from the Restrictions 1 and 2 we have put to ensure termination.

Lemma 3. *Let B be a branch of a tableau containing only a finite number of positive conditional formulas $x : \phi_0 \Rightarrow \psi_0$, $x : \phi_1 \Rightarrow \psi_1$, $x : \phi_2 \Rightarrow \psi_2$, ..., $x : \phi_{n-1} \Rightarrow \psi_{n-1}$. Then B does not contain an infinite descending chain of labels $y_1 <_x y_0$, $y_2 <_x y_1$, ..., $y_{i+1} <_x y_i$, ...*

Proof. Let B contain a descending chain of labels $y_1 <_x y_0$, $y_2 <_x y_1$, ..., $y_{i+1} <_x y_i$. This chain comes from the successive application of $(T \Rightarrow)$ and $(F\square)$ to formulas $x : \phi_i \Rightarrow \psi_i$ for $0 \le i < n$. B then contains the following formulas for $(0 \le i < n)$: $y_i : \neg\square_x\neg\phi_i$, $y_{i+1} <_x y_i$, $y_{i+1} : \phi_i$, $y_{i+1} : \square_x\neg\phi_i$.

Here $(T \Rightarrow)$ has been applied to every formula $x : \phi_i \Rightarrow \psi_i$ once and with parameter y_i previously (and newly) generated by $(F\square)$ from $y_{i-1} : \neg\square_x\neg\phi_{i-1}$. The only way to make the chain longer is by applying $(T \Rightarrow)$ a second time to one of the positive conditional formulas labelled x on B. Let this formula be $x : \phi_k \Rightarrow \psi_k$ where $0 \le k < n$. According to the restriction (RIC), $k \ne i$. Then B contains further $y_{n+1} : \phi_k$ (together with $y_n : \neg\square_x\neg\phi_k$, $y_{n+1} <_x y_n$, $y_{n+1} : \square_x\neg\phi_k$).

By the transitivity rule, we get $y_{n+1} <_x y_{k+1}$. Moreover, B contains also $y_{k+1} : \square_x\neg\phi_k$, from which we obtain by $(T\square)$ $y_{n+1} : \neg\phi_k$ which closes the branch.

Definition 4. *We say that a branch B of a tableau is regular if whenever $z <_x y \in B$ we have (i) $y <_x z \notin B$, (ii) $y \ne z$. We say that a tableau is regular if every branch is regular.*

Lemma 4. *Let T be a tableau beginning with $x : \phi$. Any expansion of T is regular.*

Proof. Given a branch B we prove something stronger, namely that the claim holds for the transitive closure $CT(B)$ of B wrt. $<_x$. Proceeding by induction on the expansion of B, we show that, if the property holds for B, then it holds also for any B' obtained from B by the application of any rule. All cases, except $(F\square)$ are trivial as $CT(B') = CT(B)$. Suppose that B' is obtained by expanding B on $u : \neg\square_v\chi$, then $CT(B') = CT(B \cup \{w <_v u\}) = CT(B) \cup \{w <_v u\} \cup \{w <_v z \mid u <_v z \in CT(B)\}$, where w does not occur in B. We leave to the reader to check that for any $a <_v b \in CT(B')$ it must be $b <_v a \notin B'$ and $b \ne a$.

Given a conditional language $\mathcal{L}$, let B be a branch. The notion of saturated branch that will be defined below expresses that all tableaux rules which could be applied to the branch have been applied to it.

Definition 5. *A branch B is* saturated *if*

1. If $x : \phi \wedge \psi \in B$ then $x : \phi \in B$ and $x : \psi \in B$.
2. If $x : \neg(\phi \wedge \psi) \in B$ then either $x : \neg\phi \in B$ or $x : \neg\psi \in B$.
3. If $x : \neg\neg\phi \in B$ then $x : \phi \in B$.
4. If $x : (\phi \Rightarrow \psi) \in B$ then for any label $y \in W_B$, either $y : \neg\phi \in B$ or $y : \neg\square_x\neg\phi \in B$ or $y : \psi \in B$.
5. If $x : \neg(\phi \Rightarrow \psi) \in B$ then there is a label y such that $y : \phi \in B$ and $y : \neg\psi \in B$ and $y : \square_x\neg\phi \in B$.
6. If $y : \square_x\phi \in B$ and $z <_x y \in B$ then $z : \phi \in B$.
7. If $y : \neg\square_x\phi \in B$ then there is a label z such that $z <_x y \in B$ and $z : \neg\phi \in B$ and $z : \square_x\phi \in B$.

We show that a saturated, open, regular branch B is satisfiable. To this purpose, we define the canonical model for B, which will satisfy all formulas on the branch.

Let $M_C = \langle W, \{\sqsubseteq_x\}_{x \in W}, I \rangle$, where: 1) $W = W_B$; 2) For each $x \in W$, $y \sqsubseteq_x z$ iff $y <_x z \in B$ or $y = z$; 3) For each $x \in W$, $I(x) = \{p \mid p \in ATM$ and $w : p \in B\}$.

Lemma 5. *Let B be an open, saturated, regular, branch which, for every label $x \in W_B$, contains only a finite number of positive conditional formulas. Then M_C is a model.*

Proof. We show the following facts:

1. $\sqsubseteq_x$ is a reflexive and transitive relation on W. Obvious.
2. Let $z \sqsubset_x y$ iff $z \sqsubseteq y$ and not $y \sqsubseteq_x z$. Then we have: $z \sqsubset_x y$ iff $z <_x y \in B$. Suppose that $z \sqsubset_x y$. Then we have $z \sqsubseteq_x y$ and not $y \sqsubseteq_x z$. From the first inequality we get $z <_x y \in B$ or $y = z$ and from 'not $y \sqsubseteq_x z$', we get $y \neq z$, thus it must be $z <_x y \in B$. Conversely, let $z <_x y \in B$, we have $z \sqsubseteq_x y$. Suppose that also $y \sqsubseteq_x z$, then we would have either $z = y$ or $y <_x z \in B$, against the fact that B is regular.
3. Let $S \subseteq W$ and $S \neq \emptyset$, then $Min_x(S) \neq \emptyset$. Suppose $Min_x(S) = \emptyset$; let $y_1 \in S \neq \emptyset$. There must be an infinite descending chain of elements of S $y_{n+1} \sqsubset_x y_n \sqsubset_x \cdots \sqsubset_x y_1$. By the previous fact we have that the infinite decreasing sequence $y_{n+1} <_x y_n <_x \cdots <_x y_1 \in B$. By lemma 3 we have a contradiction.

We now show that B is satisfiable by M_C.

Lemma 6. *Let B be an open, saturated, regular, branch which, for every label $x \in W_B$, contains only a finite number of positive conditional formulas with label x. Then B is satisfiable.*

Proof. We consider the canonical model M_C for B. Then obviously, the identity $id(x) = x$ is a **(CE)**-mapping by the construction of M_C. We show that B is satisfiable by M_C under id. We show by induction over the formulas that M_C satisfies all formulas and pseudo-formulas in B.

- The case of atomic formulas and boolean combination of formulas is easy and left to the reader.
- Let $x : \phi \Rightarrow \psi \in B$ and $y \in W_B$. Then either (i) $y : \neg\phi \in B$ or (ii) $y : \neg\Box_x\neg\phi \in B$, or (iii) $y : \psi \in B$. If (i) by the induction hypothesis, we have $M_C, y \not\models \phi$, thus $y \notin Min_x(\phi)$. If (ii) there is a label z such that $z <_x y \in B$ and $z : \phi \in B$. By the previous lemma we have $z \sqsubset_x y$ and by the induction hypothesis $M_C, z \models \phi$, thus $y \notin Min_x(\phi)$. If (iii) by the induction hypothesis we get $M_C, y \models \psi$. We have shown that if $y \in Min_x(\phi)$ then $M_C, y \models \psi$.
- Let $x : \neg(\phi \Rightarrow \psi) \in B$. Then there is y such that $y : \phi \in B$ and $y : \Box_x\neg\phi \in B$ and $y : \neg\psi \in B$. Then $M_C, y \models \phi$ and $M_C, y \models \neg\psi$ by the induction hypothesis. We show that $y \in Min_x(\phi)$. Suppose that this is not the case; since $M_C, y \models \phi$, there is $z \sqsubset_x y$ and $M, z \models \phi$. By the previous lemma,

we have $z <_x y \in B$ and by saturation $z : \neg\phi \in B$; by induction hypothesis we would have $M_C, z \not\models \phi$, a contradiction. Therefore $y \in Min_x(\phi)$, which proves $M, y \models \neg(\phi \Rightarrow \psi)$.

- Let $y : \Box_x\phi \in B$ and $v \in W$ and $v \sqsubset_x y$. By the previous lemma $v <_x y \in B$. From this it follows $v : \phi \in B$, whence $M_C, v \models \phi$.
- Let $y : \neg\Box_x\phi \in B$. Then there is a label z such that $z <_x y \in B$ and $z : \neg\phi \in B$ and $z : \Box_x\phi \in B$. From this we obtain that there is $z \in W$ and $z \sqsubset_x y$ (by the previous lemma) and $M_C, z \models \neg\phi$, from which it follows that $M_C, y \models \neg\Box_x\phi$.

Theorem 4. *If a formula ϕ is **CE**-valid then ϕ is T-provable.*

Proof. Suppose that ϕ is not T-provable. Then the tableau starting with $x : \neg\phi$ contains some open, regular, saturated branch B. Since ϕ is finite, B contains only a finite number of positive conditionals with the same label. Thus we can construct the canonical model M_C. By lemma 6, $M_C, x \models \neg\phi$. Thus ϕ cannot be valid.

5   Extensions of CE

The tableau method introduced in the previous section can be extended in order to deal with extensions of **CE** defined by combinations of the following semantic properties (and corresponding axioms):

(CV) $y \leq_x z \vee z \leq_x y$ (connectedness)
   $(\phi \Rightarrow \psi) \wedge \neg(\phi \Rightarrow \neg\chi) \rightarrow (\phi \wedge \chi \Rightarrow \psi)$
(CS) $y \leq_x x \rightarrow y = x$
   $\phi \wedge \psi \rightarrow (\phi \Rightarrow \psi)$
(MP) $y \leq_x x \rightarrow x \leq_x y$
   $((\phi \Rightarrow \psi) \wedge \phi) \rightarrow \psi$
(CEM) $y \neq z \rightarrow y <_x z \vee z <_x y$
   $(\phi \Rightarrow \psi) \vee (\phi \Rightarrow \neg\psi)$.

Theorem 1 can be extended to show that each semantic property is captured by the corresponding axiom. Let S be any subset of the above conditions/axioms, by **CE+S**-validity we mean validity in the **CE**-models satisfying the additional conditions S.

Theorem 5 ([10, 16]). *A formula ϕ is **CE+S** valid iff it is derivable in **CE** plus axioms S.*

As an example, we consider the combinations: **CE+MP**,   **CE+CV**, **CE+MP+CV+CS**.

In the extended tableau systems we have new kinds of tableau formulas, namely: $y \leq_x z$ and $y = z$. Hence, we need to change the closure conditions accordingly. Thus, for the extensions of **CE**, we say that a branch closes if it contains one of the following combinations : (i) $x : \phi$ and $x : \neg\phi$, or $x :\bot$; (ii)

$x = y$, $x : \phi$, $y : \neg\phi$; (iii) $y \leq_x z \in B$ and $u <_v w \in B$ for $y = w$, $z = u$ and $x = v$. A tableau is closed if every branch is closed. The rules we need are the following. We write $\lhd_x$ to denote either $<_x$ or $\leq_x$

$$(GenTrans) \quad \frac{y \lhd_x z \quad z \lhd_x u}{y \lhd_x u} \qquad\qquad (Trans =) \quad \frac{y = z \quad z = u}{y = u}$$

$$(Symm =) \quad \frac{y = x}{x = y}$$

where if $y \lhd_x u$ is $y <_x u$ then either $y \lhd_x z$ is $y <_x z$ or $z \lhd_x u$ is $z <_x u$.

$$(Conn) \quad \frac{}{y <_x z | z \leq_x y} \qquad\qquad (MP) \quad \frac{y <_x x}{x : \bot}$$

$$(CS) \quad \frac{y \leq_x x}{x = y}$$

We obtain the systems above by adding the above rules to those of **CE** as follows

- **CE+MP**: (MP)
- **CE+CV**: (GenTrans), (Conn)
- **CE+MP+CV+CS**: (GenTrans), (Conn), (Trans=), (Symm=), (MP),(CS)

In order to show that the rules for the extensions of **CE** are complete, we need to extend the notion of saturated branch as follows.

Definition 6. *We say that a branch is*

- *saturated with respect to* **CE+MP** *if it satisfies Definition 5 and whenever $y <_x x \in B$, then also $x :\bot \in B$*
- *saturated with respect to* **CE+CV** *if it satisfies Definition 5 and: 1) whenever $y \lhd_x z \in B$ and $z \lhd_x u \in B$, then also $y \lhd_x u \in B$; 2) for all $x, y, z \in W$, either $y <_x z \in B$ or $z \leq_x y \in B$.*
- *saturated with respect to* **CE+MP+CV+CS** *if it is saturated w.r.t.* **CE+CV** *and to* **CE+MP**, *and whenever $y \leq_x x \in B$, also $y = x \in B$, and $x = y \in B$.*

We observe that none of the rules above introduces in the tableau new labels or new formulas that create new labels when decomposed. From these observations, it follows that

Lemma 7. *Non-existence of infinite descending chains (as stated by Lemma 3) also holds for the extensions of* **CE**.

By this fact, we can show the completeness of the systems.

Theorem 6. – *Tableaux system* **CE+MP** *is complete with respect to* **CE+MP** *models.*
- *Tableaux system* **CE+CV** *is complete with respect to* **CE+CV** *models.*
- *Tableau system* **CE+MP+CV+CS** *is complete with respect to* **CE+MP+ CV+CS** *models.*

As far as **CE+MP** is concerned the proof is straightforward. As far as **CE+CV** and **CE+MP+CV+CS** are concerned, the construction of the canonical model is a bit more tricky. In particular, to cope with = in **CE+MP+CS**, we define W as a set of equivalence classes (w.r.t. =) rather than single labels. Detailed proofs are in the Appendix.

6   Conclusions

In this work we have presented a labelled calculus for **CE** and some of its extensions. The proof methods are uniform in the sense that each specific semantic condition is captured by a tableau rule. Moreover, the proof method is based on the introduction of pseudo-formulas, that are modalities in a hybrid language indexed on worlds. We have been able to obtain a terminating procedure for **CE** by performing a loop-checking. In future research we shall try to extend this approach to the other extensions of **CE** as well. Another issue that deserves investigation is whether *transitivity* is necessary for **CE**. It seems that this rule is not necessary, but we do not have conclusive evidence for it[3].

We briefly remark on some related works. De Swart [8] and Gent [12] give sequent/tableaux calculi for conditional logics VC (= **CE+CV+MP+CS**) and VCS. The kind of systems they propose are based on the entrenchment connective $\leq$, from which the conditional operator can be defined. Their systems are analytic and comprise an infinite set of rules $\leq F(n,m)$, with a uniform pattern, to decompose each sequent with m negative and n positive entrenchment formulas.

Crocco and Farinas [4] present sequent calculi for some conditional logics including minmal CK, CEM, CO (= **CE** without CA) and others. Their calculi comprise two levels of sequents: principal sequents with $\vdash_P$ corresponds to the basic deduction relation, whereas auxiliary sequents with $\vdash_a$ corresponds to the conditional operator: thus the constituents of $\Gamma \vdash_P \Delta$ are sequents of the form $X \vdash_a Y$, where X, Y are sets of formulas.

Artosi, Governatori, and Rotolo [2] develop labelled tableau for the *first-degree* fragment (i.e. without nested conditionals) of conditional logic CO (they call it CU and it corresponds to cumulative non-monotonic logics). Formulas are labelled by path of worlds containing also variable worlds. Since they adopt a selection function semantics, they have to cope with the problem of equivalent antecedents: i.e. if $[A]^M = [A']^M$ then $f(A, w) = f(A', w)$. They use an efficient unification procedure to propagate positive conditionals, and the unification procedure takes care of checking the equivalence of antecedents. Their tableau system is not analytical, as it contains a cut-rule, called PB, which is not eliminable. Moreover it is not clear how to extend it to **CE** and stronger systems on the one hand, and to nested conditionals, on the other.

---

[3] It is easy to see that the transitivity rule can be replaced by incorporating $\Box$-propagation in the $(T\Box)$-rule (as in standard rules for transitive modal logics). However, this just would shift the problem on the need of the modified $(T\Box)$-rule.

Broda, Gabbay, Lamb, and Russo develop an elegant natural deduction system for Boutillier's conditional logic of normality and some variants of it [3], [1]. Their proof system uses labels following the methodology of *Labelled Deductive Systems* [1], where the objects involved in the proofs are structured configurations of formulas, worlds, and relations thereof. In this respect their approach is rather similar to ours. However, as we already observed, Boutilier's conditional logic has a simpler semantics defined in terms of standard modal logic without world-indexed relations or modalities (and thus it cannot handle nested and iterated conditionals). Moreover, it is not evident if one can extract a decision procedure for Boutilier's logic from their natural deduction system.

Lamarre [20] presents tableaux systems for conditional logics V(= **CE**+CV), VN, VC and VW(= **CE**+CV+MP). Lamarre's method is a consistency-checking procedure which tries to build a system of sphere falsifying the input formulas. The method makes use of a subroutine to compute the *core*, that is defined as the set of formulas characterizing the minimal sphere. The computation of the core needs in turn the consistency checking procedure. Thus there is a mutual recursive definition between the procedure for checking consistency and the one to compute the core.

Gronebner and Delgrande [17] have developed a tableau method for conditional logic VN which is based on the translation of this logic into modal logic S4.3.

In [23], it is presented a labelled sequent calculus minimal normal conditional logic **CK** and some extensions of it. The calculi are based on the selection function semantics for these logics. In case of **CK**, the calculus is used to provide a polynomial-space complexity bound for this logic.

Finally, complexity results for conditional logic in the neighborhood of **CE** and its extensions have been provided in [10] the results are obtained semantically, by arguing about the size of possible countermodels.

Acknowledgements

We are grateful to the anonymous referee for their helpful suggestions and interesting observations. We are also grateful to Matteo Baldoni for his precious help in preparing the layout of the paper.

References

[1] K. Broda, D. Gabbay, L.Lamb and A. Russo. Labelled Natural Deduction for Conditional Logic of Normality. *Logic Journal of the IGPL*, Vol 10(2), 2002, 123-163, Oxford University Press.

[2] A. Artosi, G. Governatori, and A. Rotolo. Labelled tableaux for non-monotonic reasoning: Cumulative consequence relations. *Journal of Logic and Computation*, 12, 2002.

[3] C. Boutilier, Conditional logics of normality: a modal approach. *Artificial Intelligence*, 68:87–154, 1994.

[4] G. Crocco and L. Fariñas del Cerro, *Structure, Consequence relation and Logic*, volume 4, pages 239–259. Oxford University Press, 1992.

[5] G. Crocco, L. Fariñas del Cerro, and A. Herzig, *Conditionals: From philosophy to computer science*, Oxford University Press, Studies in Logic and Computation, 1995.

[6] G. Crocco and P. Lamarre, *On the Connection between Non-Monotonic Inference Systems and Conditional Logics*, In B. Nebel and E. Sandewall, editors, *Principles of Knowledge Representation and Reasoning: Proceedings of the 3rd International Conference*, pages 565-571, 1992.

[7] J. P. Delgrande, A first-order conditional logic for prototypical properties. *Artificial Intelligence*, (33):105–130, 1987.

[8] H. C.M de Swart, A Gentzen-or Beth-type system, a practical decision procedure and a constructive completeness proof for the counterfactual logics VC and VCS, *Journal of Symbolic Logic*, 48:1-20, 1983.

[9] M. Fitting, *Proof methods for Modal and Intuitionistic Logic*, vol 169 of Synthese library, D. Reidel, Dordrecht, 1983.

[10] N. Friedman and J. Halpern, On the complexity of conditional logics. In *Principles of Knowledge Representation and Reasoning: Proceedings of the 4th International Conference, KR'94*, pages 202–213.

[11] D. M. Gabbay, *Labelled Deductive Systems* (vol I), Oxford Logic Guides, Oxford University Press, 1996.

[12] I. P. Gent, A sequent or tableaux-style system for Lewis's counterfactual logic VC. *Notre Dame j. of Formal Logic*, 33(3): 369-382, 1992.

[13] M. L. Ginsberg. Counterfactuals. *Artificial Intelligence*, 30(2):35–79, 1986.

[14] L. Giordano, V. Gliozzi, and N. Olivetti. Iterated Belief Revision and Conditional Logic. *Studia Logica*, special issue on Belief Revision, Vol.70, No. 1, pp.23-47, 2002.

[15] R. Goré.Tableau Methods for Modal and Temporal Logics Rajeev Goré in *Handbook of Tableau Methods*, M D'Agostino, D Gabbay, R Haehnle, J Posegga (Eds.) Kluwer Academic Publishers, pages 297-396, 1999.

[16] G. Grahne, Updates and Counterfactuals, *in* Journal of Logic and Computation, Vol 8 No.1:87-117, 1998.

[17] C. Groeneboer and James Delgrande, A general approach for determining the validity of commonsense assertions using conditional logics. *International Journal of Intelligent Systems*, (5):505–520, 1990.

[18] S. Kraus, D. Lehmann, and M. Magidor. Nonmonotonic reasoning, preferential models and cumulative logics. *Artificial Intelligence*, 44:167–202, 1990.

[19] P. Lamarre. Etude des raisonnements non-monotones: apports des logiques des conditionnels et des logiques modales. PhD thesis, université Paul Sabatier, Toulouse 1992.

[20] P. Lamarre. A tableaux prover for conditional logics. In *Principles of Knowledge Representation and Reasoning: Proceedings of the 4th International Conference, KR'94*, pages 572–580.

[21] D. Lewis, *Counterfactuals*. Basil Blackwell Ltd, 1973.

[22] D. Nute, *Topics in Conditional Logic*, Reidel, Dordrecht, 1980.

[23] N. Olivetti and C. Schwind, A sequent calculus and a complexity bound for minimal conditional logic. In *Proc. ICTCS01 - Italian Conference on Theoretical Computer Science*, vol LNCS 2202, pages 384-404, 2001.

[24] C. B. Schwind, Causality in Action Theories. Electronic Articles in Computer and Information Science, Vol. 3 (1999), section A, pp. 27-50.

[25] L. Viganò, *Labelled Non-classical Logics*. Kluwer Academic Publishers, Dordrecht, 2000.

Appendix

Proposition 1 Let T be a satisfiable tableaux and let T' be obtained from T by applying one of the rules given for **CE** in the table above. Then T' is also satisfiable.

Proof. If T is a satisfiable tableau, then it contains at least one branch B which is satisfiable; that is there is a model $M = (W, \{\leq_x\}_{x \in W}, I)$ and a (**CE**)-mapping f such that 1 and 2 from definition 3 hold. We will show that for each tableau rule applied to T, the resulting tableau T' is still satisfiable.

(class) The case of classical connectives is easy and left to the reader.

$(T \Rightarrow)$ Let $x : \phi \Rightarrow \psi \in B$. Then T' is T with B replaced by three new branches $B_1 = B \cup \{y : \neg\phi\}$, $B_2 = B \cup \{y : \neg\Box_x\neg\phi\}$ and $B_3 = B \cup \{y : \psi\}$. If $y \notin W_B$, no B_i can be closed by the new element, since B does not contain any element labelled y. Consider the case where $y \in W_B$. We will show that at least one of these branches is satisfiable. Since B is satisfiable, there is a model $M = (W, \{\leq_x\}_{x \in W}, I)$ and a (**CE**)-mapping f such that $M, f(x) \models \phi \Rightarrow \psi$. Then for all $v \in Min_{\leq_{f(x)}}(\phi)$, $M, v \models \psi$. This is true iff $\forall v \in W, (M, v \models \neg\phi$ or $\exists v'(v' <_{f(x)} v$ and $v' \models \phi)$ or $M, v \models \psi$. Since $y \in W_B$, $f(y) \in W$, one of the following is true:

1. $M, f(y) \models \neg\phi$
2. $\exists v'(v' <_{f(x)} f(y)$ and $v' \models \phi$
3. $M, f(y) \models \psi)$

   In the first case, B_1 is satisfiable, in the second case, B_2 is satisfiable and in the third case, B_3 is satisfiable.

$(F \Rightarrow)$ Let be $x : \neg(\phi \Rightarrow \psi) \in B$. Then $M, f(x) \models \neg(\phi \Rightarrow \psi)$, i.e. there is $v \in Min_{\leq_{f(x)}}(\phi)$ such that $M, v \models \neg\psi$. If rule $(F \Rightarrow)$ is applied to $x : \neg(\phi \Rightarrow \psi)$ on T, the resulting tableau T' is T with B replaced by $B' = B \cup \{y : \phi, y : \Box_x\neg\phi, y : \neg\psi\}$, where $y \notin W_B$. Therefore, f is not defined for y. We define a new mapping $f' : W_{B'} \longrightarrow W$ by $f'(i) = f(i)$ if $i \neq y$ and $f'(y) = v$. f' is a (**CE**)-mapping from $W_{B'}$ to W: let be $w <_u w' \in B'$. Since y is new on B', $u \neq y$, $w \neq y$ and $w' \neq y$ and hence $u \in W_B$, $w \in W_B$ and $w' \in W_B$. Therefore, we have $f'(w) \leq_{f'(u)} f'(w')$. Consider the new formulas on B', y : ϕ, $y : \neg\psi$ and $\Box_x\neg\phi$. Then $M, f'(y) \models \phi$, because $f'(y) = v \in Min_{\leq_{f'(x)}}(\phi)$, i.e. $M, v \models \phi$ and $M, f'(y) \models \neg\psi$. For $y : \Box_x\neg\phi \in B'$, let be $w \in W$ and $w <_{f'(x)} f'(y)(= v)$. Since $v \in Min_{\leq_{f'(x)}}(\phi)$, $w \not\models \phi$.

$(T\Box)$ Let $y : \Box_x\phi \in B$ and $z <_x y \in B$. After application of rule $(T\Box)$, the resulting tableaux T' is T with B replaced by $B' = B \cup \{z : \phi\}$. Since B is satisfiable by M under f, we have that for every $v <_{f(x)} f(y)$, $M, v \models \phi$. Since $z <_x y \in B$, $f(z) <_{f(x)} f(y)$. Consequently, $M, f(z) \models \phi$, hence B' is also satisfiable.

$(F\Box)$ Let $y : \neg\Box_x\phi \in B$. Since B is satisfiable by M under f and $y : \neg\Box_x\phi \in B$ there is $w \in W$ and $w <_{f(x)} f(y)$ and $M, w \models \neg\phi$. Then $W(\neg\phi) \neq \emptyset$ and therefore by the limit assumption there is $v \in Min_{\leq_{f(x)}}(\neg\phi)$ and therefore $M, v \models \neg\phi$. After applying $(F\Box)$, we have $B' = B \cup \{z <_x y, z : \neg\phi, z : \Box_x\phi\}$, where $z \notin W_B$. We define a new mapping f' by $f'(u) = f(u)$ for $u \neq z$ and $f'(z) = v$. Since $f'(z) <_{f'(x)} f'(y)$, f' is a $(\mathbf{CE})$-mapping. Moreover, we have $M, f'(z) \models \neg\phi$. For $z : \Box_x\phi \in B'$ we have to show that whenever $u <_{f'(x)} f'(z)$, $M, u \models \phi$. Since $f'(z) = v \in Min_{(f'x)}(\neg\phi)$, for all $u <_{f'(x)} v$, $u \not\models \neg\phi$, i.e. $u \models \phi$.

Lemma 1 If a branch B closes, it still closes under (Restriction 1)

Proof. (sketch) Assume that the application of rule $(F \Rightarrow)$ to $z : \neg(A \Rightarrow B)$ has given rise to the new world z' and added to the branch the following formulas:

$z' : A$
$z' : \Box_z\neg A$ (i)
$z' : \neg B$

On the other hand, the application of rule $(F \Rightarrow)$ to $y : \neg(A \Rightarrow B)$ would introduce a new world y' and would add to the branch the following formulas:

$y' : A$
$y' : \Box_y\neg A$
$y' : \neg B$

We want to show that there are no inferences which can be done on y' or its descendants and cannot be done on z' or its descendants.

Both z' and y' are not related to other labels on the branch by the relation $<_w$ (for all w). The formulas on the branch which can contribute to introduce new formulas with label z' on the branch are the positive conditionals, to which the $(T \Rightarrow)$ rule can be applied with z' as parameter. All other formulas (propositional formulas, modal pseudo-formulas and the negative conditionals) which are currently on the branch with a label different from z' cannot give any contribution to introduce formulas with label z'. Among the positive conditionals on the branch we have to distinguish those with label z. Indeed, the application of the $(T \Rightarrow)$ rule to a conditional $z : \alpha \Rightarrow \beta$ with parameter z' introduces the formula $z' : \neg\Box_z\neg\alpha$ on the branch, which may interact with the formula (i) to close the branch. The same can be said for y': only positive conditionals on the branch can introduce new formulas with label y' and, in particular, the positive conditionals labeled with y play a special role, as they allow formulas of the form $y' : \neg\Box_y\neg\alpha$ to be added to the branch.

As all conditionals on the branch are equally applicable to z' and y', in order to be sure that y' cannot allow more inferences that z' we only have to require that for all those conditionals labeled with y (as $y : \alpha \Rightarrow \beta$) there is a corresponding conditional labeled with z ($z : \alpha \Rightarrow \beta$), which is given by condition (1b) of (Restriction 1).

In order to prove the lemma, we can use the observations above as follows. Given a closed branch B, we can obtain from it another branch B' by removing from B all the inferences $(F \Rightarrow)$ which do not satisfy (Restriction 1) and by replacing all the formulas labeled by y' (the world created by the application of $(F \Rightarrow)$) and all it descendants with the corresponding formulas labeled with z' and its descendants (if such formulas have not already been introduced on the branch).

Lemma 2 If a branch B closes, it still closes under (Restriction 2)

Proof. (sketch) Assume that the application of rule $(F\square)$ to $z : \neg\square_w A$ has given rise to the new world z' and added to the branch the following formulas:

$$z' <_w z$$
$$z' : \neg A$$
$$z' : \square_w A.$$

On the other hand, the application of rule $(F\square)$ to $y : \neg\square_x A$ would introduce a new world y' and would add to the branch the following formulas:

$$y' <_x y$$
$$y' : \neg A$$
$$y' : \square_x A \quad \text{(ii)}.$$

Observe that z' belongs to a descending $<_w$ chain and labels a $\square_w$ modality, while y' belongs to a descending $<_x$ chain and labels a $\square_x$ modality. The formulas on the branch that may add new formulas with label y' are the positive conditionals and the positive modalities. In particular a positive conditional labeled with x, like $x : \alpha \Rightarrow \beta$, by the $(T \Rightarrow)$ rule, can produce the addition to the branch of the formula $y' : \neg\square_x\neg\alpha$ (observe that this modality may interact with formula (ii), $y' : \square_x A$, that is also on the branch). On the other hand , if the branch contains a positive modality $y_1 : \square_x\phi$, such that either y_1 is y or $y <_x y_1$ is on the branch (that is, y_1 is on the same $<_x$ descending chain of labels as y'), by applying $(Trans)$ and $(T\square)$ we can get $y' : \phi$.

Condition (2b) guarantees that if there is a conditional formula $x : \alpha \Rightarrow \beta$ on the branch, which may introduce the formula $y' : \neg\square_x\neg\alpha$ by $(T \Rightarrow)$, then there is a corresponding formula $w : \alpha \Rightarrow \beta$, which can be used to add to the branch the formula $z' : \neg\square_w\neg\alpha$. All other labeled conditionals are equally applicable to y' and to z'.

Condition (2c) guarantees that when there is a positive modal formula $y_1 : \square_x\phi$ (occurring on the branch) that is in the $<_x$-descending chain as y' and which can introduce the formula ϕ at y', then there must be a similar labeled formula $z_1 : \square_w\phi$ on the branch, which can introduce the formula ϕ at z'

As for all inferences which can add formulas to y' the same inferences can be done on z', given a closed branch B, we can obtain from it another branch B' by removing from B all the inferences $(F\square)$ which do not satisfy (Restriction 2) and by replacing all the formulas labeled by y' (the world created by the application of $(F \Rightarrow)$) and all its descendants with the corresponding formulas labeled with z' and its descendants (if such formulas have not already been introduced on the branch).

Theorem 6

Proof. -Tableaux system **CE+MP** is complete with respect to **CE+MP** models.

Let B be an open branch, saturated w.r.t. to **CE+MP**. It is enough to consider the canonical model M_C used to prove the completeness of **CE**. M_C is a **CE**model. Furthermore, it satisfies (MP). Indeed, let $y \sqsubseteq_x x$. By construction, either $y <_x x \in B$ or $x = y$. But it cannot be $y <_x x \in B$, since by MP we would have $x :\bot \in B$. Hence, $x = y$, and by construction, $y \sqsubseteq_x x$. Reasoning by induction as we did for **CE**, we can show that B is satisfiable.

- Tableaux system **CE+CV** is complete with respect to **CE+CV** models.

Let B be a branch open and saturated w.r.t. **CE+CV**. We build a canonical model as follows. W and I are defined as in M_C, whereas $\sqsubseteq_x$ is defined as follows: $y \sqsubseteq_x z$ iff $y \leq_x z \in B$. We can easily show that $\sqsubseteq_x$ is reflexive and transitive. Obvious. The model satisfies the limit assumption, by Lemma 7 of section 5. Furthermore, the model satisfies (CV): by (Conn), either $y \leq_x z \in B$ or $z <_x y \in B$. Let $y \leq_x z \in B$. By definition of $\sqsubseteq_x$, $y \sqsubseteq_x z$. Let $z <_x y \in B$, by (GenTrans) we derive that $z \leq_x y \in B$ and hence that $z \sqsubseteq_x y$. Hence, the model is a **CE+CV** model. Moreover, we can show that $z \sqsubset_x y$ iff $z <_x y \in B$. Suppose that $z \sqsubset_x y$. We have $z \sqsubseteq_x y$ and not $y \sqsubseteq_x z$. From the first inequality we get $z \leq_x y \in B$ and from 'not $y \sqsubseteq_x z$', we get not $y \leq_x z \in B$. By (Conn), we have that $z <_x y \in B$. Conversely, let $z <_x y \in B$. By (GenTrans) we have that $z \leq_x y \in B$ and hence $z \sqsubseteq_x y$. On the other hand, we cannot have that $y \leq_x z$, since the branch is open. Therefore, $y \sqsubset_x z$.

Finally, we can reason by induction as we did in the completeness of **CE** to prove that B is satisfiable by M_C.

-Tableau system **CE+MP+CV+CS** is complete with respect to **CE+MP+CV+CS** models.

Let B be an open branch saturated w.r.t. **CE+MP+CV+CS**. We build the canonical model as follows. For all $x, y \in W_B$, let $x =_M y$ iff $x = y \in B$. Notice that $=_M$ is an equivalence relation. Indeed: it is symmetric, by (Symm=). It is reflexive: by (Conn) either $x \leq_x x \in B$ or $x <_x x \in B$. However, it cannot be that $x <_x x \in B$, since by (MP) we would have $x :\bot \in B$. Hence, for all x, $x \leq_x x \in B$, and by (CS), $x = x \in B$, hence $x =_M x$. It is transitive, by (Trans=).

We let $[x] = \{y : x = y \in B\}$ and $W = \{[w]/ =_M\}$, the set of all the equivalence classes of W_B with respect to $=_M$.

We define $I([x]) = \bigcup_{y \in [x]} \{p : p \in ATM$ and $y : p \in B\}$.

Furthermore, we let $[y] \sqsubseteq_{[x]} [z]$ iff $y \leq_x z \in B$. Notice that by $(Conn)$ and the closure condition (iii), we have that if $y \leq_x z \in B$, $x' =_M x$, $y' =_M y$ and $z' =_M z$, then also $y' \leq_{x'} z' \in B$ (otherwise by (Conn) it would be that $z' <'_x y'$, which would close the branch by condition (iii)). Hence, relation $\sqsubseteq_{[x]}$ does not depend on the choice of the representative element, and if $[x'] = [x]$, $[y'] = [y]$, $[z'] = [z]$, we have that $[y] \sqsubseteq_{[x]} [z]$ iff $[y'] \sqsubseteq_{[x']} [z']$.

It can be shown that the model is a **CE+MP+CV+CS** model: $\sqsubseteq_{[x]}$ is reflexive and transitive. Reflexive: as a consequence of (Conn) and (MP), we have that for all $x \in W_B$, $x \leq_x x \in B$. Transitive: it follows immediately from (GenTrans). Furthermore, as we did for the other logics, we can show that if $S \subseteq W$ and $S \neq \emptyset$, then $Min_x(S) \neq \emptyset$ and that it satisfies (MP) and (CV). The model is thus a **CE+MP+CV** model.

Last, we show that it satisfies (CS). Let $[y] \sqsubseteq_{[x]} [x]$. By definition, $y \leq_x x \in B$, and by (CS), $y = x \in B$. By construction of the model, $y \in [x]$ and since $[x]$ is an equivalence class, $[x] = [y]$.

Reasoning by induction on the complexity of the formulas, we show that B is satisfiable.

A General Tableau Method
for Propositional Interval Temporal Logics

Valentin Goranko[1], Angelo Montanari[2], and Guido Sciavicco[2]

[1] Department of Mathematics, Rand Afrikaans University, South Africa
vfg@rau.ac.za
[2] Dipartimento di Matematica e Informatica, University of Udine, Italy
{montana,sciavicc}@dimi.uniud.it

Abstract. Logics for time intervals provide a natural framework for representing and reasoning about timing properties in various areas of computer science. However, while various tableau methods have been developed for linear and branching time point-based temporal logics, not much work has been done on tableau methods for interval-based temporal logics. In this paper, we introduce a new, very expressive propositional interval temporal logic, called (Non-Strict) Branching CDT (BCDT$^+$) which extends most of the propositional interval temporal logics proposed in the literature. Then, we provide BCDT$^+$ with a generic tableau method which combines features of explicit tableau methods for modal logics with constraint label management and the classical tableau method for first-order logic, and we prove its soundness and completeness.

1 Introduction

Logics for time intervals provide a natural framework for representing and reasoning about timing properties in various areas of computer science. However, while various tableau methods have been developed for linear and branching time point-based temporal logics [17, 5, 15, 2], not much work has been done on tableau methods for interval-based temporal logics. One reason for this disparity is that operators of interval temporal logics are in many respects more difficult to deal with. As an example, there exist straightforward inductive definitions of the main operators of point-based temporal logics, such as the *future* operator and the *until* operator, while inductive definitions of basic interval modalities (consider, for instance, the one for the *chop* operator given in [1]) turn out to be more complex.

Various propositional and first-order interval temporal logics have been proposed in the literature. In this paper we focus our attention on propositional ones. There are two different natural semantics for interval logics, namely, a **strict** one, which excludes point-intervals, and a **non-strict** one, which includes them. The most studied propositional interval logics are Halpern and Shoham's Modal Logic of Time Intervals (HS) [9], Venema's CDT logic [16], Moszkowski's Propositional Interval Temporal Logic (PITL) [12], and Goranko, Montanari, and Sciavicco's family of Propositional Neighborhood Logics $\mathcal{PNL}$ [8].

M. Cialdea Mayer and F. Pirri (Eds.): TABLEAUX 2003, LNAI 2796, pp. 102–116, 2003.
© Springer-Verlag Berlin Heidelberg 2003

HS features four basic operators: $\langle B \rangle$ (*begin*) and $\langle E \rangle$ (*end*), and their transposes $\langle \overline{B} \rangle$ and $\langle \overline{E} \rangle$. Given a formula ϕ and an interval $[d_0, d_1]$, $\langle B \rangle \phi$ holds at $[d_0, d_1]$ if ϕ holds at $[d_0, d_2]$, for some $d_2 < d_1$, and $\langle E \rangle \phi$ holds at $[d_0, d_1]$ if ϕ holds at $[d_2, d_1]$, for some $d_2 > d_0$. A number of other temporal operators can be defined by means of the basic ones. In particular, it is possible to define the *strict after* operator $\langle A \rangle$ (and its transpose $\langle \overline{A} \rangle$) such that $\langle A \rangle \phi$ holds at $[d_0, d_1]$ if ϕ holds at $[d_1, d_2]$ for some $d_2 > d_1$; the *non-strict after* operator $\Diamond_r$ (and its transpose $\Diamond_l$) such that $\Diamond_r \phi$ holds at $[d_0, d_1]$ if ϕ holds at $[d_1, d_2]$ for some $d_2 \geq d_1$; and the *subinterval* operator $\langle D \rangle$ such that $\langle D \rangle \phi$ holds at a given interval $[d_0, d_1]$ if ϕ holds at a proper subinterval of $[d_0, d_1]$.

CDT has three binary operators C (*chop*), D, and T, which correspond to the ternary interval relations occurring when an extra point is added in one of the three possible distinct positions with respect to the two endpoints of the current interval (*before, between,* and *after*), plus a modal constant π which holds at a given interval if and only if it is a point-interval. PITL provides two modalities, namely, $\bigcirc$ (*next*) and C (the specialization of the *chop* operator for discrete structures). In PITL an interval is defined as a finite or infinite sequence of states. Given two formulas ϕ, ψ and an interval $s_0, \ldots, s_n$, $\bigcirc \phi$ holds over $s_0, \ldots, s_n$ if and only if ϕ holds over $s_1, \ldots, s_n$, while $\phi C \psi$ holds over $s_0, \ldots, s_n$ if and only if there exists i, with $0 \leq i \leq n$, such that ϕ holds over $s_0, \ldots, s_i$ and ψ holds over $s_i, \ldots, s_n$. Finally, propositional neighborhood logics in $\mathcal{PNL}$ feature two modalities for right and left interval neighborhoods, namely, $\langle A \rangle$ and $\langle \overline{A} \rangle$ in the *strict* semantics ($\mathcal{PNL}^-$ logics), and $\Diamond_r$ and $\Diamond_l$ in the *non-strict* semantics ($\mathcal{PNL}^+$ logics).

The main contributions of the paper are:

(i) Introduction of a new propositional interval logic, called (Non-Strict) Branching CDT (BCDT$^+$ for short), which features the same operators as CDT, but it is interpreted over partially ordered domains with linear intervals, and it is therefore expressive enough to include as subsystems or specializations all the above-described interval logics.

(ii) Development of an original sound and complete tableau method for BCDT$^+$, which combines features of tableau methods for modal logics with constraint label management and the classical tableau method for first-order logic. The proposed method can be adapted to variations and subsystems of BCDT$^+$, thus providing a general tableau method for propositional interval logics.

We conclude this introduction with a brief comparison between the tableaux method proposed here and other existing methods for point-based and interval-based modal and temporal logics (see [17, 5, 10]). As a preliminary remark, we note that most tableau methods for modal and temporal logics are terminating tableaux for *decidable* logics, and thus they yield decision procedures. Tableau methods for modal and (point-based) temporal logics can be classified as *explicit* or *implicit* (see [4]). Unlike implicit tableaux, explicit ones maintain the accessibility relation by means of some sort of external device. In implicit tableaux [6, 14], the accessibility relation is built-in into the rules. In particular, in linear and branching time point-based temporal logics the tableau represents a model of the satisfiable formulas (a time-line or a tree, respectively). The non-standard finite model property can then be exploited to show that the resulting tableau methods are actually decision procedures (they do not lead to infinite computations). Explicit tableau methods have been developed for several modal logics. They capture the accessibility relation by means of labeled formulas, and they provide suitable notions of closed branches and tableaux. Whenever the logic is decidable, its properties can be exploited to turn the tableau method into a decision procedure. In this respect, the tableau method for BCDT$^+$, while sharing basic features with explicit tableaux for modal logics, comes closer to the classical, possibly non-terminating tableau method for first-order logic, which only provides a semi-decision procedure for non-satisfiability. It also presents some similarities with the explicit tableau method developed for the *guarded fragment* of first-order logic (see [7]).

To the best of our knowledge, there exist very few other tableau methods for interval temporal logics (and duration calculi) in the literature. A tableau-based decision procedure for an extension of Local QPITL (a decidable fragment of PITL extended with quantification over propositional variables, which has been obtained by imposing a suitable *locality* constraint), which, besides the *chop* operator C and the modal constant π, has a projection operator *proj*, has been proposed by Kono [11] and later refined by Bowman and Thompson [1]. They introduce a normal form for the formulas of the resulting logic that allows them to exploit a classical tableau method, devoid of any mechanism for constraint label management. In [3], Chetcuti-Sperandio and Fariñas del Cerro focus on a decidable fragment of Duration Calculus (DC) which encompasses a proper subset of DC operators, namely, $\land$, $\lor$, and C. The tableau construction for the resulting logic combines application of the rules of classical tableaux with that of a suitable constraint resolution algorithm and it essentially depends on the assumption of *bounded* variability of the state variables. Finally, tableau systems for the propositional and first-order Linear Temporal Logic, which employ a mechanism for labeling formulas with temporal constraints somewhat similar to ours, have been developed respectively in [15] and [2]. The main differences between these tableau methods and ours are: (i) they are specifically designed to deal with integer time structures (i.e., linear and discrete) while ours is quite generic; (ii) the LTL is essentially point-based, and intervals only play a secondary role in it (viz., a formula it true on an interval if and only if it is true at every point in it), while in our systems intervals are primary semantic objects on which the truth definitions are entirely based; (iii) the closedness of a tableau in the cited papers is defined in terms of unsatisfiability of the associated set of temporal constraints, while in our system it is entirely syntactic.

2   Non-strict Branching CDT (BCDT$^+$)

In this section we give syntax and semantics of BCDT$^+$ and discuss its expressive power. To this end, we introduce some preliminary notions. Let $\mathbb{D}$ be a set of time points, called **domain**, and $<$ be a partial order on it. A (non-strict) **interval** on

$\mathbb{D}$ is an ordered pair $[d_0, d_1]$ such that $d_0, d_1 \in \mathbb{D}$, and $d_0 \leq d_1$. When $d_0 < d_1$ we say that the interval is **proper or strict**; when $d_0 = d_1$ it is a **point-interval**.

As in [9], we assume intervals to be *linear*, that is, for every interval $[d_0, d_1]$ and every pair of points d, d' belonging to it, namely, $d_0 \leq d \leq d_1$ and $d_0 \leq d' \leq d_1$, $d < d'$ or $d' < d$ or $d = d'$. Such an assumption keeps the temporal setting still very general, while making it fitting our intuition about the nature of time [9]. A pair $\langle \mathbb{D}, < \rangle$ is called an **interval structure**. An interval structure is: **linear** if every two points are comparable; **discrete** if every point with a successor/predecessor has an immediate successor/predecessor along every path starting from/ending at it; **dense** if for every pair of comparable (under $<$) points there exists another point in between; **unbounded** if and only if there are no points without successors (resp., predecessors); **Dedekind complete** if every non-empty and upward bounded set of points has a least upper bound. An element $d \in \mathbb{D}$ such that there are no elements $d' \in \mathbb{D}$ with $d < d'$ (resp., $d' < d$) is called **minimal** (resp., **maximal**) element.

Here we assume the non-strict semantics, but we add the modal constant π (as in [16]) that is satisfied by point-intervals only, and hence enables one to distinguish between point-intervals and proper ones.

2.1   BCDT$^+$ Syntax and Semantics

The language $\mathbf{L}^+$ for BCDT$^+$ consists of a set of propositional variables $\mathcal{AP}$, the logical connectives $\neg$ and $\wedge$, the modalities C, D, and T, and the modal constant π. The other logical connectives, as well as the logical constants $\top$ (*true*) and $\bot$ (*false*), can be defined in the usual way. BCDT$^+$ **formulas**, denoted by $\phi, \psi, \ldots$, are recursively defined as follows (where $p \in \mathcal{AP}$):

$$\phi = \pi \mid p \mid \neg\phi \mid \phi \wedge \psi \mid \phi C \psi \mid \phi D \psi \mid \phi T \psi.$$

The semantics of BCDT$^+$ is given in terms of **non-strict models**, i.e., based on non-strict interval structures, equipped with a **valuation function** for propositional variables. The valuation function is a mapping $V : \mathbb{I}(\mathbb{D})^+ \mapsto 2^{\mathcal{AP}}$, where $\mathbb{I}(\mathbb{D})^+$ is the set of all intervals in $\mathbb{D}$, such that, for any $p \in \mathcal{AP}$, p is **true** over $[d_0, d_1]$ if and only if $p \in V([d_0, d_1])$. **Truth** over an interval $[d_0, d_1]$ in a model $\mathbf{M}^+$ is defined by induction on the structure of formulas:

1. $\mathbf{M}^+, [d_0, d_1] \Vdash \pi$ iff $d_0 = d_1$;
2. $\mathbf{M}^+, [d_0, d_1] \Vdash p$ iff $p \in V([d_0, d_1])$, for all $p \in \mathcal{AP}$;
3. $\mathbf{M}^+, [d_0, d_1] \Vdash \neg\psi$ iff it is not the case that $\mathbf{M}^+, [d_0, d_1] \Vdash \psi$;
4. $\mathbf{M}^+, [d_0, d_1] \Vdash \phi \wedge \psi$ iff $\mathbf{M}^+, [d_0, d_1] \Vdash \phi$ and $\mathbf{M}^+, [d_0, d_1] \Vdash \psi$;
5. $\mathbf{M}^+, [d_0, d_1] \Vdash \phi C \psi$ iff there exists $d_2 \in \mathbb{D}$ such that $d_0 \leq d_2 \leq d_1$, $\mathbf{M}^+, [d_0, d_2] \Vdash \phi$, and $\mathbf{M}^+, [d_2, d_1] \Vdash \psi$;
6. $\mathbf{M}^+, [d_0, d_1] \Vdash \phi D \psi$ iff there exists $d_2 \in \mathbb{D}$ such that $d_2 \leq d_0$, $\mathbf{M}^+, [d_2, d_0] \Vdash \phi$, and $\mathbf{M}^+, [d_2, d_1] \Vdash \psi$;
7. $\mathbf{M}^+, [d_0, d_1] \Vdash \phi T \psi$ iff there exists $d_2 \in \mathbb{D}$ such that $d_1 \leq d_2$, $\mathbf{M}^+, [d_1, d_2] \Vdash \phi$, and $\mathbf{M}^+, [d_0, d_2] \Vdash \psi$.

Satisfiability and validity of BCDT$^+$ formulas are defined in the usual way.

2.2   Expressive Power of BCDT$^+$

Let us compare the expressive power of BCDT$^+$ with that of the above-described propositional interval logics. We say that a logic $\mathbf{L_1}$ is **at least as expressive** as a logic $\mathbf{L_2}$ if for every $\mathbf{L_2}$ formula there exists an equivalent $\mathbf{L_1}$ formula, and that $\mathbf{L_1}$ is (strictly) **more expressive** than $\mathbf{L_2}$ if and only if $\mathbf{L_1}$ is at least as expressive as $\mathbf{L_2}$, but not vice versa.

We first note that both CDT and $\mathcal{PNL}^+$ logics are interpreted over linear structures, and that the operators of $\mathcal{PNL}^+$ logics can be expressed in CDT by means of the formulas $\Diamond_r \phi := \phi T \top$ and $\Diamond_l \phi := \phi D \top$. Furthermore, it is well known that CDT does not semantically include HS in its full generality, since the latter allows the interval structure to be branching, while the former does not. On the other hand, HS is not more expressive than CDT, because it cannot express the *chop* operator (see [13]).

BCDT$^+$ generalizes Venema's CDT (and thus propositional neighborhood logics in $\mathcal{PNL}^+$) by allowing the interval structure to be non-linear, for as long as all intervals in it are linear (as in HS). Furthermore, it is strictly more expressive than HS and PITL. HS operators can be defined in BCDT$^+$ as follows: $\langle B \rangle \phi := \phi C \neg \pi$, $\langle \overline{B} \rangle \phi := \neg \pi T \phi$, $\langle E \rangle \phi := \neg \pi C \phi$, and $\langle \overline{E} \rangle \phi := \neg \pi D \phi$. Besides, the strict neighborhood operators $\langle A \rangle$ and $\langle \overline{A} \rangle$ can be defined in BCDT$^+$ by using π as follows: $\langle A \rangle \phi := (\phi \wedge \neg \pi) T \top$, and $\langle \overline{A} \rangle \phi := (\phi \wedge \neg \pi) D \top$.

By exploiting such derived operators, all conditions on the interval structure mentioned in the preliminaries can be easily expressed in BCDT$^+$. In particular, linearity can be expressed in BCDT$^+$ by means of the following formula:

$$\langle A \rangle p \rightarrow [A](p \vee \langle B \rangle p \vee \langle \overline{B} \rangle p) \wedge \langle \overline{A} \rangle p \rightarrow [\overline{A}](p \vee \langle E \rangle p \vee \langle \overline{E} \rangle p),$$

while discreteness of linear interval structures can be imposed by means of the formula:

$$\pi \vee l1 \vee (\langle B \rangle l1 \wedge \langle E \rangle l1),$$

where $l1$ stands for $\langle B \rangle \top \wedge [B][B] \bot$, together with the dual one.

As for the PITL operators, C is an operator of BCDT$^+$, while $\bigcirc$ can be defined over (linear) discrete structures as follows: $\bigcirc \phi := l1 C \phi$.

The undecidability of BCDT$^+$ with respect to a number of interval structures immediately follows from results in [9], while finding meaningful decidable fragments of BCDT$^+$ is an interesting open problem.

3   A Tableau Method for BCDT$^+$

In this section we devise a tableau method for BCDT$^+$. That method can be adapted to its strict version BCDT$^-$, and can be accordingly restricted to CDT, HS, PITL, and $\mathcal{PNL}$ logics.

First, some basic terminology. A **finite tree** is a finite directed connected graph in which every node, apart from one (the **root**), has exactly one incoming arc. A **successor** of a node $\mathbf{n}$ is a node $\mathbf{n'}$ such that there is an edge from $\mathbf{n}$ to $\mathbf{n'}$.

A **leaf** is a node with no successors; a **path** is a sequence of nodes $n_0, \ldots, n_k$ such that, for all $i = 0, \ldots, k - 1$, n_{i+1} is a successor of n_i; a **branch** is a path from the root to a leaf. The **height** of a node n is the maximum length (number of edge) of a path from n to a leaf. If n, n' belong to the same branch and the height of n is less than or equal to the height of n', we write $n \prec n'$.

Let $\langle \mathbb{C}, < \rangle$ be a finite partial order. A **labeled formula**, with label in $\mathbb{C}$, is a pair $(\phi, [c_i, c_j])$, where $\phi \in \text{BCDT}^+$ and $[c_i, c_j] \in \mathbb{I}(\mathbb{C})^+$.

For a node n in a tree, the **decoration** $\nu(n)$ is a triple $((\phi, [c_i, c_j]), \mathbb{C}, u_n)$, where $\langle \mathbb{C}, < \rangle$ is a finite partial order, $(\phi, [c_i, c_j])$ is a labeled formula, with label in $\mathbb{C}$, and u_n is a **local flag function** which associates the values 0 or 1 with every branch B containing n. Intuitively, the value 0 for a node n with respect to a branch B means that n can be expanded on B. For the sake of simplicity, we will often assume the interval $[c_i, c_j]$ to consist of the elements $c_i < c_{i+1} < \cdots < c_j$, and sometimes, with a little abuse of notation, we will write $\mathbb{C} = \{c_i < c_k, c_m < c_j, \ldots\}$. A **decorated tree** is a tree in which every node has a decoration $\nu(n)$. For every decorated tree, we define a **global flag function** u acting on pairs *(node, branch through that node)* as $u(n, B) = u_n(B)$. Sometimes, for convenience, we will include in the decoration of the nodes the global flag function instead of the local ones. For any branch B in a decorated tree, we denote by $\mathbb{C}_B$ the ordered set in the decoration of the leaf B, and for any node n in a decorated tree, we denote by $\Phi(n)$ the formula in its decoration. If B is a branch, then $B \cdot n$ denotes the result of the expansion of B with the node n (addition of an edge connecting the leaf of B to n). Similarly, $B \cdot n_1 | \ldots | n_k$ denotes the result of the expansion of B with k immediate successor nodes $n_1, \ldots, n_k$ (which produces k branches extending B). A tableau for BCDT^+ will be defined as a special decorated tree. We note again that $\mathbb{C}$ remains finite throughout the construction of the tableau.

Definition 1. *Given a decorated tree T, a branch B in T, and a node $n \in B$ such that $\nu(n) = ((\phi, [c_i, c_j]), \mathbb{C}, u)$, with $u(n, B) = 0$, the **branch-expansion rule** for B and n is defined as follows (in all the considered cases, $u(n', B') = 0$ for all new pairs (n', B') of nodes and branches).*

- *If $\phi = \neg\neg\psi$, then expand the branch to $B \cdot n_0$, with $\nu(n_0) = ((\psi, [c_i, c_j]), \mathbb{C}_B, u)$.*
- *If $\phi = \psi_0 \wedge \psi_1$, then expand the branch to $B \cdot n_0 \cdot n_1$, with $\nu(n_0) = ((\psi_0, [c_i, c_j]), \mathbb{C}_B, u)$ and $\nu(n_1) = ((\psi_1, [c_i, c_j]), \mathbb{C}_B, u)$.*
- *If $\phi = \neg(\psi_0 \wedge \psi_1)$, then expand the branch to $B \cdot n_0 | n_1$, with $\nu(n_0) = ((\neg\psi_0, [c_i, c_j]), \mathbb{C}_B, u)$ and $\nu(n_1) = ((\neg\psi_1, [c_i, c_j]), \mathbb{C}_B, u)$.*
- *If $\phi = \neg(\psi_0 C \psi_1)$ and c is the least element of $\mathbb{C}_B$, with $c_i \leq c \leq c_j$, which has not been used yet to expand the node n on B, then expand the branch to $B \cdot n_0 | n_1$, with $\nu(n_0) = ((\neg\psi_0, [c_i, c]), \mathbb{C}_B, u)$ and $\nu(n_1) = ((\neg\psi_1, [c, c_j]), \mathbb{C}_B, u)$.*
- *If $\phi = \neg(\psi_0 D \psi_1)$, c is a minimal element of $\mathbb{C}_B$ such that $c \leq c_i$, and there exists $c' \in [c, c_i]$ which has not been used yet to expand the node n on B, then take the least such $c' \in [c, c_i]$ and expand the branch to $B \cdot n_0 | n_1$, with $\nu(n_0) = ((\neg\psi_0, [c', c_i]), \mathbb{C}_B, u)$ and $\nu(n_1) = ((\neg\psi_1, [c', c_j]), \mathbb{C}_B, u)$.*

- If $\phi = \neg(\psi_0 T \psi_1)$, c is a maximal element of $\mathbb{C}_B$ such that $c_j \leq c$, and there exists $c' \in [c_j, c]$ which has not been used yet to expand the node $\mathbf{n}$ on B, then take the greatest such $c' \in [c_j, c]$ and expand the branch to $B \cdot \mathbf{n_0}|\mathbf{n_1}$, so that $\nu(\mathbf{n_0}) = ((\neg\psi_0, [c_j, c']), \mathbb{C}_B, u)$ and $\nu(\mathbf{n_1}) = ((\neg\psi_1, [c_i, c']), \mathbb{C}_B, u)$.

- If $\phi = (\psi_0 C \psi_1)$, then expand the branch to $B \cdot (\mathbf{n_i} \cdot \mathbf{m_i})|\ldots|(\mathbf{n_j} \cdot \mathbf{m_j})|(\mathbf{n_i'} \cdot \mathbf{m_i'})|\ldots|(\mathbf{n_{j-1}'} \cdot \mathbf{m_{j-1}'})$, where:

  1. for all $c_k \in [c_i, c_j]$, $\nu(\mathbf{n_k}) = ((\psi_0, [c_i, c_k]), \mathbb{C}_B, u)$ and $\nu(\mathbf{m_k}) = ((\psi_1, [c_k, c_j]), \mathbb{C}_B, u)$;
  2. for all $i \leq k \leq j - 1$, let $\mathbb{C}_k$ be the interval structure obtained by inserting a new element c between c_k and c_{k+1} in $[c_i, c_j]$, $\nu(\mathbf{n_k'}) = ((\psi_0, [c_i, c]), \mathbb{C}_k, u)$, and $\nu(\mathbf{m_k'}) = ((\psi_1, [c, c_j]), \mathbb{C}_k, u)$.

- If $\phi = (\psi_0 D \psi_1)$, then repeatedly expand the current branch, once for each minimal element c (where $[c, c_i] = \{c = c_0 < c_1 < \cdots c_i\}$), by adding the decorated sub-tree $(\mathbf{n_0} \cdot \mathbf{m_0})|\ldots|(\mathbf{n_i} \cdot \mathbf{m_i})|(\mathbf{n_1'} \cdot \mathbf{m_1'})|\ldots|(\mathbf{n_i'} \cdot \mathbf{m_i'})|(\mathbf{n_0''} \cdot \mathbf{m_0''})|\ldots|(\mathbf{n_i''} \cdot \mathbf{m_i''})$ to its leaf, where:

  1. for all c_k such that $c_k \in [c, c_i]$, $\nu(\mathbf{n_k}) = ((\psi_0, [c_k, c_j]), \mathbb{C}_B, u)$ and $\nu(\mathbf{m_k}) = ((\psi_1, [c_k, c_i]), \mathbb{C}_B, u)$;
  2. for all $0 < k \leq i$, let $\mathbb{C}_k$ be the interval structure obtained by inserting a new element c' immediately before c_k in $[c, c_i]$, and $\nu(\mathbf{n_k'}) = ((\psi_0, [c', c_i]), \mathbb{C}_k, u)$ and $\nu(\mathbf{m_k'}) = ((\psi_1, [c', c_j]), \mathbb{C}_k, u)$;
  3. for all $0 \leq k \leq i$, let $\mathbb{C}_k$ be the interval structure obtained by inserting a new element c' in $\mathbb{C}_B$, with $c' < c_k$, which is incomparable with all existing predecessors of c_k, $\nu(\mathbf{n_k''}) = ((\psi_0, [c', c_i]), \mathbb{C}_k, u)$, and $\nu(\mathbf{m_k''}) = ((\psi_1, [c', c_j]), \mathbb{C}_k, u)$.

- If $\phi = (\psi_0 T \psi_1)$, then repeatedly expand the current branch, once for each maximal element c (where $[c_j, c] = \{c_j < c_{j+1} < \cdots c_n = c\}$), by adding the decorated sub-tree $(\mathbf{n_j} \cdot \mathbf{m_j})|\ldots|(\mathbf{n_n} \cdot \mathbf{m_n})|(\mathbf{n_j'} \cdot \mathbf{m_j'})|\ldots|(\mathbf{n_{n-1}'} \cdot \mathbf{m_{n-1}'})|(\mathbf{n_j''} \cdot \mathbf{m_j''})|\ldots|(\mathbf{n_n''} \cdot \mathbf{m_n''})$ to its leaf, where:

  1. for all c_k such that $c_k \in [c_j, c]$, $\nu(\mathbf{n_k}) = ((\psi_0, [c_j, c_k]), \mathbb{C}_B, u)$ and $\nu(\mathbf{m_k}) = ((\psi_1, [c_i, c_k]), \mathbb{C}_B, u)$;
  2. for all $j \leq k < n$, let $\mathbb{C}_k$ be the interval structure obtained by inserting a new element c' immediately after c_k in $[c_j, c]$, and $\nu(\mathbf{n_k'}) = ((\psi_0, [c_j, c']), \mathbb{C}_k, u)$ and $\nu(\mathbf{m_k'}) = ((\psi_1, [c_i, c']), \mathbb{C}_k, u)$;
  3. for all $j \leq k \leq n$, let $\mathbb{C}_k$ be the interval structure obtained by inserting a new element c' in $\mathbb{C}_B$, with $c_k < c'$, which is incomparable with all existing successors of c_k, $\nu(\mathbf{n_k''}) = ((\psi_0, [c_j, c']), \mathbb{C}_k, u)$, and $\nu(\mathbf{m_k''}) = ((\psi_1, [c_i, c']), \mathbb{C}_k, u)$.

Finally, for any node $\mathbf{m}$ ($\neq \mathbf{n}$) in B and any branch B' extending B, let $u(\mathbf{m}, B')$ be equal to $u(\mathbf{m}, B)$, and for any branch B' extending B, $u(\mathbf{n}, B') = 1$, unless $\phi = \neg(\psi_0 C \psi_1)$, $\phi = \neg(\psi_0 D \psi_1)$, or $\phi = \neg(\psi_0 T \psi_1)$ (in such cases $u(\mathbf{n}, B') = 0$).

Let us briefly explain the expansion rules for $\psi_0 C \psi_1$ and $\neg(\psi_0 C \psi_1)$ (similar considerations hold for the other temporal operators). The rule for the (existential) formula $\psi_0 C \psi_1$ deals with the two possible cases: either there exists c_k in

$\mathbb{C}_B$ such that $c_i \leq c_k \leq c_j$ and ψ_0 holds over $[c_i, c_k]$ and ψ_1 holds over $[c_k, c_j]$ or such an element c_k must be added. The (universal) formula $\neg(\psi_0 C \psi_1)$ states that, for all $c_i \leq c \leq c_j$, ψ_0 does not hold over $[c_j, c]$ or ψ_1 does not hold over $[c, c_j]$. As a matter of fact, the expansion rule imposes such a condition for a single element c in $\mathbb{C}_B$ (the least element which has not been used yet), and it does not change the flag (which remains equal to 0). In this way, all elements will be eventually taken into consideration, including those elements in between c_i and c_j that will be added to $\mathbb{C}_B$ in some subsequent steps of the tableau construction.

Let us define now the notions of open and closed branch. We say that a node $\mathbf{n}$ in a decorated tree $\mathcal{T}$ is **available on a branch** B to which it belongs if and only if $u(\mathbf{n}, B) = 0$. The branch-expansion rule is **applicable** to a node $\mathbf{n}$ on a branch B if the node is available on B and the application of the rule generates at least one successor node with a new labeled formula. This second condition is needed to avoid looping of the application of the rule on formulas $\neg(\psi_0 C \psi_1)$, $\neg(\psi_0 D \psi_1)$, and $\neg(\psi_0 T \psi_1)$.

Definition 2. *A branch B is **closed** if some of the following conditions holds:*

(i) *there are two nodes $\mathbf{n}, \mathbf{n}' \in B$ such that $\nu(\mathbf{n}) = ((\psi, [c_i, c_j]), \mathbb{C}, u)$ and $\nu(\mathbf{n}') = ((\neg\psi, [c_i, c_j]), \mathbb{C}', u)$ for some formula ψ and $c_i, c_j \in \mathbb{C} \cap \mathbb{C}'$;*
(ii) *there is a node $\mathbf{n}$ such that $\nu(\mathbf{n}) = ((\pi, [c_i, c_j]), \mathbb{C}, u)$ and $c_i \neq c_j$; or*
(iii) *there is a node $\mathbf{n}$ such that $\nu(\mathbf{n}) = ((\neg\pi, [c_i, c_j]), \mathbb{C}, u)$ and $c_i = c_j$.*

*If none of the above conditions hold, the branch is **open**.*

Definition 3. *The **branch-expansion strategy** for a branch B in a decorated tree $\mathcal{T}$ is defined as follows:*

1. *Apply the branch-expansion rule to a branch B only if it is open;*
2. *If B is open, apply the branch-expansion rule to the closest to the root available node in B for which the branch-expansion rule is applicable.*

Definition 4. *A **tableau** for a given formula $\phi \in \mathrm{BCDT}^+$ is any finite decorated tree $\mathcal{T}$ obtained by expanding the three-node decorated tree built up from an empty-decoration root and two leaves with decorations $((\phi, [c_b, c_e]), \{c_b < c_e\}, u)$ and $((\phi, [c_b, c_b]), \{c_b\}, u)$, where the value of u is 0, through successive applications of the branch-expansion strategy to the existing branches.*

It is easy to show that if $\phi \in \mathrm{BCDT}^+$, $\mathcal{T}$ is a tableau for ϕ, $\mathbf{n} \in \mathcal{T}$, and $\mathbb{C}$ is the ordered set in the decoration of $\mathbf{n}$, then $\langle \mathbb{C}, < \rangle$ is an interval structure.

Definition 5. *A tableau for BCDT^+ is **closed** if and only if every branch in it is closed, otherwise it is **open**.*

As an example, consider the unsatisfiable BCDT^+ formula $\phi = pT\psi$, where $\psi = \neg(TCp)$. Here we show some steps of the construction of a closed tableau for that formula.

The initial tableau is:

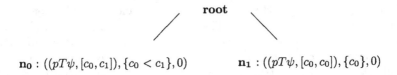

$$\mathbf{n_0} : ((pT\psi, [c_0, c_1]), \{c_0 < c_1\}, 0) \qquad \mathbf{n_1} : ((pT\psi, [c_0, c_0]), \{c_0\}, 0)$$

We suppose that the flag function is correctly updated during the construction. According to the branch-expansion strategy, by expanding $\mathbf{n_0}$ we obtain:

$$\mathbf{n_0}$$

$$\mathbf{n_2} : ((\psi, [c_0, c_1]), \{c_0 < c_1\}, 0) \qquad \mathbf{n_4} : ((\psi, [c_0, c_2]), \{c_0 < c_1 < c_2\}, 0)$$

$$\mathbf{n_3} : ((p, [c_1, c_1]), \{c_0 < c_1\}, 0) \qquad \mathbf{n_5} : ((p, [c_1, c_2]), \{c_0 < c_1 < c_2\}, 0)$$

The node $\mathbf{n_2}$ is expanded by an application of a $\neg C$ rule, attaching the decorated sub-tree

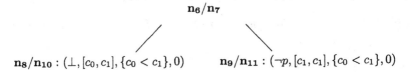

$$\mathbf{n_6} : (\bot, [c_0, c_0], \{c_0 < c_1\}, 0) \qquad \mathbf{n_7} : (\neg p, [c_0, c_1], \{c_0 < c_1\}, 0)$$

to $\mathbf{n_3}$ and the following one to each of the leaves $\mathbf{n_6}$ and $\mathbf{n_7}$:

$$\mathbf{n_6}/\mathbf{n_7}$$

$$\mathbf{n_8}/\mathbf{n_{10}} : (\bot, [c_0, c_1], \{c_0 < c_1\}, 0) \qquad \mathbf{n_9}/\mathbf{n_{11}} : (\neg p, [c_1, c_1], \{c_0 < c_1\}, 0)$$

It is straightforward to check that all branches are closed. The remaining branches can be obtained in a similar way, and they are closed as well.

3.1   Soundness and Completeness

Definition 6. *Given a set S of labeled formulas with labels in an interval structure $\langle \mathbb{C}, < \rangle$, we say that S is **satisfiable over** $\mathbb{C}$ if there exists a non-strict model $\mathbf{M}^+ = \langle \mathbb{D}, V \rangle$ such that $\langle \mathbb{D}, < \rangle$ is an extension of $\langle \mathbb{C}, < \rangle$, $\mathbf{M}^+, [c_i, c_j] \Vdash \psi$ for all $(\psi, [c_i, c_j]) \in S$.*

If S contains only one labeled formula, the notion of satisfiability of a (labeled) formula over $\mathbb{C}$ is equivalent to the notion of satisfiability given in Section 2.

Theorem 1 (Soundness). *If $\phi \in \mathrm{BCDT}^+$ and a tableau $\mathcal{T}$ for ϕ is closed, then ϕ is not satisfiable.*

Proof. We will prove by induction on the height h of a node $\mathbf{n}$ in the tableau $\mathcal{T}$ the following claim: *if every branch including* $\mathbf{n}$ *is closed, then the set* $S(\mathbf{n})$ *of all labeled formulas in the decorations of the nodes between* $\mathbf{n}$ *and the root is not satisfiable over* $\mathbb{C}$, *where* $\mathbb{C}$ *is the interval structure in the decoration of* $\mathbf{n}$.

If $h = 0$, then $\mathbf{n}$ is a leaf and the unique branch B containing $\mathbf{n}$ is closed. Then, either $S(\mathbf{n})$ contains both the labeled formulas $(\psi, [c_k, c_l])$ and $(\neg\psi, [c_k, c_l])$ for some BCDT$^+$-formula ψ and $c_k, c_l \in \mathbb{C}$, or the labeled formula $(\pi, [c_k, c_l])$ and $c_k \neq c_l$, or the labeled formula $(\neg\pi, [c_k, c_l])$ and $c_k = c_l$. Take any model $\mathbf{M}^+ = \langle \mathbb{D}, V \rangle$ where $\langle \mathbb{D}, < \rangle$ is an extension of $\langle \mathbb{C}, < \rangle$. In the first case, clearly $\mathbf{M}^+, [c_k, c_l] \Vdash \psi$ if and only if $\mathbf{M}^+, [c_k, c_l] \nVdash \neg\psi$. In the second (resp., third) case, $\mathbf{M}^+, [c_k, c_l] \Vdash \pi$ (resp., $\neg\pi$) if and only if $c_k = c_l$ (resp., $c_k \neq c_l$). Hence, $S(\mathbf{n})$ is not satisfiable over $\mathbb{C}$.

Suppose $h > 0$. Then either $\mathbf{n}$ has been generated as one of the successors, *but not the last one*, when applying the branch-expansion rule in $\wedge$, C, D, T, $\neg C, \neg D$, or $\neg T$ cases, or the branch-expansion rule has been applied to some labeled formula $(\psi, [c_i, c_j]) \in S(\mathbf{n}) -\{\varPhi(\mathbf{n})\}$ to extend the branch at $\mathbf{n}$. We deal with the latter case. The former can be dealt with in the same way. Let $\mathbb{C} = \{c_1, \ldots, c_n\}$, be the interval structure from the decoration of $\mathbf{n}$. Notice that every branch passing through any successor of $\mathbf{n}$ must be closed, so the inductive hypothesis applies to all successors of $\mathbf{n}$. We consider the possible cases for the branch-expansion rule applied at $\mathbf{n}$:

- Let $\psi = \neg\neg\xi$. Then there exists $\mathbf{n_0}$ such that $\nu(\mathbf{n_0}) = ((\xi, [c_i, c_j]), \mathbb{C}, u)$ and $\mathbf{n_0}$ is a successor of $\mathbf{n}$. Since every branch containing $\mathbf{n}$ is closed, then every branch containing $\mathbf{n_0}$ is closed. By the inductive hypothesis, $S(\mathbf{n_0})$ is not satisfiable over $\mathbb{C}$ (since $\mathbf{n_0} \prec \mathbf{n}$). Since ξ_0 and $\neg\neg\xi_0$ are equivalent, $S(\mathbf{n})$ cannot be satisfiable over $\mathbb{C}$.

- Let $\psi = \xi_0 \wedge \xi_1$. Then there are two nodes $\mathbf{n_0} \in B$ and $\mathbf{n_1} \in B$ such that $\nu(\mathbf{n_0}) = ((\xi_0, [c_i, c_j]), \mathbb{C}, u)$, $\nu(\mathbf{n_1}) = ((\xi_1, [c_i, c_j]), \mathbb{C}, u)$, and, without loss of generality, $\mathbf{n_0}$ is the successor of $\mathbf{n}$ and $\mathbf{n_1}$ is the successor of $\mathbf{n_0}$. Since every branch containing $\mathbf{n}$ is closed, then every branch containing $\mathbf{n_1}$ is closed. By the inductive hypothesis, $S(\mathbf{n_1})$ is not satisfiable over $\mathbb{C}$ since $\mathbf{n_1} \prec \mathbf{n}$. Since every model over $\mathbb{C}$ satisfying $S(\mathbf{n})$ must, in particular, satisfy $(\xi_0 \wedge \xi_1, [c_i, c_j])$, and hence $(\xi_0, [c_i, c_j])$ and $(\xi_1, [c_i, c_j])$, it follows that $S(\mathbf{n})$, $S(\mathbf{n_0})$, and $S(\mathbf{n_1})$ are equi-satisfiable over $\mathbb{C}$. Therefore, $S(\mathbf{n})$ is not satisfiable over $\mathbb{C}$.

- Let $\psi = \neg(\xi_1 \wedge \xi_2)$. Then there exist two successor nodes $\mathbf{n_0}$ and $\mathbf{n_1}$ of $\mathbf{n}$ such that $\nu(\mathbf{n_0}) = ((\xi_0, [c_i, c_j]), \mathbb{C}, u_0)$, $\nu(\mathbf{n_1}) = ((\xi_1, [c_i, c_j]), \mathbb{C}, u_1)$, $\mathbf{n_0}, \mathbf{n_1} \prec \mathbf{n}$. Since every branch containing $\mathbf{n}$ is closed, then every branch containing $\mathbf{n_0}$ and every branch containing $\mathbf{n_1}$ is closed. By the inductive hypothesis $S(\mathbf{n_0})$ and $S(\mathbf{n_1})$ are not satisfiable over $\mathbb{C}$. Since every model over $\mathbb{C}$ satisfying $S(\mathbf{n})$ must also satisfy $(\xi_0, [c_i, c_j])$ or $(\xi_1, [c_i, c_j])$, it follows that $S(\mathbf{n})$ cannot be satisfiable over $\mathbb{C}$.

- Let $\psi = \neg(\xi_0 C \xi_0)$. Suppose that $S(\mathbf{n})$ is satisfiable over $\mathbb{C}$. Then, since $(\neg(\xi_0 C \xi_1), [c_i, c_j]) \in S(\mathbf{n})$, there is a model $\mathbf{M}^+ = \langle \mathbb{D}, V \rangle$ such that $\langle \mathbb{D}, < \rangle$ is an extension of $\langle \mathbb{C}, < \rangle$ and $\mathbf{M}^+, [c_i, c_j] \Vdash \neg(\xi_0 C \xi_1)$. So, for every c_k such that $c_i \leq c_k \leq c_j$, we have that $\mathbf{M}^+, [c_i, c_k] \Vdash \neg \xi_0$ or $\mathbf{M}^+, [c_k, c_j] \Vdash \neg \xi_1$. By construction, the two immediate successors of $\mathbf{n}$ are $\mathbf{n_1}$ and $\mathbf{n_2}$ such that, for an element c_k with $c_i \leq c_k \leq c_j$, $(\neg \xi_0, [c_i, c_k])$ is in the decoration of $\mathbf{n_0}$ and $(\neg \xi_1, [c_k, c_j])$ is in the decoration of $\mathbf{n_1}$. By inductive hypothesis, since $\mathbf{n_1}, \mathbf{n_2} \prec \mathbf{n}$, $S(\mathbf{n_1})$ and $S(\mathbf{n_2})$ are not satisfiable over $\mathbb{C}$. Thus, such a model $\mathbf{M}^+$ cannot exist, and $S(\mathbf{n})$ is not satisfiable over $\mathbb{C}$.

- The cases $\psi = \neg(\xi_0 D \xi_1)$ and $\psi = \neg(\xi_0 T \xi_1)$ are analogous.
- Let $\psi = \xi_0 C \xi_1$. Assuming that $S(\mathbf{n})$ is satisfiable over $\mathbb{C}$, there is a model $\mathbf{M}^+ = \langle \mathbb{D}, V \rangle$, where $\langle \mathbb{D}, < \rangle$ is an extension of $\langle \mathbb{C}, < \rangle$, such that $\mathbf{M}^+, [c_i, c_j] \Vdash \theta$ for all $(\theta, [c_i, c_j]) \in S(\mathbf{n})$. In particular, $\mathbf{M}^+, [c_i, d] \Vdash \xi_0$ and $\mathbf{M}^+, [d, c_j] \Vdash \xi_1$ for some $c_i \leq d \leq c_j$. Consider two cases:

  1. If $d \in \mathbb{C}$, then $d = c_m$ for some $c_i \leq c_m \leq c_j$. But among the successors of $\mathbf{n}$ there are two nodes $\mathbf{n_m}, \mathbf{m_m}$ where $\nu(\mathbf{n_m}) = ((\xi_0, [c_i, c_m]), \mathbb{C}, u)$ and $\nu(\mathbf{m_m}) = ((\xi_1, [c_m, c_j]), \mathbb{C}, u)$, and since $\mathbf{n_m}, \mathbf{m_m} \prec \mathbf{n}$ (without loss of generality, suppose $\mathbf{n_m} \prec \mathbf{m_m}$), by the inductive hypothesis $S(\mathbf{n_m}) = S(\mathbf{n}) \cup \{(\xi_0, [c_i, c_m]), (\xi_1, [c_m, c_j])\}$ is not satisfiable over $\mathbb{C}$, which is a contradiction;
  2. If $d \notin \mathbb{C}$, then there is an m such that $i \leq m \leq j - 1$ and $c_m < d < c_{m+1}$. Hence, there are two successors $\mathbf{n'_m}, \mathbf{m'_m}$ of $\mathbf{n}$ such that $\nu(\mathbf{n'_m}) = ((\xi_0, [c_i, d]), \mathbb{C} \cup \{d\}, u)$, $\nu(\mathbf{m'_m}) = ((\xi_1, [d, c_j]), \mathbb{C} \cup \{d\}, u)$, and since $\mathbf{n'_m}, \mathbf{m'_m} \prec \mathbf{n}$ (without loss of generality, suppose $\mathbf{n'_m} \prec \mathbf{m'_m}$), by the inductive hypothesis $S(\mathbf{n'_m}) = S(\mathbf{n}) \cup \{(\xi_0, [c_i, d]), (\xi_1, [d, c_j])\}$ is not satisfiable over $\mathbb{C} \cup \{d\}$ which, again, is a contradiction.

  Thus, in either case $S(\mathbf{n})$ is not satisfiable over $\mathbb{C}$.
- Let $\psi = \xi_0 D \xi_1$. Assuming that $S(\mathbf{n})$ is satisfiable over $\mathbb{C}$, there is a model $\mathbf{M}^+ = \langle \mathbb{D}, V \rangle$, where $\langle \mathbb{D}, < \rangle$ is an extension of $\langle \mathbb{C}, < \rangle$, such that $\mathbf{M}^+, [c_i, c_j] \Vdash \theta$ for all $(\theta, [c_i, c_j]) \in S(\mathbf{n})$. In particular, $\mathbf{M}^+, [d, c_i] \Vdash \xi_0$ and $\mathbf{M}^+, [d, c_j] \Vdash \xi_1$ for some $d \leq c_i$. Consider 3 cases:

  1. If $d \in \mathbb{C}$, then $d = c_m$ for some $c_m \leq c_i$. But between the successors of $\mathbf{n}$ there are two nodes $\mathbf{n_m}, \mathbf{m_m}$ where $\nu(\mathbf{n_m}) = ((\xi_0, [c_m, c_i]), \mathbb{C}, u)$ and $\nu(\mathbf{m_m}) = ((\xi_1, [c_m, c_j]), \mathbb{C}, u)$, and since $\mathbf{n_m}, \mathbf{m_m} \prec \mathbf{n}$ (without loss of generality, suppose $\mathbf{n_m} \prec \mathbf{m_m}$), by the inductive hypothesis $S(\mathbf{n_m}) = S(\mathbf{n}) \cup \{(\xi_0, [c_m, c_i]), (\xi_1, [c_m, c_j])\}$ is not satisfiable over $\mathbb{C}$, which is a contradiction.
  2. If $d \notin \mathbb{C}$ and there is a minimal element $c \in \mathbb{C}$ and an index m such that $c_m, c_{m+1} \in [c, c_i]$ and $c_m < d < c_{m+1}$, then there are two successors $\mathbf{n'_m}, \mathbf{m'_m}$ of $\mathbf{n}$ such that $\nu(\mathbf{n'_m}) = ((\xi_0, [c_i, d]), \mathbb{C} \cup \{d\}, u)$ and $\nu(\mathbf{m'_m}) = ((\xi_1, [d, c_j]), \mathbb{C} \cup \{d\}, u)$, and since $\mathbf{n'_m}, \mathbf{m'_m} \prec \mathbf{n}$ (without loss of generality, suppose $\mathbf{n'_m} \prec \mathbf{m'_m}$), by the inductive hypothesis $S(\mathbf{n'_m}) = S(\mathbf{n}) \cup \{(\xi_0, [c_i, d]), (\xi_1, [d, c_j])\}$ is not satisfiable over $\mathbb{C} \cup \{d\}$ which, again, is a contradiction.
  3. If $d \notin \mathbb{C}$ and there is an index m such that $c_{m+1} \in [c, c_i]$, $d < c_{m+1}$, and d is not comparable with all predecessors of c_{m+1}, then, again, there are two successor nodes $\mathbf{n''_m}, \mathbf{m''_m}$ of $\mathbf{n}$ such that $\nu(\mathbf{n''_m}) = ((\xi_0, [c_i, d]), \mathbb{C} \cup \{d\}, u)$ and $\nu(\mathbf{m''_m}) = ((\xi_1, [d, c_j]), \mathbb{C} \cup \{d\}, u)$, and since $\mathbf{n''_m}, \mathbf{m''_m} \prec \mathbf{n}$

(without loss of generality, suppose $\mathbf{n}_m'' \prec \mathbf{m}_m''$), by the inductive hypothesis $S(\mathbf{n}_m'') = S(\mathbf{n}) \cup \{(\xi_0, [c_i, d]), (\xi_1, [d, c_j])\}$ is not satisfiable over $\mathbb{C} \cup \{d\}$ which, again, is a contradiction.

Thus, in either case $S(\mathbf{n})$ is not satisfiable over $\mathbb{C}$.

– The case of $\psi = \xi_0 T \xi_1$ is similar.

□

Definition 7. *If T_0 is the three-node tableau built up from a root with void decoration and two leaves decorated respectively by $((\phi, [c_b, c_e]), \{c_b < c_e\}, 0)$ and $((\phi, [c_b, c_b]), \{c_b\}, 0)$ for a given* BCDT$^+$*-formula ϕ, the **limit tableau** $\overline{T}$ for ϕ is the (possibly infinite) decorated tree obtained as follows. First, for all i, T_{i+1} is the tableau obtained by the simultaneous application of the branch-expansion strategy to every branch in T_i. Then, we ignore all flags from the decorations of the nodes in every T_i. Thus, we obtain a chain by inclusion of decorated trees: $T_1 \subseteq T_2 \subseteq \ldots$, and we define $\overline{T} = \bigcup\limits_{i=0}^{\infty} T_i$.*

Notice that the chain above may stabilize at some T_i if it closes, or if the branch-expansion rule is not applicable to any of its branches. If $\overline{T}$ is a limit tableau, we associate with each branch B in $\overline{T}$ the interval structure $\mathbb{C}_B = \bigcup\limits_{i=0}^{\infty} \mathbb{C}_{B_i}$, where, for all i, $\mathbb{C}_{B_i}$ is the interval structure from the decoration of the leaf of the (sub-)branch B_i of B in T_i. The definitions of closed and open branches readily apply to $\overline{T}$.

Definition 8. *A branch in a (limit) tableau is **saturated** if there are no nodes on that branch to which the branch-expansion rule is applicable on the branch. A (limit) tableau is **saturated** if every open branch in it is saturated.*

Now we will show that the set of all labeled formulas on an open branch in a limit tableau has the saturation properties of a Hintikka set in first-order logic.

Lemma 1. *Every limit tableau is saturated.*

Proof. Given a node $\mathbf{n}$ in a limit tableau $\overline{T}$, we denote by $d(\mathbf{n})$ the distance (number of edges) between $\mathbf{n}$ and the root of $\overline{T}$. Now, given a branch B in $\overline{T}$, we will prove by induction on $d(\mathbf{n})$ that after every step of the expansion of that branch at which the branch-expansion rule becomes applicable to $\mathbf{n}$ (because $\mathbf{n}$ has just been introduced, or because a new point has been introduced in the interval structure on B) that rule is subsequently applied on B to that node.

Suppose the inductive hypothesis holds for all nodes with distance to the root less than l. Let $d(\mathbf{n}) = l$ and the branch-expansion rule has become applicable to $\mathbf{n}$. If there are no nodes between the root (incl. the root) and $\mathbf{n}$ (excl. $\mathbf{n}$) to which the branch-expansion rule is applicable at that moment, the next application of the branch-expansion rule on B is to $\mathbf{n}$. Otherwise, consider the closest-to-$\mathbf{n}$-node $\mathbf{n}^*$ between the root and $\mathbf{n}$ to which the branch-expansion rule is applicable or will become applicable on B at least once thereafter. (Such a node exists because there are only finitely many nodes between $\mathbf{n}$ and the root.) Since $d(\mathbf{n}^*) < d(\mathbf{n})$, by the inductive hypothesis the branch-expansion rule has been subsequently applied to $\mathbf{n}^*$. Then the next application of the branch-expansion rule on B must have been to $\mathbf{n}$ and that completes the induction. Now, assuming that a branch in a limit tableau is not saturated, consider the closest to the root node $\mathbf{n}$ on that branch B to which the branch-expansion rule is applicable on that branch. If $\Phi(\mathbf{n})$ is none of the cases $\neg C, \neg D$, and $\neg T$, then the branch-expansion rule has become applicable to $\mathbf{n}$ at the step when $\mathbf{n}$ is introduced, and by the claim above, it has been subsequently applied, at which moment the node has become unavailable thereafter, which contradicts the assumption. Suppose that $\Phi(\mathbf{n}) = \neg(\psi_0 C \psi_1)$. Then an application of the rule on B would create two successors with labels $(\neg\psi_0, [c_i, c])$ and $(\neg\psi_1, [c, c_j])$, at least one of them new on B. But c_i, c_j, c have already been introduced at some (finite) step of the construction of B and at the first step when the three of them, as well as $\mathbf{n}$, have appeared on the branch, the branch-expansion rule has become applicable to $\mathbf{n}$, hence is has been subsequently applied on B and that application must have introduced the labels $(\psi_0, [c_i, c])$ and $(\psi_1, [c, c_j])$ on B, which again contradicts the assumption. The same holds if $\Phi(\mathbf{n}) = \neg(\psi_0 D \psi_1)$ or $\Phi(\mathbf{n}) = \neg(\psi_0 D \psi_1)$. □

Corollary 1. *Let ϕ be a BCDT^+-formula and $\overline{T}$ be the limit tableau for ϕ. For every open branch B in $\overline{T}$, the following closure properties hold.*

- *If there is a node $\mathbf{n} \in B$ such that $\nu(\mathbf{n}) = ((\neg\neg\psi, [c_i, c_j]), \mathbb{C}, u)$, then there is a node $\mathbf{n_0} \in B$ such that $\nu(\mathbf{n_0}) = ((\psi, [c_i, c_j]), \mathbb{C}, u_0)$.*
- *If there is a node $\mathbf{n} \in B$ such that $\nu(\mathbf{n}) = ((\psi_0 \wedge \psi_1, [c_i, c_j]), \mathbb{C}, u)$, then there is a node $\mathbf{n_0} \in B$ such that $\nu(\mathbf{n_0}) = ((\psi_0, [c_i, c_j]), \mathbb{C}, u_0)$ and a node $\mathbf{n_1} \in B$ such that $\nu(\mathbf{n_1}) = ((\psi_1, [c_i, c_j]), \mathbb{C}, u_1)$.*
- *If there is a node $\mathbf{n} \in B$ such that $\nu(\mathbf{n}) = ((\neg(\psi_0 \wedge \psi_1), [c_i, c_j]), \mathbb{C}, u)$, then there is a node $\mathbf{n_0} \in B$ such that $\nu(\mathbf{n_0}) = ((\neg\psi_0, [c_i, c_j]), \mathbb{C}, u_0)$ or a node $\mathbf{n_1} \in B$ such that $\nu(\mathbf{n_1}) = ((\neg\psi_1, [c_i, c_j]), \mathbb{C}, u_1)$.*
- *If there is a node $\mathbf{n} \in B$ such that $\nu(\mathbf{n}) = ((\psi_0 C \psi_1, [c_i, c_j]), \mathbb{C}, u)$, then, for some $c \in \mathbb{C}_B$ such that $c_i \leq c \leq c_j$ there are two nodes $\mathbf{n}', \mathbf{m}' \in B$ such that $\nu(\mathbf{n}') = ((\psi_0, [c_i, c]), \mathbb{C}', u')$ and $\nu(\mathbf{m}') = ((\psi_1, [c, c_j]), \mathbb{C}', u')$.*
- *Similarly for every node $\mathbf{n}$ with $\Phi(\mathbf{n}) = \psi_0 D \psi_1$ or $\Phi(\mathbf{n}) = \psi_0 T \psi_1$.*
- *If there is a node $\mathbf{n} \in B$ such that $\nu(\mathbf{n}) = ((\neg(\psi_0 C \psi_1), [c_i, c_j]), \mathbb{C}, u)$, then for all $c \in \mathbb{C}_B$ such that $c_i \leq c \leq c_j$, there is a node $\mathbf{n}' \in B$ such that $\nu(\mathbf{n}') = ((\neg\psi_0, [c_i, c]), \mathbb{C}', u')$ or a node $\mathbf{m}' \in B$ such that $\nu(\mathbf{m}') = ((\neg\psi_1, [c, c_j]), \mathbb{C}', u')$.*
- *Similarly for every node $\mathbf{n}$ with $\Phi(\mathbf{n}) = \neg(\psi_0 D \psi_1)$ or $\Phi(\mathbf{m}) = \neg(\psi_0 T \psi_1)$.*

Lemma 2. *If the limit tableau for some formula $\phi \in \mathrm{BCDT}^+$ is closed, then some finite tableau for ϕ is closed.*

Proof. Suppose the limit tableau for ϕ is closed. Then every branch closes at some finite step of the construction and then remains finite. Since the branch-expansion rule always produces finitely many successors, every finite tableau is finitely branching, and hence so is the limit tableau. Then, by König's lemma, the limit tableau, being a finitely branching tree with no infinite branches, must be finite, hence its construction stabilizes at some finite stage. At that stage a closed tableau for ϕ is constructed. □

Theorem 2 (Completeness). *Let $\phi \in BCDT^+$ be a valid formula. Then there is a closed tableau for $\neg\phi$.*

Proof. We will show that the limit tableau $\overline{\mathcal{T}}$ for $\neg\phi$ is closed, whence the claim follows by the previous lemma.

By contraposition, suppose that $\overline{\mathcal{T}}$ has an open branch B. Let $\mathbb{C}_B$ be the interval structure associated with B and $S(B)$ be the set of all labeled formulas on B. Consider the model $\mathbf{M}^+ = \langle \mathbb{C}_B, V \rangle$ where, for every $[c_i, c_j] \in \mathbb{I}(\mathbb{C}_B)^+$ and $p \in \mathcal{AP}$, $p \in V([c_i, c_j])$ iff $(p, [c_i, c_j]) \in \Phi(B)$. We show by induction on ψ that, for every $(\psi, [c_i, c_j]) \in S(B)$, $\mathbf{M}^+, [c_i, c_j] \Vdash \psi$.

We reason by induction on the complexity of ψ:

- Let $\psi = \pi$ (resp., $\psi = \neg\pi$). Since $(\pi, [c_i, c_j]) \in S(B)$ (resp., $(\neg\pi, [c_i, c_j]) \in S(B)$) and B is open, then $c_i \neq c_j$ (resp., $c_i = c_j$). Hence $\mathbf{M}^+, [c_i, c_j] \Vdash \pi$ (resp., $\mathbf{M}^+, [c_i, c_j] \Vdash \neg\pi$).
- Let $\psi = p$ or $\psi = \neg p$ where $p \in \mathcal{AP}$. Then the claim follows by definition, because if $(\neg p, [c_i, c_j]) \in S(B)$ then $(p, [c_i, c_j]) \notin S(B)$ since B is open.
- Let $\psi = \neg\neg\xi$. Then by Corollary 1, $(\xi, [c_i, c_j]) \in S(B)$, and by inductive hypothesis $\mathbf{M}^+, [c_i, c_j] \Vdash \xi$. So $\mathbf{M}^+, [c_i, c_j] \Vdash \psi$.
- Let $\psi = \xi_0 \wedge \xi_1$. Then by Corollary 1, $(\xi_0, [c_i, c_j]) \in S(B)$ and $(\xi_1, [c_i, c_j]) \in S(B)$. By inductive hypothesis, $\mathbf{M}^+, [c_i, c_j] \Vdash \xi_0$ and $\mathbf{M}^+, [c_i, c_j] \Vdash \xi_1$, so $\mathbf{M}^+, [c_i, c_j] \Vdash \psi$.
- Let $\psi = \neg(\xi_0 \wedge \xi_1)$. Then by Corollary 1, $(\neg\xi_0, [c_i, c_j]) \in S(B)$ or $(\neg\xi_1, [c_i, c_j]) \in S(B)$. By inductive hypothesis $\mathbf{M}^+, [c_i, c_j] \Vdash \neg\xi_0$ or $\mathbf{M}^+, [c_i, c_j] \Vdash \neg\xi_1$, so $\mathbf{M}^+, [c_i, c_j] \Vdash \psi$.
- Let $\psi = \xi_0 C \xi_1$. Then by Corollary 1, $(\xi_0, [c_i, c]) \in S(B)$ and $(\xi_1, [c, c_i]) \in S(B)$ for some $c \in \mathbb{C}_B$ such that $c_i \leq c \leq c_j$. Thus, by inductive hypothesis, $\mathbf{M}^+, [c_i, c] \Vdash \xi_0$ and $\mathbf{M}^+, [c, c_j] \Vdash \xi_1$, and thus $\mathbf{M}^+, [c_i, c_j] \Vdash \psi$.
- Similarly for $\psi = \xi_0 D \xi_1$ and $\psi = \xi_0 T \xi_1$.
- Let $\psi = \neg(\xi_0 C \xi_1)$. Then by Corollary 1, for all $c \in \mathbb{C}_B$ such that $c_i \leq c \leq c_j$, $(\neg\xi_0, [c_i, c]) \in S(B)$ and $(\neg\xi_1, [c, c_j]) \in S(B)$. Hence, by the inductive hypothesis, $\mathbf{M}^+, [c_i, c] \Vdash \neg\xi_0$ and $\mathbf{M}^+, [c, c_j] \Vdash \neg\xi_1$, for all $c_i \leq c \leq c_j$. Thus, $\mathbf{M}^+, [c_i, c_j] \Vdash \psi$.
- Similarly for $\psi = \neg(\xi_0 D \xi_1)$ and $\psi = \neg(\xi_0 T \xi_1)$.

This completes the induction. In particular, we obtain that $\neg\phi$ is satisfied in $\mathbf{M}^+$, which is in contradiction with the assumption that ϕ is valid. □

Concluding Remark: The main natural continuation of this work would be to identify cases (fragments of the logic, or classes of interval structures) when the tableau will terminate and therefore provide a decision procedure.

Acknowledgments

The authors would like to thank the Italian Ministero degli Affari Esteri and the National Research Foundation of South Africa for the research grant, under the Joint Italy/South Africa Science and Technology Agreement, that they received for the project: "Theory and applications of temporal logics to computer science and artificial intelligence". We also thank the anonymous referees for some useful remarks and references.

References

[1] H. Bowman and S. Thompson. A decision procedure and complete axiomatization of finite interval temporal logic with projection. *Journal of Logic and Computation*, 13(2):195–239, 2003.

[2] S. Cerrito, M. Cialdea Mayer, and S. Praud. First-order linear temporal logic over finite time structures. In H. Ganzinger, D. McAllester, and A. Voronkov, editors, *6th Int. Conf. on Logic for Programming and Automated Reasoning (LPAR)*, pages 62–76. LNAI 1705, Springer, 1999.

[3] N. Chetcuti-Serandio and L. Fariñas del Cerro. A mixed decision method for duration calculus. *Journal of Logic and Computation*, 10:877–895, 2000.

[4] M. D'Agostino, D. Gabbay, R. Hähnle, and J. Posegga, editors. *Handbook of Tableau Methods*. Kluwer, Dordrecht, 1999.

[5] E. A. Emerson. Temporal and modal logic. *Handbook of Theoretical Computer Science*, B:996–1072, 1990.

[6] M. Fitting. Proof methods for modal and intuitionistic logics. In Holland D. Reidel, Dordrecht, editor, *Synthese Library*, volume 169, 1983.

[7] E. Grädel, C. Hirsch, and M. Otto. Back and forth between guarded and modal logics. *ACM Trans. on Computational Logics*, 3(3):418–463, 2002.

[8] V. Goranko, A. Montanari, and G. Sciavicco. On propositional interval neighborhood temporal logics. *Journal of Universal Computer Science. To appear*, 2003.

[9] J. Y. Halpern and Y. Shoham. A propositional modal logic of time intervals. *Journal of the ACM*, 38(4):935–962, 1991.

[10] R. Kontchakov, C. Lutz, F. Wolter, and M. Zakharyashev. Temporalising tableaux. *Submitted for publication*, 2002.

[11] S. Kono. A combination of clausal and non-clausal temporal logic programs. In M. Fisher and R. Owens, editors, *Executable Modal and Temporal Logics*, number 897 in LNCS, pages 40–57. Springer, 1995.

[12] B. Moszkowski. *Reasoning about Digital Circuits*. PhD thesis, Stanford University, Stanford, CA, 1983.

[13] Maarten Marx and Yde Venema. *Multi-Dimensional Modal Logic*. Kluwer Academic Publishers, Dordrecht, 1997.

[14] W. Rautenberg. Modal tableau calculi and interpolation. *Journal of Philosophical Logics*, 12:403–423, 1983.

[15] P. H. Schmitt and J. Goubault-Larrecq. A tableau system for linear-time temporal logic. In E. Brinksma, editor, *3rd Workshop on Tools and Algorithms for the Construction and Analysis of Systems (TACAS)*, pages 130–144. LNCS 1217, Springer, 1997.

[16] Y. Venema. A modal logic for chopping intervals. *Journal of Logic and Computation*, 1(4):453–476, 1991.

[17] P. Wolper. The tableau method for temporal logic: an overview. *Logique et Analyse*, 28(110–111):119–136, June–September 1985.

Universal Variables in Disconnection Tableaux

Reinhold Letz[1] and Gernot Stenz[2]

[1] Institut für Informatik, Technische Universität München
D-80290 Munich, Germany
letz@in.tum.de
[2] Max-Planck-Institut für Informatik
Stuhlsatzenhausweg 85, 66123 Saarbrücken, Germany
stenz@mpi-sb.mpg.de

Abstract. The disconnection calculus since its original conception has been developed into one of the most successful tableau methods ever devised. Still, deductions in the disconnection calculus can suffer from redundancies inherent to the tableau framework. Even though the calculus can decide the Bernays-Schönfinkel class of formulae it is in many cases inferior to combinations of a grounding mechanism with a Davis-Putnam prover system. In this paper we address two enhancements of the disconnection calculus that are intended to reduce some of the redundancies typical for tableau methods. First, we investigate the use of local variables, a syntactically detectable form of universal variables. These variables can be used to relax the ∀-closure condition and introduce partial unification for branch closures. However, the use of such variables has certain ramifications we will also discuss. Then, we examine the extended use of context lemmas during proof search by allowing the use of context lemmas for subsumption of new tableau clauses. We also show limitations to this method. Both techniques described in this paper are being implemented as part of the DCTP disconnection tableau prover.

1 Introduction

For many years, automated deduction in classical first-order logic was dominated by resolution-based systems. In the last years, however, a number of generally successful systems have been developed, which are belonging to the tableau paradigm. Because tableau deductions have a richer structure, a number of strongly differing tableau calculi have been developed like connection tableaux [13, 12], hypertableaux [2], or disconnection tableaux [5, 14, 19]. Currently, the most powerful tableau-oriented theorem prover is the DCTP system [15, 18], which is based on the disconnection tableau approach. The main advantage of disconnection tableaux is that free variables in tableaux are not treated in a *rigid* manner as in Fitting's free-variable tableaux [6] or in connection tableaux. There, any free variable is just treated as a name for a single yet unknown (ground) term t, and free variables need to be instantiated during the deduction process. Since free variables in general are distributed over the entire

M. Cialdea Mayer and F. Pirri (Eds.): TABLEAUX 2003, LNAI 2796, pp. 117–133, 2003.
© Springer-Verlag Berlin Heidelberg 2003

tableau, an instantiation has strong global effects. As a consequence of this *destructive* method, a central virtue of Smullyan's original tableau calculus is lost, viz., the possibility of saturating a branch up to a Hintikka set, which represents a model. Also, the decision power of such free-variable tableau systems is significantly weakened.

With the disconnection approach, this weakness is remedied, since variables are kept local to clauses. In fact, the approach can be viewed as the first integration of unification into tableaux which preserves the property of branch saturation. For certain types of formulae, this can be exploited for model generation [14]. Also, disconnection tableaux offer the possibility for the integration of a number of methods that increase the performance dramatically. Examples are techniques like the special treatment of unit clauses (including simplification and subsumption), folding up (which implements a controlled integration of the cut rule), and an efficient equality handling by using an equivalent of ordered paramodulation [1]. In summary, the disconnection approach seems to be most promising for future developments.

In this paper, we concentrate on an important weakness of the current method concerning the treatment of variables. In certain cases, we may be able to deduce a clause c containing a literal l with one or more *universal* or *local* variables $x_1, \ldots, x_n$, i.e., c is of the form $(\forall x_1 \cdots x_n l) \vee c'$ where c' is the rest of the clause. As we will demonstrate, universal variables can be treated differently and may permit a significant reduction of the proof length and the search space, in certain cases just one such clause may speed up the search by magnitudes.

Obviously, the application of this method depends on the number of occurrences of universal variables. Since in practical examples, universal variables occur very rarely only, we have to think about methods which support the generation of clauses with universal variables. Fortunately, here we can observe an important synergetic effect of different methods. When using the methods of folding up and unit simplification, the number of generated universal variables significantly increases.

The paper is organised as follows. Following this introduction, Section 2 defines the basic notation and describes the disconnection calculus. In Section 3 we illustrate a fundamental weakness of the current method and show how this weakness can be remedied by treating universal variables appropriately. In Section 4 we consider the modifications that have to be made when units and the folding up rule are used. Then, we describe an implementation of the method in the system DCTP and give results of an experimental evaluation of the new system. We conclude with an assessment of this work and address future perspectives.

2   The Disconnection Tableau Calculus

The disconnection tableau calculus was first developed in [5], the method works on sets of clauses. As usual, a *literal* is an atomic formula or a negated atomic formula. A *clause* is a disjunction of literals; occasionally, we will treat clauses as sets of literals. A *literal occurrence* is any pair $\langle c, l \rangle$ where c is a clause, l is a literal, and $l \in c$. Throughout this paper, we assume that all clauses in the clause sets and in the constructed tableau are variable-disjoint.

Definition 1 (Link, Linking Instance). *Given two variable-disjoint clauses c and c' and two literal occurrences $\langle c, l \rangle$ and $\langle c', \neg l' \rangle$, if there is a unifier σ of l and $\neg l'$, then the set $\ell = \{\langle c, l \rangle, \langle c', \neg l' \rangle\}$ is called a* connection *or* link *(between the clauses c and c'). The clause $c\sigma$ is a* linking instance *of c wrt. the connection ℓ.*

The generation of linking instances is related with the *clause linking method* developed in [8]. In contrast to Plaisted's and Lee's approach, however, the disconnection method embeds the process of generating linking instances into a tableau guided control structure. We need the following further notions.

Definition 2 (Path). *A path through a clause set S is any total mapping $P : S \rightarrow \bigcup S$ with $P(c) \in c$, i.e., a set containing exactly one literal occurrence $\langle c, l \rangle$ for every $c \in S$. With the set of literals of P we mean the set of literals in the range of P. A path P is* complementary *if it contains two literal occurrences of the form $\langle c, l \rangle$ and $\langle c', \neg l \rangle$, otherwise P is called* consistent *or* open.

Definition 3 (Tableau). *A tableau is a (possibly infinite) downward tree with literal labels at all tree nodes except the root. Given a clause set S, a tableau for S is a tableau in which, for every tableau node N, the set of literals $c = l_1, \ldots, l_m$ at the immediate successor nodes $N_1, \ldots, N_m$ of T is an instance of a clause in S; for every N_i ($1 \leq i \leq m$), c is called the clause of N_i. With every tableau node N_i the literal occurrence $\langle c, l_i \rangle$ will be associated. Furthermore, a branch of a tableau T is any maximal sequence $B = N_1, N_2, N_3, \ldots$ of nodes in T such that N_1 is an immediate successor of the root node and any N_{i+1} is an immediate successor of N_i. In the tableaux that we will consider, no clause will occur more than once on a branch, so with every branch B we can associate a path P, viz., the set of literal occurrences associated with the nodes in B.*

The disconnection tableau calculus consists of a single inference rule, the so-called *linking rule*.

Definition 4 (Linking Rule). *Given a tableau branch B with leaf node N and two ancestor nodes N_1 and N_2 with literals l and $\neg l'$ and variable-disjoint clauses c_1 and c_2, respectively, if $\ell = \{\langle c, l \rangle, \langle c', \neg l' \rangle\}$ is a connection with unifier σ, then*

1. *expand the branch with a linking instance wrt. ℓ of one of the two clauses, say, with $c\sigma$,*
2. *below the node labeled with $l\sigma$, expand the branch with a linking instance wrt. ℓ of the other clause, i.e., $c'\sigma$.*
3. *Afterwards, rename the variables in the new tableau clauses, i.e., substitute all variables with new different variables not yet seen in the deduction process.*

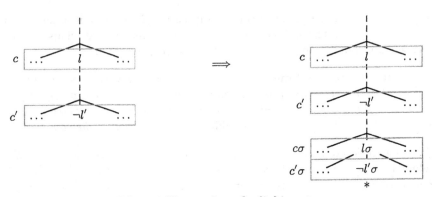

Fig. 1. Illustration of a linking step

In other terms, we perform a clause linking step (in the terminology of the clause linking method) and attach the coupled linking instances below the leaf N of the current tableau branch. In order to be able to start the tableau construction, we must provide an initial set of connections from which to choose the first links. For this we may take an arbitrary *initial path* P_S through the set S of input clauses; P_S remains fixed during the entire tableau construction.

As branch closure condition, the standard tableau closure condition is not sufficient, but the same notion as employed in the clause linking method can be used.

Definition 5 (∀-Closure). *A tableau branch B is ∀-closed if it contains two literals l and $\neg k$ such that $l\theta = k\theta$ where θ is a substitution mapping all variables to the same new constant.*[1]

So by the very nature of a linking step, in the tableau in Fig. 1, the middle branch must be ∀-closed, as indicated with an asterisk.

The *disconnection tableau calculus* then simply consists of the rule for the selection of an initial path (applicable merely once at the beginning of the proof construction) and the linking rule. The calculus is refutationally sound and complete for any initial path selection [14]. The most important completeness-preserving refinement of the calculus is the following.

Definition 6 (Variant-Freeness). *A disconnection tableau T is variant-free if, for no node N with clause c in T, there exists an ancestor node N' in T with clause c' such that c and c' are variants of each other, i.e., c can be obtained from c' by renaming its variables. (Note, however, that this restriction does not extend to the initial path.)*

With this refinement it is automatically achieved that a link can be used only once on each branch, which permits decision procedures for certain formulae

---

[1] However, this substitution is not actually applied to the tableau, as opposed to rigid-variable tableaux.

classes, most notably the Bernays-Schönfinkel class (which cannot be decided by resolution). As shown in [14], we may also extract a model from a finitely failed open branch.

Since the disconnection method constructs just a single tableau as in the original approach of Smullyan, this tableau also represents the entire search space including all useless attempts to produce closed sub-tableaux. Consequently, large parts of the tableau are irrelevant for closure, i.e. many branch literals are not used for branch closures in the dominated sub-tableau and can therefore be pruned together with the respective tableau clauses.

Definition 7 (Relevance). *A tableau node N dominating[2] a closed sub-tableau T is relevant in T if N is used for a $\forall$-closure in T. A clause in a tableau is relevant if all its nodes are relevant in their respective sub-tableaux.*

As usual, a tableau is constructed in a depth-first manner by exploring one branch at a time until it can be closed, and afterwards backtracking[3] to the next selected open branch with maximal depth. This method is similar to the working of SAT-solvers following the DPLL approach. Accordingly, we can also apply dependency-directed backtracking as used in most of those SAT-solvers. In order to identify whether a branch literal is relevant, we use the method of *relevance sets*, which is described in [11].

3 Universal Variables and Their Potential

In a tableau proof a branch formula F containing free variables has often to be used more than once, but with different instances for some of the variables, say, $u_1, \ldots, u_n$. This may lead to the multiple occurrence of similar subproofs. Such multiple occurrences can be avoided if one can show that the respective variables are universal wrt. the formula F, i.e., when $\forall u_1 \cdots u_n F$ holds on the branch. A general description of this property is given, e.g., in [3]. Since proving this property is undecidable in general, efficient sufficient conditions for universality have been developed, e.g., the concept of locality in [10]. We will use a similar approach.

Definition 8 (Local Variable). *A variable u is called* local *in a literal l of a clause $l \vee c$ if u does occur in l but not in c.*

Obviously, since all clauses in disconnection tableau are variable-disjoint, any local variable in a literal on a tableau branch is universal wrt. the literal.

In order to illustrate a fundamental weakness of the current system and the potential of the use of universal variables, we discuss a very simple example. As a matter of fact, in practice, such clauses will normally not occur in the

---

[2] A node dominates a sub-tableau if it is on the branch from the root of the entire tableau to the root of the sub-tableau, including both.

[3] Note that this does not mean that we backtrack over *different* tableaux as in the connection tableau method or in Fitting's free-variable tableaux.

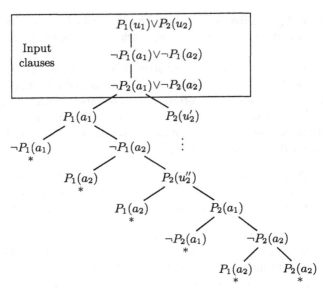

Fig. 2. Tableau for the clause set from Example 1

respective input formulae, but, as our experiments show, clauses with universal variables may be dynamically generated during the proof process.

Example 1. Let S be a set consisting of the three clauses $P_1(u_1) \vee P_2(u_2)$, $\neg P_1(a_1) \vee \neg P_1(a_2)$, and $\neg P_2(a_1) \vee \neg P_2(a_2)$ where the u_i denote variables and the a_i constants.

A part of a minimal closed disconnection tableau for this set is displayed in Fig. 2. Note that for the entire proof thirteen linking steps are needed.

The redundancy in the proof is obvious. Since the variables u_i in the first clause are universal, one should treat the respective branches separately, i.e., not mix the instantiations resulting from the independent parts, and exploit this for obtaining a shorter proof and a smaller search space. Different methods have been developed in order to achieve a better behaviour for such formulae.

One approach to to avoid that the product of the instantiations is formed is to split a clause $c \vee c'$, where c and c' are variable-disjoint and refute the sub-clauses separately. This problem splitting is, for example, used in the SPASS prover system [23, 24]. Another approach is to replace the first clause with two new clauses $D_1 \vee c$ and $\neg D_1 \vee c'$ where D_1 is a new predicate symbol. This method of clause splitting is used quite successfully in resolution systems like Vampire [16]. Unfortunately, both methods have to be used carefully. In resolution, for example, we have to avoid that the original clause is reintroduced by a resolution step. In the tableau framework, the introduction of new formulae which are not sub-formulae of the original ones, even contradicts the basic working paradigm and can lead to other problems. The first approach is problematic because all optimisations of the system such as pruning and the reuse of subproofs would have to adapted or would not be possible at all.

Therefore, we propose to use a more natural method which is also more general in that it applies to some cases where the clause parts are not entirely variable-disjoint. We extend the closure rule in such a manner that a literal containing universal variables can be used for different branch closures. We require, however, that one of the clause parts is a literal.

In order to avoid the forming of instantiation products, we simply forbid that universal literals be instantiated when the linking instances are generated. This procedure reduces the search space to an extent which cannot even be achieved by the clause splitting approach.

Definition 9 (U-Closure). *A u-substitution is a substitution on universal variables only. Two terms/literals k, l are u-unifiable if there is a u-substitution σ with $k\sigma = l\sigma$. A tableau branch is u-closed if it contains two literals $k, \neg l$ and a u-substitution σ such that $k\sigma$ and $l\sigma$ are equal under variable identification.*

In order to capture the restriction that universal variables must not be further instantiated in linking steps, the linking rule has to be modified, as follows.

Definition 10 (U-Linking Rule). *Let $c\sigma$ be a linking instance according to the standard linking rule (Definition 1). Then instead of putting $c\sigma$ on the tableau, u-linking puts $c\tau$ on the tableau where τ is the restriction of σ to the non-universal variables of c.*

With the new rules, we achieve the disconnection tableau proof displayed in Fig. 3. Furthermore, since the u-linking rule forbids that clauses are put on the tableau in which universal variables are instantiated, no further instance of the first clause can appear in the tableau.

Proposition 1 (Soundness of u-Closure). *If there is a u-closed disconnection tableau T for a set of clauses S, then S is unsatisfiable.*

Proof. We show the soundness by transforming T into a tableau which can be closed according to general tableau rules. First, we replace every literal l

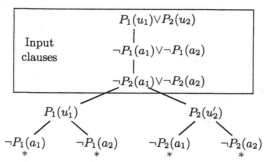

Fig. 3. Closed tableau for the clause set from Example 1 using local variables

containing the universal variables $u_1, \ldots, u_n$ with the finite conjunction C_l of all instances $l\theta_i$ of l in which l is used in u-closure steps; this is sound, because every pair of a variable assignment and an interpretation which satisfies $\forall u_1 \cdots u_n l$ also satisfies C_l. Now, we can simulate any u-closure step using literals k and l with two applications of the α-rule selecting the respective literals $k\theta$ and $l\theta'$ from the conjunctions C_k and C_l, respectively, followed by a standard $\forall$-closure using $k\theta$ and $l\theta'$. □

To show the completeness of the method is more difficult. The problem is that the very paradigm of the disconnection method is to produce clause instances whereas the u-linking rule blocks this for universal variables. Although in many examples the new method will lead to a shortening of proofs, it may also lead to a lengthening of proofs, as can be seen with the following example in Fig. 4. The normal instantiated tableau on the left can be closed quicker than the u-linking tableau on the right, as the instantiation of the non-local subgoal $Q(z)$ is delayed and thus an additional u-linking step has to be performed to ensure the u-closedness of the tableau.

So with the new calculus it is not possible to simulate the old one step by step, which would significantly facilitate the completeness proof. Instead one should proof completeness by extending the argument given in [14] to the new calculus, which, however, is out of the scope of this paper.

4   Universal Variables and Unit Lemmas

By far the most significant improvement of the performance of the basic disconnection method was achieved by the integration of a special handling of unit clauses. In the resolution context, unit resolution is a very favourable strategy

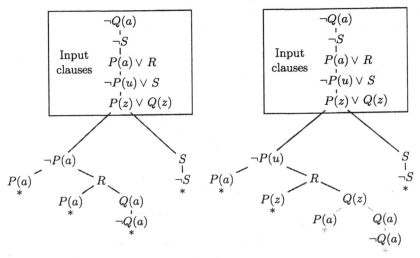

Fig. 4. Proof lengthening by the use of universal variables

for Horn problems, and even though unit resolution is incomplete for non-Horn problems, unit preference is a very successful method in general. Inferences with unit clauses can also improve the search behaviour of hyper-linking calculi [9]. In the tableau context, unit clauses are favourable because they avoid a branching of the tableau.

The treatment of unit clauses in our system DCTP is completely different from the handling of non-unit clauses. On the one hand, no instances of unit clauses are put on the tableau. On the other hand, the closure for unit clauses is u-closure. So in DCTP unit clauses are already treated in a universal manner, since all their variables are trivially universal. The concept of universal variables put forward in this paper can therefore be viewed as a generalisation of the special handling of units to the non-unit case. With the u-linking rule and the u-closure condition, we have achieved a general framework in which the up to now specialised treatment of units becomes standard.

Two further features of unit clauses are *unit subsumption* and *unit simplification*. While the deletion of subsumed clauses on a branch in general is incomplete for the disconnection method, unit subsumption can be safely applied, i.e., any clause can be deleted which is *subsumed* by (i.e., contains a literal which is an instance of) a unit clause. Unit simplification applies to a clause c if a unit clause *subsumes* the complement of a literal l in c. The result is that l is removed from the clause. Obviously, this may eventually lead to the generation of a new unit clause, which can be added to the data base.

So, in the disconnection framework, we can also dynamically generate new unit clauses. There is a further method for generating new unit clauses, which originates from a completely different source, this is so-called *folding up* procedure. Folding up was introduced in the context of tableaux with rigid variables like connection tableaux [12]. The method generalises Shostak's *c-reduction rule* [17] and provides an efficient way of producing *bottom-up lemmas* that can be used in other parts of the proof. However, in the context of the disconnection method with its more general closure condition, folding up has to be adapted appropriately to preserve soundness. The method in the case of the standard disconnection method with standard linking and ∀-closure is as follows.

Definition 11 (Folding Up). *For every branch literal l at the root of a closed sub-tableau T such that l is relevant in T, we can formulate a so-called context lemma $k = \neg l\tau$ where τ is a substitution which identifies all variables in l. The unit k can than be used for branch closure on all branches which are dominated by the lowest relevant literal r in T above l.*

This can be implemented by adding k to the edge immediately above r. As a matter of fact, one has to take care that the relevance information remains correct when context lemmas are used for branch closure. The usefulness of these methods can be already seen at Example 1, with folding up even the standard disconnection procedure can reduce the minimal proof length to five linking steps instead of thirteen.

We will discuss now what happens when u-closure and u-linking are used together with folding up. The problem here is that a naive combination of both features destroys the soundness of the disconnection method, which can be seen from the following example.

Example 2. Consider the following two clauses $c_1 = P(u, x) \vee P(x, x)$ and $c_2 = \neg P(a, y) \vee \neg P(b, y)$.

Assume we start the tableau construction with a u-linking step on the literals $P(u, x)$ and $\neg P(a, y)$. Then the original clauses are put on the tableau, assume first c_1 and below $P(u, x)$ the clause c_2. According to the u-linking rule the universal variable u is not instantiated. Next, we u-close the $\neg P(b, y)$-branch. Now we can walk back and fold up the literal $P(u, x)$ and generate a context unit lemma of the form $\neg P(u, x)\tau$. This situation is depicted in Fig. 5.

But what is the correct substitution τ? If the standard procedure would be used, we would have to generate a context unit lemma of the form $\neg P(x, x)$ and insert it at the root of the tableau. Afterwards it can be used to $\forall$-close the remaining open branch and hence the entire tableau. Unfortunately, the set of the two input clauses is satisfiable. What went wrong is that we did not take into account that the stronger the used closure rule, the weaker the resulting context lemmas.

In order to develop the correct generalisation of folding up in the presence of u-closure, we have to recall that folding up is just an efficient encoding of clauses derivable from certain sub-tableaux, as pointed out in [12]. Let T be a closed sub-tableau with relevant root literal l and S_T the set of clauses occurring in the sub-tableau T. Assume further, $\neg l_1, \ldots, \neg l_n$ be the leaf literals in T whose branches are closed using l and P the disjunction of the leaf literals in T whose branches are closed using literals above l. Then we can conclude that $S_T \models (\neg l_1 \vee \cdots \vee \neg l_n \vee P)\theta$ where θ is a substitution identifying all non-universal variables. Now, in order to achieve that $(\neg l_1 \vee \cdots \vee \neg l_n)\theta$ becomes unit its literals must be unifiable by a u-substitution τ, and only if such a unifier exists, we may generate a context unit. We give now a more implementation-oriented reformulation of this method.

Definition 12 (Folding Up with Universal Variables). *Let T be a closed sub-tableau with relevant root literal l and $\sigma_1, \ldots, \sigma_n$ be the u-substitutions in which l is used in u-closure steps in T. Let further θ be the substitution which maps all non-universal to the same variable x not occurring in the tableau. Then*

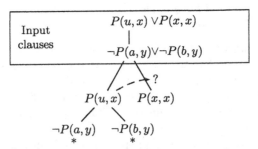

Fig. 5. Folding up with universal variables as in Example 2

we can perform folding up for l only if there is a unifier τ for $l\sigma_1\theta, \ldots, l\sigma_n\theta$ and the generated context unit lemma is $\neg l\tau$ whose literals are all treated as local except for x. The point to which the context unit lemma is folded up is the same as in the standard procedure.

With this method, more and logically stronger context units can be generated (note that in the old approach, context units may only be used for $\forall$-closure and not for u-closure as context-free units).

Furthermore, when context units contain only universal variables, then we can use them also for pruning similar to context-free units. This amounts to a generalisation of the well-known regularity condition [14].

5   Implementation: The DCTP Theorem Prover

The disconnection calculus has been implemented and continuously refined in the theorem prover DCTP. The design and implementation of DCTP is described briefly in [18] and in greater detail in [19]. While the underlying inference mechanism of DCTP has remained unchanged since version 1.2, significant alterations have been made to several aspects of the system for the current version 1.3. These changes include a complete redesign of the preprocessing mechanism, where the connection graph computation has been made redundant by an index unification algorithm. The mechanism for resolving so-called isolated connections [4] has been improved to allow more resolution steps, while still ensuring termination of this technique. The priority for branch closures has been reversed from local to global closures in order to facilitate the generation of more general lemmas.

6   Motivation: The CASC-18 System Competition

The current state of the art with regard to automated theorem proving is demonstrated by the annual CASC prover competitions [21]. After first participating at CASC-JC in 2001, an improved version 1.2 of DCTP took part in CASC-18 in Copenhagen in 2002, along with a strategy parallel version DCTP-10.1p that employed differently parameterised versions of DCTP using the technology of the e-SETHEO system [20]. Certain aspects of CASC-18 outcome deserve attention. The disconnection calculus is able to decide the problems presented in the EPR class of the competition. But even though DCTP was the key strategy used by the winning e-SETHEO system in that class, DCTP itself did not perform quite as well as expected, in particular on the unsatisfiable problems of this class. Many problems that could readily be solved by resolution provers and trivially be solved with the use of propositional Davis-Putnam provers were far beyond the capabilities of DCTP. Also, quite a few of the satisfiable problems of the EPR class could not be solved by any of the participating systems due to the sheer size of those problems of several megabytes each.

In the all-important MIX class of the most general problems, DCTP performed reasonably well on the problems containing equality literals, despite the fact that the equational reasoning component of DCTP is not on the same level of refinement as the top systems for equational problems. On the other hand, DCTP did not perform too well on the set of Horn problems without equality, a problem class that is widely considered as comparatively easy.

The discrepancies between the theoretical expectations and the practical results in both the EPR and MIX classes of CASC-18 strongly suggest that improvements in the implementation of DCTP are possible on a fundamental level. This was the primary motivation for the development of the new features described in this paper.

7   Enhancements for Version 1.3

Preprocessing and the connection graph. During formula preprocessing, DCTP initially used a connection graph on the input set for finding pure clauses and isolated connections. However, for very large problems it is infeasible to compute such a connection graph even once. To improve the scalability of DCTP, the entire concept of the connection graph was dropped. Instead, as part of a complete re-implementation of the formula preprocessing stage, a unification index was introduced.

Factorisation. Clause factoring is a standard technique in resolution systems. Tableau provers, on the other hand, can only incorporate limited versions of factorisation into the proof search, such as the folding up of solved subgoals. It is, however, possible to have full factorisation of the clause set as a part of the formula preprocessing stage. This factorisation is combined with full clause subsumption to limit the number of generated clauses.

Isolated connections. Isolated connections are links in the clause set where one of the linked subgoals has just this one link in all the clauses. This link then can deterministically be resolved upon to reduce the number of input clauses and increase the instantiatedness of the clauses. But resolving one isolated connection can lead to new isolated connections, and if the mechanism is employed to the full it can lead to non-terminating loops of resolution steps. A mechanism was included that guarantees termination while at the same time making near-optimal use of isolated connections. Alternatively, in cases where the resolution of isolated connections is not wanted, the information present can still be used to propagate instantiations.

Better closure heuristics. It is possible that more than one path subgoal is available at a time during proof search to be used for branch closure with a leaf subgoal. Older versions of DCTP used the strategy of selecting the lowest possible subgoal for branch closure. This method had been imported from model elimination and was supposed to keep branch closures as local as possible and therefore provide frequent possibilities for folding up subgoals. But then tests showed that it is far more profitable to use the uppermost closing subgoal on the branch. This way fewer subgoals can be folded up, but they can be folded up to higher points on the tableau and thus are available for more widespread use. Additionally, this *global* selection of closing subgoals better fits into the framework of a proof confluent calculus as it allows more redundant tableau clauses to be pruned.

8   Future Extensions of DCTP

Model extraction. As mentioned above, the disconnection calculus allows the extraction of models from failed proof attempts. The theory of model extraction for the non-equality case has been discussed in detail in [19], also with regard to the effects of calculus refinements on the extraction of models. An implementation of such a model extraction mechanism is still pending.

Relaxation of the active path restriction. The proper selection of a suitable initial active path as explained in Section 2 is crucial for the success of a proof attempt. In order to gain greater independence from this selection process we intend to devise a way that allows us to select arbitrary links from the input set while at the same time limiting the search space growth.

Decidable classes. There are several decidable subclasses of first-order logic, e.g. the monadic fragment. We want to identify further classes of formulae that the disconnection calculus can decide and what special restrictions to the inference process are necessary in order to achieve this goal. This also includes the integration of separate decision procedures into DCTP. Currently we investigate the possibility of extending the disconnection calculus by certain kinds of literal orderings and selection functions to provide a decision procedure for the guarded fragment.

9   Integration of the New Features

Both concepts described in this paper, u-linking and the extended use of context lemmas, have very recently been implemented as part of the theorem prover DCTP [18]. These implementation efforts have required a major reconstruction of large parts of the inference mechanism.

In older versions of DCTP, branch closure was realised as either ∀-closure or *t*-closure (as described in [18, 19]). Both could be implemented in a lean way by membership checks of branch subgoal indexes. The introduction of u-closure on the other hand made it necessary to partially unify branch subgoals. In order to do this efficiently, the entire linking and branch closure mechanism was changed to use a unification index.

Another problem was presented by the way in which DCTP performs linking steps. When a subgoal to be extended and a link to be applied have been selected, DCTP unifies the linked subgoals to create temporarily instantiated versions of the original linked clauses. Only when a linking step has been found not to be redundant, i.e. at least one new clause is placed on the tableau as a result of said linking step, new and variable-disjoint clause copies are created to be placed on the tableau. Finally, the effects of the linking instantiation are undone.

This manner of performing linking steps was introduced to prevent the excessive creation of redundant clause copies.

However, in combination with u-linking a number of problems arise. Consider a tableau clause $c = P(u_1, u_1, x) \vee Q(x)$ with branch literal $P(u_1, u_1, x)$, which contains the local variable u_1 and another tableau clause $d = \neg P(a, u_2, u_2) \vee R(y) \vee S(y)$ with branch literal $\neg P(a, u_2, u_2)$ with u_2 local. If a u-linking step is applied to the link between c and d, the unification temporarily produces the clauses $c' = P(a, a, a) \vee Q(a)$ and $d' = \neg P(a, a, a) \vee R(y) \vee S(y)$. The application of the unifier is necessary to properly instantiate the shared variable x in c. But as local variables are not instantiated, the final clauses resulting from the linking step are $c'' = P(u_1, u_1, a) \vee Q(a)$ and $d'' = \neg P(a, u_2, u_2) \vee R(y) \vee S(y)$. Note that d'' can be deleted as a variant of d while d' cannot. This example illustrates the necessity to make significant changes to the linking mechanism. Either all potential linking instances are immediately produced in their final versions or all procedures for variant deletion, regularity failure or subsumption checks must be properly adapted to handle intermediate instantiations of local variables.

Even though both local variables and extended context lemmas have been included in DCTP in a working fashion, the implementation is not entirely completed. In order to fully exploit the potential of these new calculus enhancements is necessary to integrate them properly into all heuristics and guidance functions.

10   Experiments

We were able to identify a number of test problems from the TPTP problem library [22] where the use of local variables proved advantageous. Table 1 shows the proof times (in seconds) and proof search inferences for these example problems. All tests were conducted on a Linux PC with 512 MB of memory and clocked at 2.4 GHz. The maximum allowed time for each proof attempt was 200 seconds. DCTP was used with a fixed parameterisation largely identical to the one used by DCTP-1.2 at the CASC-18 competition in 2002.

The example of LCL225-1 demonstrates that the number of inferences need not necessarily be reduced to achieve shorter proof times. The LCL type problems often feature very large terms with term depths greater than 10. In these cases it can be useful to employ local variables to avoid creating larger terms. This way even a greater number of simpler inferences can be handled in much less time.

The problem SYN036-1, also known as Andrew's Challenge, had long been unsolvable for DCTP. It first became feasible when clause factoring on the set of input clauses was introduced to the prover. With the inclusion of clause factoring the differences between the number of inferences becomes even greater. With factoring, it takes DCTP 1402 successful inferences to prove SYN036-1 (along with 30951 unsuccessful ones) without local variables, while with local variables the proof can be completed within three inferences. This dramatic reduction in proof search is a direct consequence of the fact that clause factoring, along with clause splitting, is a heuristic that favours the creation of universal variables.

Table 1. Performance of DCTP on selected problems with and without the use of local variables (without the extended use of context lemmas)

Problem	w. local variables		w/o local variables	
	Time	Inferences	Time	Inferences
GRP003-2	1.03	949	9.08	3371
GRP048-2	38.43	18333	timeout	-
LCL004-1	51.19	1970	timeout	-
LCL218-1	3.49	5917	53.42	13496
LCL223-1	21.96	13964	timeout	-
LCL224-1	21.88	13961	timeout	-
LCL225-1	31.83	16322	192.43	15123
LCL230-1	60.04	19997	timeout	-
SYN036-1	2.18	3372	13.49	6432
SYN897-1	0.77	214	14.21	7923

We are well aware of the fact that the experimental results presented above are not fully comprehensive and, theoretically, it is possible to find favourable examples for even the most pathological of refinement techniques. However, the implementation of the concepts described in this paper in the context of the DCTP theorem prover is still of a prototypical kind. The correct and efficient integration of local variables into the proof search does not only require a major reconstruction of many data structures, but also a global adaptation of all pruning and clause deletion techniques and, even more important, the development of new guidance heuristics. This extensive reconstruction of DCTP has not fully been completed yet, but we hope to conclude this work in the forseeable future.

With respect to the other new features of DCTP 1.3, let us take another look at the results of CASC-18 in the light of the refinements described in the previous sections.

There was a number of unsatisfiable ALC problems [7] (e.g. SYN440-1) that DCTP could not solve during the competition. With the use of improved preprocessing, global closure and adapted subgoal selection, all of these ALC problems can be solved by DCTP 1.3. The EPR class also contained a number of very large satisfiable translated QBF problems of more than 10 megabytes each (e.g. SYN852-1). None of these could be solved by either of the participating systems. Due to the improved and scalable preprocessing (an initial number of 3240 clauses can be reduced to 26 clauses) combined with the inherent decision power of the disconnection calculus, all of these problems now can be solved within 80 seconds on a 2.4 GHz PC. A number of similar problems also were featured as part of the SAT class of the competition for satisfiable problems (e.g. SYN904-1). These problems, too, can now be solved by DCTP within 20 seconds.

11   Conclusion and Further Work

In this paper, we have integrated the concept of universal variables into the disconnection tableau calculus. We have also developed an improved method for using context lemmas in order to reduce the proof length and the search space. Both of these enhancements have been implemented as part of the DCTP disconnection tableau prover. Even though the current implementation is not the most efficient possible, in many examples, a significant speed-up could be achieved. Also, the results do not yet reflect the full potential of the new method, since not everything is implemented. So when all techniques are integrated and adapted to the proof search mechanism, we can expect further significant improvements.

Once the integration of local variables and extended context lemmas has been completed we see a number of topics for future work. First, we intend to generalise the concept of the extended context lemmas to a more general form of branch regularity. Then, we want to find a way of extending the use of local variables to the equational case. Until now, local variables are used only in problems without equality literals. The integration of extended context lemmas and local variables into the equational reasoning mechanism of the disconnection calculus presents additional problems, as then local variables can become non-local in eq-instances and eq-instances of context lemmas may have to be placed on the tableau in non-unit clauses. The proper handling of context information in the equational case is an interesting and challenging problem.

References

[1] L. Bachmair and H. Ganzinger. Equational reasoning in saturation-based theorem proving. In *Automated Deduction: A Basis for Applications. Volume I, Foundations: Calculi and Methods*, pages 353–398. Kluwer, Dordrecht, Dordrecht, 1998.

[2] P. Baumgartner. Hyper tableau — the next generation. In *Proceedings, TABLEAUX-98, Oisterwijk, The Netherlands*, volume 1397 of *LNAI*, pages 60–76, 1998.

[3] B. Beckert and R. Hähnle. Analytic tableaux. In *Automated Deduction — A Basis for Applications*, volume I: Foundations, pages 11–41. Kluwer, Dordrecht, 1998.

[4] W. Bibel, R. Letz, and J. Schumann. Bottom-up enhancements of deductive systems. *Artifial Intelligence and Information-Control Systems of Robots*, pages 1–9, 1987.

[5] J.-P. Billon. The disconnection method: a confluent integration of unification in the analytic framework. In *Proceedings, 5th TABLEAUX*, volume 1071 of *LNAI*, pages 110–126, Berlin, 1996. Springer.

[6] Melvin C. Fitting. *First-Order Logic and Automated Theorem Proving*. Springer, second revised edition, 1996.

[7] U. Hustadt, R. Schmidt, and C. Weidenbach. Optimised functional translation and resolution. In Harrie de Swart, editor, *Proceedings, TABLEAUX-98*, volume 1397 of *LNAI*, pages 36–37, Berlin, May 5–8 1998. Springer.

[8] S.-J. Lee and D. Plaisted. Eliminating duplication with the hyper-linking strategy. *Journal of Automated Reasoning*, pages 25–42, 1992.

[9] R. Letz. LINUS: A link instantiation prover with unit support. *Journal of Automated Reasoning*, 18(2):205–210, April 1997.

[10] R. Letz. Clausal tableaux. In Wolfgang Bibel and Peter H. Schmitt, editors, *Automated Deduction — A Basis for Applications*, volume I: Foundations, pages 43–72. Kluwer, Dordrecht, 1998.

[11] R. Letz. Lemma and model caching in decision procedures for quantified boolean formulas. In *Proceedings, TABLEAUX-2002*, volume 2381 of *LNCS*, pages 160–175. Springer-Verlag, July 30–August 1 2002.

[12] R. Letz, K. Mayr, and C. Goller. Controlled integration of the cut rule into connection tableau calculi. *Journal of Automated Reasoning*, 13(3):297–337, December 1994.

[13] R. Letz, J. Schumann, S. Bayerl, and W. Bibel. SETHEO: A high-performance theorem prover. *Journal of Automated Reasoning*, 8(2):183–212, 1992.

[14] R. Letz and G. Stenz. Proof and Model Generation with Disconnection Tableaux. In Andrei Voronkov, editor, *Proceedings, 8th LPAR, Havanna, Cuba*, pages 142–156. Springer, Berlin, December 2001.

[15] Reinhold Letz and Gernot Stenz. Integration of Equality Reasoning into the Disconnection Calculus. In *Proceedings, TABLEAUX-2002, Copenhagen, Denmark*, volume 2381 of *LNAI*, pages 176–190. Springer, Berlin, 2002.

[16] A. Riazanov and A. Voronkov. Vampire 1.1 (System Description). In *Proceedings, 1st IJCAR, Siena, Italy*, volume 2083 of *LNAI*, pages 376–380. Springer, 2001.

[17] R. E. Shostak. Refutation graphs. *Artificial Intelligence*, 7:51–64, 1976.

[18] G. Stenz. DCTP 1.2 – System Abstract. In *Proceedings, TABLEAUX-2002, Copenhagen, Denmark*, volume 2381 of *LNAI*, pages 335–340. Springer, Berlin, 2002.

[19] G. Stenz. *The Disconnection Calculus*. Logos Verlag, Berlin, 2002. Dissertation, Fakultät für Informatik, Technische Universität München.

[20] Gernot Stenz and Andreas Wolf. E-SETHEO: An Automated[3] Theorem Prover – System Abstract. In Roy Dyckhoff, editor, *Proceedings of TABLEAUX-2000, St. Andrews, Scotland*, volume 1847 of *LNAI*, pages 436–440. Springer, Berlin, July 2000.

[21] G. Sutcliffe and C. Suttner. The CADE-14 ATP system competition. Technical Report JCU-CS-98/01, Department of Computer Science, James Cook University, 3 March 1998.

[22] G. Sutcliffe, C. Suttner, and T. Yemenis. The TPTP problem library. In A. Bundy, editor, *Proceedings, 12th CADE, Nancy, France*, LNCS 814, pages 708–722. Springer, 1994.

[23] C. Weidenbach. Spass - version 0.49. *Journal of Automated Reasoning*, 18(2):247–252, April 1997.

[24] C. Weidenbach. *Handbook of Automated Reasoning*, chapter Combining Superposition, Sorts and Splitting. Elsevier, 1999.

A Tableau Algorithm for Reasoning about Concepts and Similarity

Carsten Lutz[1], Frank Wolter[2], and Michael Zakharyaschev[3]

[1] Institut für Theoretische Informatik, TU Dresden
Fakultät Informatik, 01062 Dresden, Germany
lutz@tcs.inf.tu-dresden.de
[2] Department of Computer Science, University of Liverpool
Liverpool L69 7ZF, UK
frank@csc.liv.ac.uk
[3] Department of Computer Science, King's College London
Strand, London WC2R 2LS, UK
mz@dcs.kcl.ac.uk

Abstract. We present a tableau-based decision procedure for the fusion (independent join) of the expressive description logic $\mathcal{ALCQO}$ and the logic $\mathcal{MS}$ for reasoning about distances and similarities. The resulting 'hybrid' logic allows both precise and approximate representation of and reasoning about concepts. The tableau algorithm combines the existing tableaux for the components and shows that the tableau technique can be fruitfully applied to fusions of logics with nominals—the case in which no general decidability transfer results for fusions are available.

1 Introduction

Undoubtedly, there will come a day when, to attract submissions, organisers will be trying to annotate their conference sites with machine readable information. Imagine, for instance, that we want to do this now for *Tableaux 2003*. Choosing a formalism for representation of and reasoning about the terminology used in the Tableaux 2003 site, we may naturally try the description logic $\mathcal{ALCQO}$ underlying the DAML+OIL language of the semantic web [7, 1]. Then we start with a definition of tableau-style algorithms and, as a first attempt, write something like this:

$$\text{Tableau\_style\_algorithm} = \text{Algorithm} \sqcap \exists \text{comprises.Rule}, \qquad (1)$$

saying that tableau-style algorithms are precisely those algorithms that are equipped with rules. Well, it seems unlikely that any potential participant of *Tableaux 2003* would be happy with this provocative definition (according to which almost all reasoning procedures may be called tableau-based). Then how to improve it? Do we really have a good, clear and concise definition (which is better than 'lots of rules, but few axioms')? How can we represent in $\mathcal{ALCQO}$ many other 'vague' concepts from the site, such as 'related techniques,' 'related methods,' 'new calculi,' etc.?

M. Cialdea Mayer and F. Pirri (Eds.): TABLEAUX 2003, LNAI 2796, pp. 134–149, 2003.
© Springer-Verlag Berlin Heidelberg 2003

One of the possible solutions to these problems is to introduce a *similarity measure* between the objects of the application domain—in our case the reasoning procedures (which can be based on common sense, or defined by an expert, or automatically generated using certain algorithms). Then, by taking a role name similar_to_degree ≤ 1, we could say, for instance, that tableau-style algorithms are similar to degree ≤ 1 to at least one of the prototypical tableau algorithms $ta_1, \ldots, ta_7$. However, this approach is in conflict with the expressive capabilities of standard description logics (DLs) such as $\mathcal{ALC}$ or DAML+OIL because usually similarity measures are supposed to satisfy a number of natural axioms like the axioms of *metric spaces*, in particular, a sort of 'triangular inequality' which is *not expressible in standard DLs*.

The main idea of this paper is not to extend the family of DLs by introducing a new one, but rather to *combine* the existing knowledge representation formalisms, viz.,

- the standard description logic $\mathcal{ALCQO}$—i.e., the basic DL $\mathcal{ALC}$ extended with qualified number restrictions, nominals and general TBoxes [5], and
- the logic $\mathcal{MS}$ [13] for reasoning about metric spaces[1]

in order to achieve the desirable expressivity.

To illustrate the expressive power of the resulting 'hybrid' logic $sim\text{-}\mathcal{ALCQO}$, we show how one can further 'approximate' the definition of tableau-style algorithms. First, we add to the right-hand side of (1) the conjunct

$$\mathsf{E}^{\leq 1}(ta_1 \sqcup \cdots \sqcup ta_7)$$

which is an $\mathcal{MS}$-formula saying that tableau-style algorithms should be similar to degree ≤ 1 to at least one of $ta_1, \ldots, ta_7$. If this 'positive information' is still not enough, one can add some 'negative' bit. For example, it may be natural to say that tableau-style algorithms are neither similar to degree ≤ 0.5 to a certain Hilbert-style algorithm ha, nor similar to degree ≤ 0.5 to any resolution-based decision procedure:

$$\neg \mathsf{E}^{\leq 0.5} ha \sqcap \neg \mathsf{E}^{\leq 0.5} \mathsf{Resolution\_based\_algorithm}.$$

Of course, the individual algorithms such as ha can also be described by means of concepts, possibly involving similarity measures:

$$ha : \mathsf{Algorithm} \sqcap \neg \exists \mathsf{feature}.\mathsf{Termination} \sqcap \mathsf{A}^{\leq 0.5}(\exists \mathsf{comprises}.\mathsf{Modus\_ponens})$$

(i.e., ha does not necessarily terminate and all ≤ 0.5 similar algorithms use a kind of *modus ponens* as one of their inference rules). It may seem more natural to specify similarity in terms of a finite set of symbolic similarity measures such as 'close' and 'far' rather than in terms of rational numbers as above. In our

---

[1] This metric logic differs considerably from the metric logics investigated in [9]. Here we quantify over open and closed 'balls,' while in [9] over closed balls and their complements. The expressive power of the two languages is, therefore, incomparable.

approach, however, the user is free to choose either option: one may fix a rational number for each symbolic similarity measure, say, 1 for 'close' and 10 for 'far' (or the other way round), and then work with the symbolic names.

In this paper, we provide a tableau-style decision procedure for the new logic *sim-ALCQO*. Technically, this logic is the *fusion* (or *independent join*) [8, 3] of *ALCQO* and *MS*. We believe that this is a reasonable starting point, since many similarity measures are indeed metric, and our approach without any problems can be adapted to similarity measures which do *not* satisfy all of the axioms of metric spaces. Moreover, we can easily extend *sim-ALCQO* and the tableau algorithm with additional similarity measures (say, between inference rules).

In our opinion, *sim-ALCQO* provides just the right compromise between expressive power and computational cost:

(1) In *sim-ALCQO*, we can mix constructors of *ALCQO* and *MS* in order to define concepts based on similarity measures as illustrated above. Moreover, as our tableau algorithm shows, reasoning in *sim-ALCQO* is decidable. It is of interest to contrast this with the fact that a tighter coupling of *ALCQO* and *MS* leads to undecidability: as we also show, the extension of *MS* with qualified number restrictions such as 'there exists at most 1 point x with property P within distance ≤ 1' results in an undecidable logic. Therefore, the fusion of the two formalisms seems to be a good starting point for investigating the interaction between concepts and similarity measures.

(2) Although there exists a number of general results regarding the transfer of decidability from the components of a fusion to the fusion itself [8, 3, 12, 2, 11], these results do not apply to logics with nominals (atomic concepts interpreted as singleton sets) such as *ALCQO*. In fact, no transfer result is available from which we could derive the decidability of *sim-ALCQO* using the decidability of both *ALCQO* and *MS*. Despite the fact that they are not applicable, it is of interest to note that our algorithm has an important advantage over general approaches to proving decidability: structurally, it is very similar to the tableau algorithms for *SHIQ* and *SHOQ* proposed in [6, 5]. Since these algorithms have turned out to be implementable in efficient reasoning systems, we do hope that our algorithm shares this attractive property as well.

The paper is organised as follows: in Section 2, we introduce the description logic *sim-ALCQO*. In Section 3, we describe the tableau algorithm for deciding the satisfiability of *sim-ALCQO*-knowledge bases, whose correctness is then proved in Section 4. Section 5 is concerned with the undecidability of *MS* extended with qualifying number restrictions. A version of this paper with detailed proofs is available at http://www.csc.liv.ac.uk/~frank.

2   The Logic *sim-ALCQO*

In this section, we introduce the combined logic *sim-ALCQO*. The *alphabet* for forming concepts and assertions consists of the following elements:

- a countably infinite list of *concept names* $A_1, A_2, \ldots$;
- a countably infinite list of *object names* $\ell_1, \ell_2, \ldots$;
- binary *distance* (δ), *equality* ($=$) and *membership* ($:$) *predicates*;
- the *Boolean operators* $\sqcap$, $\sqcup$, $\neg$;
- two *distance quantifiers* $\mathsf{E}^{<a}$, $\mathsf{E}^{\leq a}$ and their duals $\mathsf{A}^{<a}$, $\mathsf{A}^{\leq a}$, for every positive rational number a (i.e., $a \in \mathbb{Q}^+$);
- *role names* $R_1, R_2, \ldots$;
- *qualified number restrictions* $(\leq nR.C)$ and $(\geq nR.C)$, for every natural n, every role name R, and every concept C.

Using this alphabet, *sim-$\mathcal{ALCQO}$-concepts* are defined by the formation rule:

$$C ::= A_i \mid \ell_i \mid \neg C \mid C_1 \sqcap C_2 \mid C_1 \sqcup C_2 \mid \mathsf{E}^{<a}C \mid \mathsf{E}^{\leq a}C \mid \mathsf{A}^{<a}C \mid$$
$$\mid \mathsf{A}^{\leq a}C \mid (\leq nR_i.C) \mid (\geq nR_i.C).$$

As usual, we write $\top$ as an abbreviation for an arbitrary propositional tautology, $\bot$ for $\neg\top$, $\exists R.C$ for $(\geq 1R.C)$, and $\forall R.C$ for $(\leq 0R.\neg C)$. At first sight, it may seem strange to have both strict and non-strict versions of the E and A constructors available for talking about similarity measures. Note, however, that this allows us to define the concept $\mathsf{E}^{\leq a}C \sqcap \neg\mathsf{E}^{<a}C$ which states that the most similar object from C is located precisely at distance a. Object names occurring in concepts will also be called *nominals*.

Now we define *sim-$\mathcal{ALCQO}$-assertions* as expressions of the following forms:

- $\ell : C$, where ℓ is an object name and C a concept;
- $C_1 = C_2$, where C_1 and C_2 are concepts;
- $\delta(k, \ell) < a$, $\delta(k, \ell) \leq a$, $\delta(k, \ell) > a$, $\delta(k, \ell) \geq a$, where k, ℓ are object names and $a \in \mathbb{Q}^+$.

Assertions of the third form are called *distance assertions*. A *sim-$\mathcal{ALCQO}$-knowledge base* is a finite set of *sim-$\mathcal{ALCQO}$-assertions*.

Observe that knowledge bases subsume both general TBoxes and ABoxes. In particular, the rather common ABox assertions of the form $(\ell_1, \ell_2) : R$, where ℓ_1 and ℓ_2 are object names and R a role name, can be viewed as abbreviations for $\ell_1 : \exists R.\ell_2$.

The semantics of *sim-$\mathcal{ALCQO}$-concepts* is a blend of the semantics of the logic of metric spaces [13] and the usual set-theoretic semantics of description logics. A *concept-distance model* (a *CD-model*, for short) is a structure of the form

$$\mathfrak{B} = \langle W, d, A_1^{\mathfrak{B}}, A_2^{\mathfrak{B}}, \ldots, R_1^{\mathfrak{B}}, R_2^{\mathfrak{B}}, \ldots, \ell_1^{\mathfrak{B}}, \ell_2^{\mathfrak{B}} \ldots \rangle,$$

where $\langle W, d \rangle$ is a *metric space* with a *distance function* d satisfying, for all $x, y, z \in W$, the axioms

$$d(x, y) = 0 \quad \text{iff} \quad x = y, \tag{2}$$
$$d(x, z) \leq d(x, y) + d(y, z), \tag{3}$$
$$d(x, y) = d(y, x), \tag{4}$$

the $A_i^{\mathfrak{B}}$ are subsets of W, the $R_i^{\mathfrak{B}}$ are binary relations on W, and the $\ell_i^{\mathfrak{B}}$ are singleton subsets of W such that $i \neq j$ implies $\ell_i^{\mathfrak{B}} \neq \ell_j^{\mathfrak{B}}$.

The extension $C^{\mathfrak{B}}$ of a $sim\text{-}\mathcal{ALCQO}$-concept C is computed inductively:

$$(C_1 \sqcap C_2)^{\mathfrak{B}} = C_1^{\mathfrak{B}} \cap C_2^{\mathfrak{B}}, \quad (C_1 \sqcup C_2)^{\mathfrak{B}} = C_1^{\mathfrak{B}} \cup C_2^{\mathfrak{B}}, \quad (\neg C)^{\mathfrak{B}} = W - C^{\mathfrak{B}},$$

$$(\mathsf{E}^{\leq a}C)^{\mathfrak{B}} = \{x \in W \mid \exists y \in W \; (d(x,y) \leq a \; \wedge \; y \in C^{\mathfrak{B}})\},$$
$$(\mathsf{E}^{<a}C)^{\mathfrak{B}} = \{x \in W \mid \exists y \in W \; (d(x,y) < a \; \wedge \; y \in C^{\mathfrak{B}})\},$$
$$(\mathsf{A}^{\leq a}C)^{\mathfrak{B}} = \{x \in W \mid \forall y \in W \; (d(x,y) \leq a \; \rightarrow \; y \in C^{\mathfrak{B}})\},$$
$$(\mathsf{A}^{<a}C)^{\mathfrak{B}} = \{x \in W \mid \forall y \in W \; (d(x,y) < a \; \rightarrow \; y \in C^{\mathfrak{B}})\},$$
$$(\leq nR.C)^{\mathfrak{B}} = \{x \in W \mid |\{y \in W \mid (x,y) \in R^{\mathfrak{B}} \wedge y \in C^{\mathfrak{B}}\}| \leq n\},$$
$$(\geq nR.C)^{\mathfrak{B}} = \{x \in W \mid |\{y \in W \mid (x,y) \in R^{\mathfrak{B}} \wedge y \in C^{\mathfrak{B}}\}| \geq n\}.$$

We still have to specify when a CD-model satisfies a $sim\text{-}\mathcal{ALCQO}$-assertion: the *truth-relation* $\models$ between CD-models $\mathfrak{B}$ and assertions φ is defined as follows:

- $\mathfrak{B} \models \ell : C$ iff $\ell^{\mathfrak{B}} \subseteq C^{\mathfrak{B}}$,
- $\mathfrak{B} \models C_1 \doteq C_2$ iff $C_1^{\mathfrak{B}} = C_2^{\mathfrak{B}}$,
- $\mathfrak{B} \models \delta(k,\ell) \leq a$ iff $d(k^{\mathfrak{B}}, \ell^{\mathfrak{B}}) \leq a$,
- $\mathfrak{B} \models \delta(k,\ell) < a$ iff $d(k^{\mathfrak{B}}, \ell^{\mathfrak{B}}) < a$, and similar for $\geq$ and $>$.

Finally, a $sim\text{-}\mathcal{ALCQO}$-knowledge base Σ is called *satisfiable* if there exists a CD-model $\mathfrak{B}$ such that $\mathfrak{B} \models \varphi$ for all $\varphi \in \Sigma$. In this case we write $\mathfrak{B} \models \Sigma$.

Note that we make the *unique name assumption* (*UNA*), i.e., different object names denote distinct domain elements. The sole purpose of this assumption is to allow a clearer presentation of our tableau algorithm. It is, however, easily seen that the UNA has no influence on decidability, and that our tableau algorithm can be extended to deal with $sim\text{-}\mathcal{ALCQO}$ without UNA.

3 The Tableau Algorithm

Now we present a sound, complete and terminating algorithm for checking the satisfiability of $sim\text{-}\mathcal{ALCQO}$-knowledge bases. In fact, it is a (labelled) tableau algorithm that generalises the existing tableau algorithms for metric logics [13] and for the description logic $\mathcal{ALCQO}$ [5]. Before formulating the algorithm and proving its correctness, we introduce some notations and auxiliary definitions.

Suppose we are given a $sim\text{-}\mathcal{ALCQO}$-knowledge base Σ. Denote by $con(\Sigma)$ the set of concepts occurring in Σ (including all subconcepts), by $rol(\Sigma)$ the set of role names occurring in Σ, by $par(\Sigma)$ the set of rational numbers occurring in Σ (either in E/A concepts or in distance assertions), and by $ob(\Sigma)$ we denote the set of object names occurring in Σ. Without loss of generality, we may assume that neither $par(\Sigma)$ nor $ob(\Sigma)$ are empty: if this is not the case, we can always add an assertion $\ell : \mathsf{A}^{<a}\top$ with a fresh object name ℓ. To simplify presentation, it is convenient to make three assumptions:

(1) A concept C is in *negation normal form* (*NNF*) if negation occurs only in front of concept names and nominals. Each concept can be transformed into an equivalent one in NNF by pushing negation inwards: for example, $\neg E^{<a} C$ is equivalent to $A^{<a} \neg C$. So, without loss of generality, we may assume that all concepts are in NNF. In what follows, we use $\dot{\neg} C$ to denote the NNF of $\neg C$.

(2) We may also assume that knowledge bases contain only assertions of the form $\ell : C$ and $C \doteq \top$. To see this, note first that distance assertions can be expressed using nominals and distance quantifiers:

$$\delta(k, \ell) < a \text{ is equivalent to } k : E^{<a} \ell, \quad \delta(k, \ell) \le a \text{ is equivalent to } k : E^{\le} \ell,$$
$$\delta(k, \ell) > a \text{ is equivalent to } k : A^{\le a} \neg \ell, \quad \delta(k, \ell) \ge a \text{ is equivalent to } k : \neg A^{<a} \neg \ell.$$

Assertions of the form $C_1 \doteq C_2$ can be rewritten as $(C_1 \sqcap C_2) \sqcup (\dot{\neg} C_1 \sqcap \dot{\neg} C_2) \doteq \top$.

(3) Without loss of generality, we may assume that $par(\Sigma)$ contains only natural numbers: given a knowledge base Σ with $par(\Sigma) \subseteq \mathbb{Q}^+$, we may replace every element q of $par(\Sigma)$ with $q \cdot x$, where x is the least common multiple of the denominators of all elements of $par(\Sigma)$. It is then straightforward to show that any CD-model of the resulting knowledge base can be converted into a CD-model of Σ and vice versa.

We use α_Σ to denote the largest natural number that occurs in $par(\Sigma)$ and $M[\Sigma]$ to denote the smallest set satisfying the following conditions:

- $par(\Sigma) \subseteq M[\Sigma]$;
- if $a, b \in M[\Sigma]$ and $a + b < \alpha_\Sigma$, then $a + b \in M[\Sigma]$;
- if $a, b \in M[\Sigma]$ and $a - b > 0$, then $a - b \in M[\Sigma]$.

Having started on the input knowledge base Σ (in the form described above), the tableau algorithm considers only certain 'relevant' concepts. More precisely, we define the *closure* $cl(\Sigma)$ of Σ to be the (finite) set of concepts

$$con(\Sigma) \cup \{\dot{\neg} C \mid C \in con(\Sigma)\} \cup$$
$$\{A^{<a} C, A^{\le a} C \mid a \in M[\Sigma] \text{ and } \exists b \ge a \{A^{\le b} C, A^{<b} C\} \cap con(\Sigma) \ne \emptyset\}.$$

Similar to the set $cl(\Sigma)$ of relevant concepts, $M[\Sigma]$ describes the set of relevant numbers. However, the numbers in $M[\Sigma]$ are not enough: to distinguish between '$\le a$' and '$< a$,' we require some additional symbols that will be used in the same way as numbers, namely, $M[\Sigma]^- = \{a^- \mid a \in M[\Sigma]\}$. Define a strict linear order $\prec$ on $M[\Sigma] \cup M[\Sigma]^-$ by setting

$$a_1^- \prec a_1 \prec a_2^- \prec a_2 \prec \cdots \prec a_n^- \prec a_n,$$

where $a_1 < a_2 < \cdots < a_n$.

We are in a position now to describe our tableau algorithm. Starting with Σ, it operates on *constraint systems* $\mathcal{S} = \langle T, <, L, S, E \rangle$, where

- $\langle T, < \rangle$ is a forest whose set of roots coincides with $ob(\Sigma)$;
- S is a *node labelling function* which associates with each $x \in T$ a set

$$S(x) \subseteq cl(\Sigma) \cup \{(R, \ell), (a, \ell), (a^-, \ell) \mid \ell \in ob(\Sigma), R \in rol(\Sigma), a \in M[\Sigma]\};$$

- L is a *labelling function* which associates with each pair $x, y \in T$ such that $x < y$ either a role name or a number from $M[\Sigma]$, or a symbol from $M[\Sigma]^-$;
- E is a set of inequalities between members of T.

Intuitively, we have $x < y$ if either x and y are related by some role R or the distance between x and y is known to be smaller than some value from $M[\Sigma]$. The purpose of the extra elements (R, ℓ) and (a, ℓ) in node labels is to represent additional edges that lead to nominals (roots in the forest), and whose explicit representation would destroy the forest structure.

The algorithm starts with $\mathcal{S}_0 = \langle T_0, <_0, L_0, S_0, E_0 \rangle$, the *initial constraint system* for Σ, where

- $T_0 = ob(\Sigma)$,
- $S_0(\ell) = \{\ell\} \cup \{C \mid \ell : C \in \Sigma\}$, for every $\ell \in ob(\Sigma)$,
- $E_0 = \{\ell \neq \ell' \mid \ell \neq \ell', \ \ell, \ell' \in ob(\Sigma)\}$, and
- $<_0 = L_0 = \emptyset$.

Before describing the completion rules, we introduce some simplifying notation required to deal with edges represented via node labels. We write $\overline{L}(x, y) = a$ to express that either $x < y$ and $L(x, y) = a$ or that a is the $\prec$-minimum of $\{c \mid (c, y) \in S(x)\}$.[2] To account for the fact that, for some rules, it is not important whether a node is a predecessor or a successor, we write $L^o(\{x, y\}) = a$ if a is the $\prec$-minimum of $\{\overline{L}(x, y), \overline{L}(y, x)\}$. Finally, for a role name R, we say that y is an R-successor of x if either $x < y$ and $L(x, y) = R$ or $(R, y) \in S(x)$.

The *completion rules* are shown in Fig. 1. Constraint systems obtained by applying the completion rules to the initial constraint system for Σ will be called *constraint systems for* Σ. The terms 'blocked' and 'indirectly blocked' in the rule premises refer to a cycle detection mechanism that is needed to ensure termination of the algorithm. Before discussing the completion rules in more detail, let us formally introduce this mechanism. The general idea is that we stop the expansion of node labels if a node is labelled with exactly the same set of concepts as one of its $<$-ancestors. This simple approach works perfectly well, but it is not the most sensible thing we can do: the problem is that, due to the 'extra' concepts $A^{<^a}C$ and $A^{\leq^a}C$, the size of $cl(\Sigma)$ is *exponential* in the size of Σ rather than polynomial, and thus paths of the forest may grow to a length doubly exponential in Σ before the blocking occurs. Fortunately, this worst case can be avoided. When comparing node labels to check for a blocking situation, it is not necessary to take into account *all* of the extra $A^{<^a}C$ and $A^{\leq^a}C$ concepts: if, for example, we find $A^{\leq^a}C \in S(x)$, then it is clear that the object x also satisfies the concepts $A^{\leq^b}C$ for all $b \leq a$, even if they do not explicitly appear

---

[2] This gives a well-defined value for $\overline{L}(x, y)$, as $(c, y) \in S(x)$ implies that y is a root.

in the node label $S(x)$. This observation leads to the following, refined variant of blocking.

For a node $x \in T$, we use $S^*(x)$ to denote the set of concepts $C \in S(x)$ such that one of the following conditions is satisfied:

1. C is not of the form $\mathsf{A}^{<a}D$ or $\mathsf{A}^{\leq a}D$;
2. C is of the form $\mathsf{A}^{\leq a}D$ and there is no $b > a$ such that $\mathsf{A}^{\leq b}D \in S(x)$;
3. C is of the form $\mathsf{A}^{<a}D$ and there is no $b > a$ such that $\mathsf{A}^{<b}D \in S(x)$.

Denote by $<^+$ the transitive closure of $<$. We say that a node $x \in T$ is *directly blocked by a node* y if $y <^+ x$, $S^*(x) = S^*(y)$, but for no distinct $u <^+ x$ and $v <^+ x$ do we have $S^*(u) = S^*(v)$. The $<^+$-successors of directly blocked nodes are called *indirectly blocked*. All directly or indirectly blocked nodes comprise the set of *blocked nodes*. Observe that the elements (R, ℓ) and (a, ℓ) of node labels are not taken into account for blocking.

Note that this blocking condition can be refined even further by taking into account implications between $\mathsf{A}^{\leq a}C$ and $\mathsf{A}^{<b}C$ concepts. We prefer to work with the above variant, since it suffices to restrict paths in forests to exponential length, and the more elaborate version makes proofs rather unreadable due to many additional case distinctions.

Let us now return to the completion rules. In what follows we assume that a rule can be applied to a tableau only if the tableau is changed. Such a rule will be called *applicable* to the tableau. The tableau algorithm applies the rules until either the obtained constraint system contains an obvious contradiction or no more rules are applicable. To be more precise, say that a constraint system S *contains a clash* if it contains a node x such that one of the following conditions hold:

1. $\{A, \neg A\} \subseteq S(x)$, for some concept name A;
2. $\{\ell, \neg \ell\} \subseteq S(x)$ for some object name ℓ;
3. $\ell' \in S(\ell)$ for some object names $\ell' \neq \ell$;
4. $(x \neq x) \in E$;
5. for some R, $(\leq nR.C) \in S(x)$ and there are $n+1$ R-successors $y_0, \ldots, y_n$ of x with $C \in L(y_i)$, for each $0 \leq i \leq n$ and $y_i \neq y_j \in E$ for each $0 \leq i < j \leq n$.

A constraint system S is *complete* if it either contains a clash or none of the rules in Fig. 1 is applicable to S.

4   Termination, Soundness and Completeness

We show now that the tableau algorithm above always *terminates*, is *sound* (i.e., if there is a complete and clash-free constraint system for Σ, then Σ is satisfiable), and *complete* (i.e., if Σ is satisfiable, then the tableau algorithm eventually succeeds in finding a complete and clash-free complete system).

$R_\sqcap$ If $C_1 \sqcap C_2 \in S(x)$ and x is not indirectly blocked,
then set $S(x) := S(x) \cup \{C_1, C_2\}$.

$R_\sqcup$ If $C_1 \sqcup C_2 \in S(x)$ and x is not indirectly blocked,
then set either $S(x) := S(x) \cup \{C_1\}$ or $S(x) := S(x) \cup \{C_2\}$.

$R_=$ If $C = \top \in \Sigma$ and x is not indirectly blocked, then set $S(x) := S(x) \cup \{C\}$.

R_A If $A^{<a}C \in S(x)$ or $A^{\leq a}C \in S(x)$ and x is not indirectly blocked,
then set $S(x) := S(x) \cup \{C\}$.

$R_{A<}$ Let $A^{<a}C \in S(x)$ and x is not indirectly blocked. Then:
if $L^o(\{y, x\}) = a^-$, then set $S(y) := \{C\} \cup S(y)$;
if $L^o(\{y, x\}) = b < a$, then set $S(y) := \{A^{<a-b}C\} \cup S(y)$;
if $L^o(\{y, x\}) = b^-$ with $b < a$, then set $S(y) := \{A^{\leq a-b}C\} \cup S(y)$.

$R_{A\leq}$ Let $A^{\leq a}C \in S(x)$, $L^o(\{y, x\}) \in \{b, b^-\}$ and x is not indirectly blocked. Then:
if $b = a$, then set $S(y) := \{C\} \cup S(y)$;
if $b < a$, then set $S(y) := \{A^{\leq a-b}C\} \cup S(y)$.

$R_{E<}$ If $E^{<a}C \in S(x)$, x is not blocked, and
$\overline{L}(x, y) \notin \{b \mid b < a\} \cup \{b^- \mid b \leq a\}$ for any y with $C \in S(y)$,
then create a new node $y > x$ and set $L(x, y) := a^-$ and $S(y) := \{C\}$.

$R_{E\leq}$ If $E^{\leq a}C \in S(x)$, x is not blocked and
$\overline{L}(x, y) \notin \{b \mid b \leq a\} \cup \{b^- \mid b \leq a\}$ for any y with $C \in S(y)$,
then create a new node $y > x$ and set $L(x, y) := a$ and $S(y) := \{C\}$.

R_{ch} If $\{(\geq nR.C), (\leq nR.C)\} \cap S(x) \neq \emptyset$, x is not blocked and y is an
R-successor of x, then set $S(y) := S(y) \cup \{C\}$ or $S(y) = S(y) \cup \{\neg C\}$.

$R_\geq$ If $(\geq nR.C) \in S(x)$, x is not blocked, and there are no R-successors $y_1, \ldots, y_n$
with $C \in S(y_i)$ and $y_i \neq y_j \in E$, for all $i \neq j$, then take new $y_1 > x, \ldots, y_n > x$
and set $L(x, y_i) := R$, $S(y_i) := \{C\}$, $E := E \cup \{y_i \neq y_j \mid 1 \leq i < j \leq n\}$.

$R_\leq$ If $(\leq nR.C) \in S(x)$, x is not blocked, has $n + 1$ R-successors $y_0, \ldots, y_n$
with $C \in S(y_i)$ for all i, and, for some $i, j \leq n$, $y_i \neq y_j \notin E$ and $y_j \notin ob(\Sigma)$,
then set $E := E \cup \{y \neq y_i \mid y \neq y_j \in E\}$, $S(y_i) := S(y_i) \cup S(y_j)$,
$S(x) := S(x) \cup \{(R', \ell) \mid R' = L(x, y_j)\}$, if $y_i = \ell \in ob(\Sigma)$,
and finally delete y_j and all z with $y_j <^+ z$ from T.

R_ℓ If $\ell \in S(x)$, $x \notin ob(\Sigma)$, and x is not indirectly blocked,
Then set $S(\ell) := S(\ell) \cup S(x)$, and, for every y,
$S(y) := S(y) \cup \{(c, \ell) \mid c = \overline{L}(y, x)$ or $c = R$ a role and x an R-successor of $y\}$,
$E := E \cup \{y \neq \ell \mid y \neq x \in E\}$, and delete x and all z with $x <^+ z$ from T.

Fig. 1. Tableau rules

Termination

Theorem 1. *Any sequence of applications of tableau rules to the initial constraint system for Σ terminates after finitely many steps.*

Proof. Let $m_0 = |con(\Sigma)|$ and m_q be the maximal number occurring in qualified number restrictions of Σ. Termination follows from the following five observations.

(1) Each rule except $R_\leq$ and R_ℓ strictly extends the constraint system. Moreover, neither R_ℓ nor $R_\leq$ removes concepts from nodes.

(2) None of the generating rules $R_{E<}$, $R_{E\leq}$, $R_\geq$ can be applied more than once to a given node and a given concept.

Suppose that $R_{E<}$ is applied to a node x, generates y with $x < y$ and updates $L(x,y) = a^-$ and $S(y) = \{C\}$. The only reason why $R_{E<}$ could be applied once again to x and $E^{<a}C$ is that later on y is removed by an application of $R_\leq$ or R_ℓ. However, unless x is removed (in this case the claim is trivial) y cannot be removed by an application of $R_\leq$ because we do not find a z and a role R with $R = L(z,y)$. Suppose y is removed by an application of R_ℓ because $\ell \in S(y)$. Then, after the application of R_ℓ, we have $(a^-, \ell) \in S(x)$ and $C \in S(\ell)$, since $a^- = L(x,y)$. But then, since a node of the form ℓ is never removed, the rule $R_{E<}$ is not applicable to x and $E^{<a}C$ afterwards. The rule $R_{E\leq}$ is considered analogously.

Suppose that $R_\geq$ is applied to a node x, generates $y_1, \ldots, y_n$ with $x < y_i$ and updates $L(x,y_i) = R$, $S(y_i) = \{C\}$, and $E = E \cup \{y_i \neq y_j \mid 1 \leq i < j \leq n\}$. Now, whenever some y_j is removed by $R_\geq$ or R_ℓ and x is not removed, after the removal of y_j we still have n R-successors $z_1, \ldots, z_n$ of x such that $C \in S(z_i)$, $E \supseteq \{z_i \neq z_j \mid 1 \leq i < j \leq n\}$. So, $R_\geq$ is not applied to x after such a removal.

(3) The out-degree of the forest constructed using the tableaux rules is bounded by $m_0 + m_q \cdot m_0$. This follows from (2) and the fact that nodes are labelled with subsets of the set

$$cl(\Sigma) \cup \{(R, \ell), (a, \ell), (a^-, \ell) \mid \ell \in ob(\Sigma), R \in rol(\Sigma), a \in M[\Sigma]\}.$$

(4) If a node x is removed, then all z with $x <^+ z$ are removed as well

(5) No $<$-branch in any constraint system for Σ can ever be of length exceeding $2^{m_0} \cdot |M[\Sigma]|^2$, since no node introducing rule can be applied to a node x such that $S^*(y) = S^*(z)$ for two distinct $y, z \leq x$.

Soundness

Before proving the soundness of the tableau algorithm, we introduce a relational semantics for $sim\text{-}\mathcal{ALCQO}$. This semantics comprises, for each $a \in M[\Sigma]$, additional binary relations R_a and S_a such that, intuitively, we have uR_av if the distance between u and v is at most a, and uS_av if the distance between u and v is less than a. Formally, a *Kripke model for Σ* is a structure of the form

$$\mathfrak{M} = \langle W, A_1^{\mathfrak{M}}, \ldots, R_1^{\mathfrak{M}}, \ldots, (R_a)_{a \in M[\Sigma]}, (S_a)_{a \in M[\Sigma]}, \ell_1^{\mathfrak{M}}, \ldots \rangle$$

satisfying, for all $u, v, w \in W$ and all $a, b \in M[\Sigma]$, the following conditions:

(S1$_R$) if uR_av and $a \leq b$, then uR_bv,

(S2$_R$) uR_av iff vR_au,

(S3$_R$) uR_au,

(S4$_R$) if uR_av, vR_bw and $a + b \in M[\Sigma]$, then $uR_{a+b}w$,

(S1$_S$) if uS_av and $a \leq b$, then uS_bv;

(S2$_S$) uS_av iff vS_au;

(S3$_S$) uS_au,

(S4$_S$) if uS_av, vS_bw and $a + b \in M[\Sigma]$, then $uS_{a+b}w$,

(C1) if uS_av then uR_av,

(C2) if uR_av and $a < b$, then uS_bv,

(C3) if uR_av, vS_bw and $a + b \in M[\Sigma]$, then $uS_{a+b}w$,

(C4) if uS_av, vR_bw and $a + b \in M[\Sigma]$, then $uS_{a+b}w$.

The *value* $C^{\mathfrak{M}}$ of a concept C in $\mathfrak{M}$ and the *truth-relation* $\mathfrak{M} \models C_1 \doteq C_2$ are defined in almost the same way as for CD-models: we only replace $\mathfrak{B}$ with $\mathfrak{M}$ and define the clauses for the distance quantifiers as follows:

$$(\mathsf{E}^{\leq a}C)^{\mathfrak{M}} = \{x \in W \mid \exists y \in W \left(xR_ay \wedge y \in C^{\mathfrak{M}}\right)\},$$

$$(\mathsf{E}^{<a}C)^{\mathfrak{M}} = \{x \in W \mid \exists y \in W \left(xS_ay \wedge y \in C^{\mathfrak{M}}\right)\},$$

$$(\mathsf{A}^{\leq a}C)^{\mathfrak{M}} = \{x \in W \mid \forall y \in W \left(xR_ay \rightarrow y \in C^{\mathfrak{M}}\right)\},$$

$$(\mathsf{A}^{<a}C)^{\mathfrak{M}} = \{x \in W \mid \forall y \in W \left(xS_ay \rightarrow y \in C^{\mathfrak{M}}\right)\}.$$

The next theorem ensures that the alternative Kripke semantics is 'equivalent' to the original one.

Theorem 2. *The knowledge base Σ is satisfiable in a CD-model iff it is satisfiable in a Kripke model for Σ.*

Proof. ($\Rightarrow$) Suppose that Σ is satisfied in a CD-model

$$\mathfrak{B} = \langle W, d, A_1^{\mathfrak{B}}, \ldots, R_1^{\mathfrak{B}}, \ldots, \ell_1^{\mathfrak{B}}, \ldots \rangle.$$

Define a Kripke model

$$\mathfrak{M} = \langle W, A_1^{\mathfrak{M}}, \ldots, R_1^{\mathfrak{M}}, \ldots, (R_a)_{a \in M[\Sigma]}, (S_a)_{a \in M[\Sigma]}, \ell_1^{\mathfrak{M}}, \ldots \rangle$$

for Σ by taking, for $a \in M[\Sigma]$,

- $A_i^{\mathfrak{M}} = A_i^{\mathfrak{B}}$, $\ell_i^{\mathfrak{M}} = \ell_i^{\mathfrak{B}}$, and $R_i^{\mathfrak{M}} = R_i^{\mathfrak{B}}$;
- xR_ay iff $d(x, y) \leq a$;
- xS_ay iff $d(x, y) < a$.

It is not difficult to see that $\mathfrak{M}$ is a Kripke model for Σ and to prove by induction that $C^{\mathfrak{M}} = C^{\mathfrak{B}}$, for all $C \in cl(\Sigma)$. It follows that $\mathfrak{M}$ satisfies Σ.

($\Leftarrow$) Suppose now that Σ is satisfied in a Kripke model

$$\mathfrak{M} = \langle W, A_1^{\mathfrak{M}}, \ldots, R_1^{\mathfrak{M}}, \ldots, (R_a)_{a \in M[\Sigma]}, (S_a)_{a \in M[\Sigma]}, \ell_1^{\mathfrak{M}}, \ldots \rangle$$

for Σ. Let $M[\Sigma] = \{a_1, \ldots, a_N\}$ with $0 < a_1 < a_2 < \cdots < a_N$. Choose a rational number $\gamma_\Sigma > a_N$ in such a way that there are no $a_1, a_2 \in M[\Sigma]$ with $a_N < a_1 + a_2 \leq \gamma_\Sigma$. Let D be the minimal number in the set

$$M[\Sigma] \cup \{a_1 + a_2 - \gamma_\Sigma \mid a_1, a_2 \in M[\Sigma] - \{\gamma_\Sigma\} \ \& \ a_1 + a_2 > \gamma_\Sigma\}.$$

Take some positive $\epsilon < \frac{D}{2^{N+1}}$. Define a function $d : W \times W \to \mathbb{R}$ by taking $d(u, v) = 0$ if $u = v$ and otherwise

$$d(u, v) = \begin{cases} \gamma_\Sigma, & \text{if } \neg \exists a \in M[\Sigma] \ uR_a v, \\ a, & \text{if } \exists a \in M[\Sigma] \ (uR_a v \wedge \neg uS_a v), \\ a_i - 2^i \cdot \epsilon, & \text{if } \exists a_i \in M[\Sigma] \ (uS_{a_i} v \wedge \forall j \ (0 < j < i \to \neg uR_{a_j} v)). \end{cases}$$

Consider the model

$$\mathfrak{B} = \langle W, d, A_1^{\mathfrak{B}}, \ldots, R_1^{\mathfrak{B}}, \ldots, \ell_1^{\mathfrak{B}}, \ldots \rangle.$$

where $A_i^{\mathfrak{B}} = A_i^{\mathfrak{M}}$, $R_i^{\mathfrak{B}} = R_i^{\mathfrak{M}}$, and $\ell_i^{\mathfrak{B}} = \ell_i^{\mathfrak{M}}$ for all i. One can show now that $\mathfrak{B}$ is a CD-model satisfying Σ.

Thus, it suffices to prove soundness with respect to Kripke semantics.

Theorem 3. *If there exists a complete and clash-free constraint system for Σ, then Σ is satisfiable in a Kripke model for Σ.*

Proof. Suppose that $\mathcal{S} = \langle T, <, S, L, E \rangle$ is a complete and clash-free constraint system for Σ that is obtained by repeatedly applying completion rules from Fig. 1 to the initial constraint system $\langle T_0, <_0, S_0, L_0, E_0 \rangle$. We use this constraint system to construct a Kripke model

$$\mathfrak{M} = \langle W, A_1^{\mathfrak{M}}, \ldots, R_1^{\mathfrak{M}}, \ldots, (R_a)_{a \in M[\Sigma]}, (S_a)_{a \in M[\Sigma]}, \ell_1^{\mathfrak{M}}, \ldots \rangle$$

satisfying Σ. Denote by T^i the set of nodes from T that are not indirectly (but possible directly) blocked. The domain W of $\mathfrak{M}$ consists of all sequences of the form $\langle \ell, x_1, \ldots, x_k \rangle$, where $\ell \in ob(\Sigma)$ and $x_1, \ldots, x_k \in T^i$ (with $k \geq 0$) such that $\ell < x_1$ and, for $1 \leq i < k$, either (i) x_i is unblocked and $x_i < x_{i+1}$ or (ii) there is a z such that z directly blocks x_i and $z < x_{i+1}$. Role names R are interpreted by setting

- $(\langle \ell_1, x_1, \ldots, x_k \rangle, \langle \ell_2 \rangle) \in R^{\mathfrak{M}}$ iff x_k is not blocked and $(R, \ell_2) \in S(x_k)$, or there exists z which directly blocks x_k such that $(R, \ell_2) \in S(z)$;
- $(\langle \ell, x_1, \ldots, x_k \rangle, \langle \ell, x_1, \ldots, x_{k+1} \rangle) \in R^{\mathfrak{M}}$ iff one of the following holds:
  - x_i is not blocked, $x_k < x_{k+1}$, and $L(x_k, x_{k+1}) = R$;
  - there is z which directly blocks x_k, $z < x_{k+1}$ and $L(z, x_{k+1}) = R$.

Given $\overline{x} = \langle \ell, x_1, \ldots, x_k \rangle \in W$, let $S(\overline{x})$ denote $S(x_k)$. We now define the relations R_a and S_a. Let R_a be the set of pairs $(\overline{x}, \overline{y}) \in W \times W$ such that, for $\{\overline{u}, \overline{v}\} = \{\overline{x}, \overline{y}\}$, the following conditions are satisfied:

(a) $A^{\leq a}C \in S(\overline{u})$ implies $C \in S(\overline{v})$;

(b) $A^{\leq b}C \in S(\overline{u})$ and $b > a$ imply that $A^{\leq c}C \in S(\overline{v})$ for some $c \geq b - a$;

(c) $A^{<b}C \in S(\overline{u})$ and $b > a$ imply that $A^{<c}C \in S(\overline{v})$ or $A^{\leq c}C \in S(\overline{v})$ for some $c \geq b - a$.

Similarly, S_a is comprised of the pairs $(\overline{x}, \overline{y}) \in W \times W$ such that, for $\{\overline{u}, \overline{v}\} = \{\overline{x}, \overline{y}\}$, the following conditions are satisfied:

(d) $A^{<a}C \in S(\overline{u})$ implies $C \in S(\overline{v})$;

(e) $A^{\leq b}C \in S(\overline{u})$ and $b > a$ imply that $A^{\leq c}C \in S(\overline{v})$ for some $c \geq b - a$;

(f) $A^{<b}C \in S(\overline{u})$ and $b > a$ imply that $A^{<c}C \in S(\overline{v})$ or $A^{\leq c}C \in S(\overline{v})$ for some $c \geq b - a$.

For all $\ell \in ob(\Sigma)$, we set $\ell^{\mathfrak{M}} = \{\langle \ell \rangle\}$. This is well-defined, since no nominal is removed from the tableau. Finally, for all concept names A_i and $\overline{x} \in W$, we set $\overline{x} \in A_i^{\mathfrak{M}}$ iff $A_i \in S(\overline{x})$. $\mathfrak{M}$ is a Kripke models for Σ which Σ. A proof of this claim can be found in the full version of this paper.

Completeness

Let us say that a model $\mathfrak{B} = \langle W, d, A_1^{\mathfrak{B}}, \ldots, \ell_1^{\mathfrak{B}}, \ldots \rangle$ *realises* a constraint system $\langle T, <, L, S, E \rangle$ for Σ if $\mathfrak{B} \models \Sigma$ and there exists a map $\rho : T \to W$ such that

- $C \in S(x)$ implies $\rho(x) \in C^{\mathfrak{B}}$;
- $L^{o}(\{x, y\}) = a \in M[\Sigma]$ implies $d(\rho(x), \rho(y)) \leq a$;
- $L^{o}(\{x, y\}) = a^{-} \in M[\Sigma]^{-}$ implies $d(\rho(x), \rho(y)) < a$;
- $x \neq y \in E$ implies $\rho(x) \neq \rho(y)$;
- if y is an R-successor of x, then $(\rho(x), \rho(y)) \in R^{\mathfrak{B}}$.

The following lemma is an immediate consequence of the definitions:

Lemma 1. *If a knowledge base Σ is satisfied in a CD-model $\mathfrak{B}$, then the initial constraint system for Σ is realisable in $\mathfrak{B}$.*

Lemma 2. *Suppose that $\mathfrak{B}$ realises a constraint system $S = \langle T, <, L, S, E \rangle$ for Σ and a completion rule R is applicable to S. Then R can be applied in such a way that $\mathfrak{B}$ realises the resulting constraint system $S' = \langle T', <', S', L', E' \rangle$ as well.*

Proof. Let $\mathfrak{B} = \langle W, d, A_1^{\mathfrak{B}}, \ldots, \ell_1^{\mathfrak{B}}, \ldots \rangle$ realise S by means of a map $\rho : T \to W$ and let S' be obtained from S using some rule R. We consider only two rules, $R = R_{E\leq}$ and $R = R_{A<}$, and and leave the remaining cases to the reader.

$R_{E\leq}$: Suppose that $E^{\leq a}C \in S(x)$, $T' = T \cup \{y\}$, $L'(\{x, y\}) = a$, $<' = < \cup \{(x, y)\}$, and $S(y) = \{C\}$. We know that $\rho(x) \in (E^{\leq a}C)^{\mathfrak{B}}$. So we can find $v \in W$ such that $d(\rho(x), v) \leq a$ and $v \in C^{\mathfrak{B}}$. Define a map $\rho' : T' \to W$ by taking $\rho'(z) = \rho(z)$ for all $z \in T$ and $\rho'(y) = v$. It should be clear that $\mathfrak{B}$ realises S' my means of ρ'.

$R_{A<}$: Let $A^{<a}C \in S(x)$, $x \in T$. Suppose that the rule is applied to some $y \in T$. Consider three possible cases.

(i) If $L^o(\{x,y\}) = a^-$ then $d(\rho(x), \rho(y)) < a$ and $S(y) = \{C\} \cup S(y)$. We need to show that $\rho(y) \in C^{\mathfrak{B}}$. But this follows immediately from $\rho(x) \in (A^{<a}C)^{\mathfrak{B}}$.

(ii) If $L^o(\{y,x\}) = b < a$ then $d(\rho(x), \rho(y)) \leq b$ and $S(y) = \{A^{<a-b}C\} \cup S(y)$. To show that $\rho(y) \in (A^{<a-b}C)^{\mathfrak{B}}$, take any $v \in W$ such that $d(\rho(y), v) < a - b$. By the triangular inequality, we then have $d(\rho(y), v) < a$ and so $v \in C^{\mathfrak{B}}$.

(iii) The case of $L^o(\{y,x\}) = b^-$ and $b < a$ is considered similarly to (ii).

As a consequence of these two lemmas and Theorem 1 we obtain

Theorem 4. *If Σ is satisfiable, then there exists a complete clash-free constraint system for Σ.*

5   Undecidability

We show now that a rather natural and closer integration of distance quantifiers and qualified number restrictions results in an undecidable logic. Denote by sim_f the language with the following concept formation rule:

$$C ::= A_i \mid \ell_i \mid \neg C \mid C_1 \sqcap C_2 \mid C_1 \sqcup C_2 \mid E^{\leq a}C \mid (\leq_a^1 .C),$$

where $(\leq_a^1 .C)$ is interpreted in concept distance models $\mathfrak{B}$ as follows

$$(\leq_a^1 .C)^{\mathfrak{B}} = \{x \in W \mid |\{y \mid d(x,y) \leq a, y \in C^{\mathfrak{B}}\}| \leq 1\}.$$

Theorem 5. *The satisfiability problem for sim_f-knowledge bases in concept distance models is undecidable.*

Proof. (sketch) We can simulate the undecidable $\mathbb{N} \times \mathbb{N}$-tiling problem in almost the same way as in the undecidability proof of [9] for the language $\mathcal{MS}_1$ with the operators $A^{\leq a}$, $A^{\geq 0}_{\leq a}$ and their duals: just replace everywhere in the proof of Theorem 3.1 the concept $A^{\geq 0}_{\leq 80}\neg\chi_{i,j}$ by the concept $(\leq_{80}^1 .\chi_{i,j})$.

6   Conclusion

We have introduced the description-metric logic sim-$\mathcal{ALCQO}$ for defining concepts based on similarity measures, and have proposed a tableau algorithm for deciding the satisfiability of sim-$\mathcal{ALCQO}$-knowledge bases. This algorithm unifies the tableau algorithms for $\mathcal{SHOQ}$ (a superlogic of $\mathcal{ALCQO}$) presented in [5] and for the logic of metric spaces $\mathcal{MS}$ as defined in [13]. It is of interest to note that, in contrast to what is done in [13], we need a different soundness proof, since the presence of number restrictions prohibits the use of filtration techniques.

We regard the presented logic only as a first step towards DLs that allow definitions of concepts based on similarity measures. Although we believe that the expressive power provided by sim-$\mathcal{ALCQO}$ is quite natural and useful, an in-depth investigation of the expressive means that are useful for defining vague concepts are in order. Some possible extensions of $sim\text{-}\mathcal{ALCQO}$ are the following:

(1) New constructors $\mathsf{E}^{<a}R.C$ and $\mathsf{A}^{<a}R.C$, where the former expresses that there exists an R-successor at distance smaller than a satisfying C, and the latter is its dual. Such constructors would, e.g., allow us to say that a person is very similar to his father: $\mathsf{E}^{<0.5}$parent.Male. The presented algorithm should be extendable to this case without any problems.

(2) New constructors $\mathsf{E}^{>a}C$ and $\mathsf{E}^{\geq a}C$ (and their duals) with the obvious semantics. Although these constructors do not seem to be so natural as the variants based on $<$ and $\leq$, they could, e.g., be used to express that a propotypical tableau algorithm pta is very close to *all* other tableau algorithms: pta : $\mathsf{A}^{>0.5}\neg$Tableau_algorithm. While [9] proves the decidability of the metric logic with the operators $\mathsf{E}^{\leq a}C$ and $\mathsf{E}^{>a}C$ (and their duals), nothing is currently known about the extension of $\mathcal{MS}$ with all four possible constructors.

Acknowledgements

The work of the second author was supported by Deutsche Forschungsgemeinschaft (DFG) grant Wo583/3-3. The work of the third author was partially supported by U.K. EPSRC grants no. GR/R45369/01 and GR/R42474/01.

References

[1] F. Baader, D. Calvanese, D. McGuinness, D. Nardi, and P. Patel-Schneider, editors. *The Description Logic Handbook*. Cambridge University Press, 2003.

[2] F. Baader, C. Lutz, H. Sturm, and F. Wolter. Fusions of description logics and abstract description systems. *J. of Artificial Intelligence Research*, 16:1–58, 2002.

[3] K. Fine and G. Schurz. Transfer theorems for stratified modal logics. In J. Copeland, editor, *Logic and Reality, Essays in Pure and Applied Logic. In memory of Arthur Prior*, pages 169–213. Oxford University Press, 1996.

[4] I. Horrocks and P. Patel-Schneider. The generation of DAML+OIL. In C. Goble, D. McGuinness, R. Möller, and P. Patel-Schneider, editors, *Proceedings of the International Workshop in Description Logics 2001 (DL2001)*, number 49 in CEUR-WS (http://ceur-ws.org/), pages 30–35, 2001.

[5] I. Horrocks and U. Sattler. Ontology reasoning in the $\mathcal{SHOQ}(D)$ description logic. In B. Nebel, editor, *Proceedings of the 17th International Joint Conference on Artificial Intelligence (IJCAI'01)*, pages 199–204. Morgan Kaufmann, 2001.

[6] I. Horrocks, U. Sattler, and S. Tobies. Reasoning with individuals for the description logic $\mathcal{SHIQ}$. In D. MacAllester, editor, *Proc. of the 17th International Conference on Automated Deduction (CADE-17)*, number 1831 in LNCS. Springer, 2000.

[7] I. Horrocks, P. Patel-Schneider, and F. van Harmelen. Reviewing the design of DAML+OIL: An ontology language for the semantic web. In *Proceedings of the 18th National Conference on Artificial Intelligence (AAAI 2002)*, 2002.

[8] M. Kracht and F. Wolter. Properties of independently axiomatizable bimodal logics. *J. Symbolic Logic*, 56:1469–1485, 1991.

[9] O. Kutz, H. Sturm, N.-Y. Suzuki, F. Wolter, and M. Zakharyaschev. Logics of metric spaces. *ACM Transactions on Computational Logic*, 2003. In print.

[10] M. Schmidt-Schauß and G. Smolka. Attributive concept descriptions with complements. *Artificial Intelligence*, 48:1–26, 1991.

[11] E. Spaan. *Complexity of Modal Logics*. PhD thesis, Department of Mathematics and Computer Science, University of Amsterdam, 1993.

[12] F. Wolter. Fusions of modal logics revisited. In M. Kracht, M. De Rijke, H. Wansing, and M. Zakharyaschev, editors, *Advances in Modal Logic*, volume 1, pages 361–379. CSLI, Stanford, 1997.

[13] F. Wolter and M. Zakharyaschev. Reasoning about distances. To appear in Proc. IJCAI 2003.

XPath and Modal Logics of Finite DAG's

Maarten Marx*

ILLC, Universiteit van Amsterdam, The Netherlands
marx@science.uva.nl

Abstract. XPath, CTL and the modal logics proposed by Blackburn et al, Palm and Kracht are variable free formalisms to describe and reason about (finite) trees. XPath expressions evaluated at the root of a tree correspond to existential positive modal formulas. The models of XPath expressions are finite ordered trees, or in the presence of XML's ID/IDREF mechanism graphs. The ID/IDREF mechanism can be seen as a device for naming nodes. Naming devices have been studied in hybrid logic by nominals. We add nominals to the modal logic of Palm and interpret the language on directed acyclic graphs. We give an algorithm which decides the consequence problem of this logic in exponential time. This yields a complexity result for query containment of the corresponding extension of XPath.

1 Introduction

This paper is about reasoning in languages interpreted on finite trees and directed acyclic graphs (DAGs). These finite structures are the core interest in both theoretical linguistics (parsing a sentence leads to a finite tree or DAG) and in the world of XML databases (an XML document is modeled as a finite tree, or in the presence of ID/IDREF attributes as a finite graph). In the field of XML databases, a key problem is the equivalence or containment of XPath expressions possibly in the presence of a Document Type Definition (DTD). This problem can be seen as an instance of the consequence problem in logic. We study the complexity of this problem in the setting in which the relevant structures are finite trees or DAGs. The language used to describe these structures is the modal tree language proposed by Palm [20].This is a fragment of Propositional Dynamic Logic (PDL) with four basic programs corresponding to the four basic movements in finite ordered trees: mother, daughter, left sister and right sister.

The novelty of this paper is the addition of nominals to this language in order to simulate XML's ID/IDREF mechanism, and the generalization of the class of models from trees to rooted DAGs. The main result is that the satisfiability problem interpreted on rooted DAGs is in EXPTIME.

We started our work by building a tableau system for a fragment of the language. But this system seemed to be horribly inefficient, as it had to build a tree from the root. It is straightforward to devise for every natural number n,

---

* Research funded by NWO grant 612.000.106.

M. Cialdea Mayer and F. Pirri (Eds.): TABLEAUX 2003, LNAI 2796, pp. 150–164, 2003.

a satisfiable formula of size $O(n^2)$ whose minimal model is a binary branching tree of depth 2^n, a structure with 2^{2^n} many nodes [5]. The known lower bound of the satisfiability problem was EXPTIME, so somewhere there was something wrong.

The decision algorithm presented here searches for a pseudomodel of the formula to be satisfied. The pseudomodel is such that it can be transformed into a finite structure in which the original formula is still satisfied. The pseudomodel on the other hand has size bound by just a single exponential in the input. Algorithms using pseudomodels are the natural alternative when either tableaux do not (or not easy) terminate or when they take more time or space than needed to solve the problem. Pseudomodels are often used in decision procedures in modal and temporal logic, but also in e.g., the family of guarded fragments, see [5] for a number of examples.

Our work is related to many areas in logic. We mention the most relevant. The EXPTIME lower bound of PDL [12] transfers to finite models and our similarity type. Upper bounds for (converse) PDL do not transfer to our case, since we are working on *finite* trees and DAGs, whence have a different logic. For instance, versions of Löb's axiom hold on finite trees and DAGs, but fail for PDL. Our EXPTIME algorithm uses several features of the one for PDL by Pratt [21]. The novelty of this paper is the adaptation of Pratt's method to the finite case: the new part in the algorithm prevents building models with cycles or infinite paths. Alternative EXPTIME lower bounds can be extracted from results about XPath query containment under DTD's by Neven and Schwentick [18]. For unordered finite trees, complexity results for part of the language can be obtained by an interpretation into CTL* [11]. The connection between CTL and XPath is first made in [17]. The addition of nominals to the language places this work in the tradition of hybrid logic [4]. The algorithm presented here uses as a subroutine (cf. Figure 4) the decision procedure for Palms language on finite trees from [6]. Alechina, de Rijke and Demri [2] analyze path constraints for semistructured data and obtain complexity results by an embedding into converse PDL with nominals. The difference with the present work is twofold. Firstly, they consider arbitrary graphs as models. Secondly, they consider edge labeled structures, while we are interested in node labeled structures (like XML documents). This shows in the difference in signatures: we consider just the four basic moves in a tree and allow whatever node label (i.e., propositional variable); [2] has no propositional variables, but arbitrary edge labels (i.e., atomic programs).

Organization. The next section presents two modal logics of finite trees and establishes the relation to first and second order logic of trees. Then follow two sections about XML motivating our work. These two sections are not needed to understand the technical part of the paper. After that we concentrate on the decision algorithm and its correctness proof. We conclude with a number of open problems.

2   Modal Logic of Finite Trees

We first recall the modal logic of finite trees proposed by Marcus Kracht in [14, 15]. The language will be called $\mathcal{L}_K$. $\mathcal{L}_K$ is a propositional modal language identical to Propositional Dynamic Logic (PDL) [13] over four basic programs: $\leftarrow$, $\rightarrow$, $\uparrow$ and $\downarrow$ which explore the left-sister, right-sister, mother-of and daughter-of relations. Recall that PDL has two sorts of expressions: programs and propositions. We suppose we have fixed a non-empty, finite or countably infinite, set of atomic symbols A whose elements are typically denoted by p. $\mathcal{L}_K$'s syntax is as follows, writing π for programs and ϕ for propositions:

$$\pi ::= \leftarrow | \rightarrow | \uparrow | \downarrow | \pi; \pi | \pi \cup \pi | \pi^* | ?\phi$$
$$\phi ::= p | \top | \neg\phi | \phi \wedge \phi | \langle\pi\rangle\phi.$$

We employ the usual boolean abbreviations and write $[\pi]\phi$ instead of $\neg\langle\pi\rangle\neg\phi$.

$\mathcal{L}_K$ is interpreted on *finite ordered trees* whose nodes are *labeled* with symbols drawn from A. We assume that the reader is familiar with finite trees and such concepts as 'daughter-of', 'mother-of', 'sister-of', 'root-node', 'terminal-node', and so on. If a node has no sister to the immediate right we call it a last node, and if it has no sister to the immediate left we call it a first node. Note that the root node is both first and last. The root node will always be called *root*. A labeling of a finite tree associates a subset of A with each tree node.

Formally, we present finite ordered trees as tuples $\mathbf{T} = (T, R_\rightarrow, R_\downarrow)$. Here T is the set of tree nodes and $R_\rightarrow$ and $R_\downarrow$ are the immediate right-sister and daughter-of relations respectively. A pair $\mathfrak{M} = (\mathbf{T}, V)$, where $\mathbf{T}$ is a finite tree and $V : A \longrightarrow Pow(T)$, is called a *model*, and we say that V is a *labeling function* or a *valuation*. Given a model $\mathfrak{M}$, we simultaneously define a set of relations on $T \times T$ and the interpretation of the language $\mathcal{L}_K$ on $\mathfrak{M}$:

$$
\begin{array}{ll}
R_\uparrow = R_\downarrow^{-1} & R_{\pi\cup\pi'} = R_\pi \cup R_{\pi'} \\
R_\leftarrow = R_\rightarrow^{-1} & R_{\pi;\pi'} = R_\pi \circ R_{\pi'} \\
R_{\pi^*} = R_\pi^* & R_{?\phi} = \{(t,t) \mid \mathfrak{M}, t \models \phi\}
\end{array}
$$

$$
\begin{array}{lll}
\mathfrak{M}, t \models p & \text{iff} & t \in V(p), \text{ for all } p \in A \\
\mathfrak{M}, t \models \top & \text{iff} & t \in T \\
\mathfrak{M}, t \models \neg\phi & \text{iff} & \mathfrak{M}, t \not\models \phi \\
\mathfrak{M}, t \models \phi \wedge \psi & \text{iff} & \mathfrak{M}, t \models \phi \text{ and } \mathfrak{M}, t \models \psi \\
\mathfrak{M}, t \models \langle\pi\rangle\phi & \text{iff} & \exists t' (t R_\pi t' \text{ and } \mathfrak{M}, t' \models \phi).
\end{array}
$$

For any formula ϕ, if there is a model $\mathfrak{M}$ such that $\mathfrak{M}, root \models \phi$, then we say that ϕ is *satisfiable*.

We note that we could have generated the same language by taking $\downarrow$ and $\rightarrow$ as primitive programs and closing the set of programs under converses. We use a number of formulas as abbreviations:

$$t \models root \iff t \models \neg\langle\uparrow\rangle\top \iff t \text{ is the root}$$
$$t \models leaf \iff t \models \neg\langle\downarrow\rangle\top \iff t \text{ is a terminal node}$$
$$t \models first \iff t \models \neg\langle\leftarrow\rangle\top \iff t \text{ is a first node}$$
$$t \models last \iff t \models \neg\langle\rightarrow\rangle\top \iff t \text{ is a last node}$$

We now discuss the expressivity of this and related languages. First two examples: (1) says that every a node has a b and a c daughter, in that order, and no other daughters; and (2) says that every a node has a b first daughter followed by some number of c daughters, and no other daughters.

(1) $a \rightarrow \langle\downarrow\rangle(first \wedge b \wedge \langle\rightarrow\rangle(c \wedge last))$

(2) $a \rightarrow \langle\downarrow\rangle(\neg\langle\leftarrow\rangle\top \wedge b \wedge \langle(\rightarrow; ?c)^*\rangle last)$.

$\mathcal{L}_K$ can express properties beyond the power of the first order logic of ordered labeled trees[1] For example, it can express the property of having an odd number of daughters: $\langle\downarrow\rangle(first \wedge \langle(\rightarrow; \rightarrow)^*\rangle last)$.

Palm [20, 19] proposed a fragment of $\mathcal{L}_K$ which is functionally complete with respect to first order logic of ordered labeled trees (an extension of results by Schlingloff [24]). There are two equivalent formulations (cf., [6]) of this language, which we both denote by $\mathcal{L}_P$. The first is by restricting the set of programs to

$$\pi ::= \leftarrow |\rightarrow| \uparrow | \downarrow | ?\phi; \pi | \pi^*.$$

The second is more economic in its modal operators and resembles temporal logic: let $\mathcal{L}_P$ be the modal language with the following four binary modal operators: for $\pi \in \{\leftarrow, \rightarrow, \uparrow, \downarrow\}$, $\mathfrak{M}, t \models Until_\pi(\phi, \psi)$ iff there exists a t' such that $tR_{\pi^+}t'$ and $\mathfrak{M}, t' \models \phi$ and for all t'' such that $tR_{\pi^+}t''R_{\pi^+}t'$ it holds that $\mathfrak{M}, t'' \models \psi$.

Present proposals for XPath [8] don't go beyond first order expressivity. For that reason we focus on $\mathcal{L}_P$ from now on. We study the complexity of the consequence problem: $\Gamma \models \phi$ (is ϕ true at every state on each finite ordered tree on which all of Γ is true at every state). For finite Γ, this reduces to the satisfiability problem because $\Gamma \models \phi$ if and only if it is not the case that $[\downarrow^*]\Gamma \wedge \langle\downarrow^*\rangle\neg\phi$ is satisfiable. We will improve on the following theorem:

Theorem 1 ([6]). *The satisfiability problem for $\mathcal{L}_P$ is in EXPTIME.*[2]

We note the remarkable fact that the satisfiability for the equally expressive first order logic on finite trees is decidable but with a non–elementary lower bound [23].

---

[1] That is first order logic in the signature with binary $R_\uparrow^*, R_\rightarrow^*$ and countably many unary predicates, interpreted on labeled ordered trees.

[2] EXPTIME is the class of all problems solvable in exponential time. A problem is solvable in exponential time if there is a deterministic exponentially time bounded Turing machine that solves it. A deterministic Turing machine is exponentially time bounded if there is a polynomial $p(n)$ such that the machine always halts after at most $2^{p(n)}$ steps, where n is the length of the input.

3   XML and XPath

XML is a new standard adopted by the World Wide Web Consortium (W3C) to complement HTML for data exchange on the web. In its simplest form XML looks just like HTML. The main difference is that the user can define its own tags and specifies a Document Type Definition (DTD) which serves as a grammar for the underlying XML document. Figure 1 contains a DTD and an XML document that conforms to it. An XML document is most naturally viewed as a finite ordered node–labeled tree. The tag–names form the labels of the non–terminal nodes and the terminals are labeled with the data (in our example of type CDATA). We assume familiarity with these concepts, for an introduction cf. e.g., [1]. XPath is a simple language for navigating an XML tree and selecting a set of element nodes [7]. It's grammar resembles the file selection mechanism in UNIX. As an example, the XPath expression /a//b[*/c]/g selects nodes labeled with g (g–nodes for short) that are children of b–nodes, which have an c–node as a grandchild and which are themselves descendants of the root a–node. A clear explanation of the semantics of XPath is given in [3]. XPath queries starting with the root symbol / can easily be translated into expressions in the positive existential fragment of $\mathcal{L}_P$. /a//b[*/c]/g selects the same nodes as

$$g \wedge \langle \uparrow \rangle (b \wedge \langle \downarrow \rangle \langle \downarrow \rangle c \wedge \langle \uparrow^* \rangle (a \wedge \ root)).$$

```
1. <!ELEMENT Collection (Painter+)>
2. <!ELEMENT Painter  (Name, Painting*)>
3. <!ELEMENT Name   CDATA  >
4. <!ELEMENT Painting CDATA>

<Collection>
  <Painter>
    <Name> Rembrandt </Name>
    <Painting> de Nachtwacht </Painting>
    <Painting> de Staalmeesters </Painting>
  </Painter>
  <Painter>
    <Name> Vermeer </Name>
    <Painting> het Melkmeisje </Painting>
  </Painter>
</Collection
```

Fig. 1. An XML DTD and document

DTD's can also be translated into $\mathcal{L}_K$. The DTD from Figure 1 translates into[3]

$Collection \rightarrow \langle \downarrow; ?first; ?Painter; (\rightarrow; ?Painter)^* \rangle last.$

$Painter \quad \rightarrow \langle \downarrow; ?first; ?Name; (\rightarrow; Painting)^* \rangle last.$

$Name \quad \rightarrow \langle \downarrow; ?first; ?CDATA \rangle last.$

$Painting \quad \rightarrow \langle \downarrow; ?first; ?CDATA \rangle last.$

These translations (which can be performed in polynomial time) make that the containment problem of XPath expressions under a DTD can be reduced to the consequence problem in the modal logic of finite ordered trees. This problem has received quite some attention lately [25, 9, 17, 18]. The situation here is quite comparable to that in description logic: an effort is made to map out the complexity landscape for a great number of XPath fragments. The result which is of interest here is that query containment under a DTD is EXPTIME hard for XPath expressions in which one can use /, //, | or /, //, [], *. When the non-deterministic operators //, |, * are left out (leaving only /, []) the problem is complete for CO–NP. Both results are in [18].

Theorem 1 now yields a matching upper bound for a large extension of these XPath fragments. Note that XPath statements correspond to existential positive modal formulas. But Theorem 1 works for the whole modal language which is of course closed under full negation, but also can express until–like constructions. We can now consider queries like

- select all A that only have B children;
- select all couples with a completely Greek descendant line (in a genealogy tree in which nationality is coded).

The first uses negation, the second the until construction.

The part of the landscape that has been investigated until now views XML documents as trees. But in the presence of XML's ID/IDREF mechanism they are really graphs. We turn to these models in the next section.

4 From Trees to DAGs

So far we have discussed XML documents as if they were trees. But XML contains a mechanism for defining and using references and, hence for describing graphs rather than trees. XML allows the association of unique identifiers to elements as the value of a certain attribute. These are attributes of type ID, and the referencing is done with an attribute of type IDREF. How this is done exactly is not important for our discussion. Figure 2 contains a DTD[4] using this

---

[3] This DTD translates to $\mathcal{L}_P$. But we need $\mathcal{L}_K$ to translate a rule like `<!ELEMENT Collection (Painter,Painting)+>`. For lack of space we cannot give the translation algorithm. For that see [16].

[4] Instead of the official but rather cumbersome, `<!ELEMENT Countries (State,City*)>` we simply write the equivalent context free grammar rule `Countries → (State,City*)`.

```
countries  → (state(id)*,city*)
state      → name
city       → (name,capitol_of)
capitol_of → state(idref)
name       → CDATA.
```

```
<countries>
   <state ID=A1>
      <name Holland />
   </state>
   <city>
      <name Amsterdam />
      <capitol_of IDREF=A1 />
   </city>
</countries>
```

Fig. 2. A DTD and an XML document with ID/IDREF

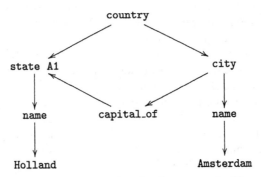

Fig. 3. Acyclic graph for the document in Figure 2

mechanism, and a document which conforms to it. The corresponding acyclic graph is drawn in Figure 3. More abstractly, the models we consider are node labeled graphs in which the nodes may have a unique name besides their label. In modal logic names for states are known as *nominals* and modal logics containing names are referred to as *hybrid logics* [4]. In a modal language a nominal is nothing but a special propositional variable which can only be true at exactly one state. Modal languages with the Difference operator D can express that p behaves like a nominal by stating (here $E\phi$ abbreviates $\phi \vee D\phi$):

$E(p \wedge \neg Dp)$.

The difference operator D is defined by $\mathfrak{M}, t \models D\phi$ iff there exists a $t' \neq t$ such that $\mathfrak{M}, t' \models \phi$. On finite ordered trees $D\phi$ is term definable as

$$D\phi \equiv \langle\downarrow^+\rangle\phi \vee \langle\uparrow^+\rangle\phi \vee \langle\uparrow^*; (\leftarrow^+ \cup \rightarrow^+); \downarrow^*\rangle\phi.$$

So in a sense, we have nominals in our modal language. But a referencing mechanism on trees is not very interesting nor can we make the connection with the XML graph models. Instead of interpreting the modal language of trees on arbitrary graphs we decided to make a smaller move, remaining as close to trees as possible. Here's the definition. We call a directed acyclic graph (DAG) $(N, R_\downarrow)$ *rooted* if there exists an $r \in N$ without ancestors and r is the ancestor of each $n \in N$. Note that a rooted DAG is a tree if every node except the root has exactly one parent. It is useful to distinguish nodes with one parent from nodes with multiple parents. The latter correspond to nodes with a name. For $NOM \subseteq N$, we call a structure $(N, R_\downarrow, NOM)$ a nominalized rooted DAG (NDAG for short) if $(N, R_\downarrow)$ is a rooted DAG and all elements in $N \setminus (NOM \cup \{root\})$ have exactly one parent.

The restriction to NDAG's is rather natural from an XML point of view, and similar in spirit to the restriction to trees encountered in the literature. Depending on certain syntactic properties of the DTD, we can restrict the class of models accordingly. If the DTD contains no ID/IDREF only trees have to be considered. If the DTD does not specify a *cycle* of naming and referencing only NDAG's need to be considered. This we will do in the next section.

We note that on DAGs there are strictly less validities than on trees. For instance, $\langle\downarrow;\uparrow\rangle\phi \to \phi$ is valid on trees but not on DAGs. Moreover $\mathcal{L}_K$ is not strong enough to capture all first order properties of NDAG's. For instance, $\exists yzw(y \ne z \wedge xR_\downarrow y \wedge yR_\downarrow w \wedge xR_\downarrow z \wedge zR_\downarrow w)$ is not expressible by an $\mathcal{L}_K$ formula as an easy bisimulation argument shows. Of course on trees, this formula is not satisfiable, whence simply expressible by $\bot$.

5   Deciding the Modal Logic of Finite Rooted Nominalized DAGs

At present we do not know the complexity of the full PDL language nor of Palms fragment with nominals on ordered DAGs. We make a restriction common in the literature on XPath query containment and remove the two sister axis[5] from the language $\mathcal{L}_P$. To this language we add a modal constant *id* and nominals and interpret it on NDAG's. Formally, there is a special set of propositional variables called nominals. On an NDAG $(N, R_\downarrow, NOM)$ each nominal is interpreted as a singleton subset of NOM. The interpretation of the modal constant *id* is exactly the set NOM. We call the resulting logic $\mathcal{L}_{DAG}$[6]. The $\mathcal{L}_{DAG}$ consequence problem consists of all pairs (Γ, χ) with $\Gamma \cup \{\chi\}$ a finite set of $\mathcal{L}_{DAG}$ formulas such that $\Gamma \models \chi$ on finite NDAG's.

The following three validities are noteworthy. (3) states that all nominals are interpreted in the set NOM; (4) that there are no cycles and (5) that nodes which are not in the set NOM have at most one parent.

(3) $i \to id$ for i a nominal

(4) $i \to \neg\langle\downarrow^+\rangle i$ for i a nominal

(5) $\neg id \to (\langle\uparrow\rangle\phi \to [\uparrow]\phi)$.

---

[5] On ordered DAGs the interpretation of the sister relation is problematic: should they share one or all parents? In the former case the tree validity $\langle\uparrow\rangle\phi \to [\leftarrow^* \cup \to^*]\langle\uparrow\rangle\phi$ does not hold. In the latter, we cannot mark first and last nodes anymore. We note that without the sister axis DTD's cannot be expressed anymore. Thus the present result only yields a decision procedure for XPath root queries without a DTD.

[6] Hybrid logics usually have besides nominals also the satisfaction operator @. Here we do not add it because $@_i\phi$ is term definable as $\langle\uparrow^*\rangle(root \wedge \langle\downarrow^*\rangle(i \wedge \phi))$.

Theorem 2. *The $\mathcal{L}_{DAG}$ consequence problem is in EXPTIME.*

The proof consists of a linear reduction and a decision algorithm. The reduction removes the transitive closure operation by adding new propositional symbols. Similar techniques are employed in [22, 10] for obtaining normalized monadic second order formulas. The reduction is most simply presented in the formulation of Palms language using the until operators.

Let $\chi \in \mathcal{L}_{DAG}$. Let $Cl(\chi)$ be the smallest set of formulas containing all subformulas of χ, the constants *root* and *leaf*, and which is closed under taking single negations and under the rule: $Until_\pi(\phi, \psi) \in Cl(\chi) \Rightarrow \psi \wedge Until_\pi(\phi, \psi) \in Cl(\chi)$.

We associate a formula $\nabla(\chi)$ with χ as follows. We create for each $\phi \in Cl(\chi)$, a new propositional variable q_ϕ. Now $\nabla(\chi)$ "axiomatizes" these new variables as follows:

$$q_p \quad\quad\quad \leftrightarrow p$$
$$q_{\neg\phi} \quad\quad\quad \leftrightarrow \neg q_\phi$$
$$q_{\phi\wedge\psi} \quad\quad\quad \leftrightarrow q_\phi \wedge q_\psi$$
$$q_{Until_\pi(\phi,\psi)} \leftrightarrow \langle\pi\rangle q_\phi \vee \langle\pi\rangle q_{(\psi\wedge Until_\pi(\phi,\psi))} \quad \text{for } \pi \in \{\downarrow, \uparrow\}.$$

Lemma 1. *(i) For every model $\mathfrak{M}$ which validates $\nabla(\chi)$, for every node n and for every subformula $\phi \in Cl(\chi)$, $\mathfrak{M}, n \models q_\phi$ iff $\mathfrak{M}, n \models \phi$.*
(ii) Thus for all $\gamma, \chi \in \mathcal{L}_{DAG}$, it holds that $\gamma \models \chi \iff \nabla(\gamma \wedge \chi), q_\gamma \models q_\chi$.

The proof is by induction on the structure of the formula, and for the left to right direction of the until case by induction on the depth of direction of π. Note that it is crucial that the models are finite and acyclic. Also note that this reduction does not work (at least not directly) for formulas of the form $\langle(\uparrow; \downarrow)^*\rangle\phi$ or even $\langle(\downarrow^*)^*\rangle\phi$.

Finally note that the right hand side of the statement in Lemma 1.(ii) contains only diamonds of the form $\langle\uparrow\rangle$ and $\langle\downarrow\rangle$. As the reduction is linear we can thus decide the consequence problem for this restricted language.

We will now give an EXPTIME algorithm that on input formulas γ, χ decides whether there exists a model $\mathfrak{M}$ in which γ is true everywhere and χ is true at the root. To this the consequence problem reduces because $\gamma \not\models \chi$ iff there exists a model in which $\gamma \wedge (p \leftrightarrow \neg\chi \vee \langle\downarrow\rangle p)$ is true everywhere and p is true at the root. Here p is a new propositional variable whose intended meaning is $\langle\downarrow^*\rangle\neg\chi$.

Preliminaries. The next notion is well known. Hintikka sets are used to label nodes of models with a set of formulas which are supposed to be true at that node. The first condition ensures that γ and $\top$ are true in every node. The other two ensure the correct behaviour of the Booleans.

Definition 1 (Hintikka Set). *Let $A \subseteq Cl(\{\gamma, \chi\})$. We call A a Hintikka Set if A satisfies the following conditions:*

1. $\gamma \in A$ and $\top \in A$.
2. If $\phi \in Cl(\{\gamma, \chi\})$ then $\phi \in A$ iff $\neg\phi \notin A$.
3. If $\phi \wedge \psi \in Cl(\{\gamma, \chi\})$ then $\phi \wedge \psi \in A$ iff $\phi \in A$ and $\psi \in A$.

Let $HS(\gamma, \chi)$ denote the set of all Hintikka Sets which are a subset of $Cl(\gamma, \chi)$. Note that $|HS(\gamma, \chi)| \le 2^{|Cl(\gamma, \chi)|}$.

For H a set of Hintikka sets, let $l : H \longrightarrow \{0, 1, \ldots, |H|\}$ be a function assigning to each $A \in H$ a level. We call a structure (H, l) an ordered set of Hintikka sets. For notational convenience we introduce a binary relation on Hintikka sets specifying that it is not directly inconsistent that the two sets stand in the parent relation in the tree: For A, B Hintikka sets, A child B holds if

1. $l(A) > l(B)$
2. for all $\langle \downarrow \rangle \psi \in Cl(\phi)$, if $\psi \in B$, then $\langle \downarrow \rangle \psi \in A$;
3. for all $\langle \uparrow \rangle \psi \in Cl(\phi)$, if $\psi \in A$, then $\langle \uparrow \rangle \psi \in B$;
4. if $id \notin B$ then also for all $\langle \uparrow \rangle \psi \in Cl(\phi)$, $\langle \uparrow \rangle \psi \in B$ implies $\psi \in A$.

The definition of saturation is the crucial one in any mosaic style proof. Informally it states that a set of Hintikka Sets is large enough to build a model from. In the temporal logic literature, the diamond formulas in Hintikka set are called *unfulfilled eventualities.*

Definition 2 (Saturation). *Let (H, l) be an ordered set of Hintikka sets. We call (H, l) down- saturated if for all $A \in H$, $\langle \downarrow \rangle \phi \in A$ only if there exists a $B \in H$ such that $\phi \in B$ and A child B.*

We call (H, l) up- saturated if for all $A \in H$ containing id, $\langle \uparrow \rangle \phi \in A$ only if there exists a $B \in H$ such that $\phi \in B$ and B child A.

We call (H, l) saturated if it is both up and down saturated.

The next definition specifies when an ordered saturated set of Hintikka sets can be turned into an NDAG.

Definition 3. *We call an ordered saturated set of Hintikka sets (H, l) rooted and nominalized if*

1. *There is exactly one $A \in H$ with root $\in H$, and for every nominal $i \in CL(\{\gamma, \chi\})$ there exists exactly one $A \in H$ such that $i \in A$.*
2. *(everyone has a predecessor) For every B in H there is a path $C_0, \ldots, C_k$ of Hintikka sets in H with $B = C_k$ such that*
   (a) root $\in C_0$
   (b) C_j child C_{j+1}.

We can now make the connection between satisfiability and the existence of certain sets of Hintikka sets.

Lemma 2. *The following are equivalent:*

1. *There exists a model over a finite NDAG in which γ is true everywhere and χ is true at the root;*

2. *There exists a rooted nominalized saturated ordered set of Hintikka Sets (H, l), with $H \subseteq HS(\gamma, \chi)$ and there is an $A \in H$ with $\{root, \chi\} \subseteq A$.*

Proof. First assume $\mathfrak{M}$ is a model over a finite NDAG in which γ is true everywhere and χ is true at the root. For each node t define $A_t = \{\psi \in Cl(\gamma, \chi) \mid \mathfrak{M}, t \models \psi\}$. Obviously each A_t is a Hintikka set and there is an A with $\{root, \chi\} \subseteq A$. Let H be the set of all such A_t. Inductively define the level function on H. First define which Hintikka Sets are of level 0: $l(A) = 0$ if *leaf* $\in A$. Next, suppose the j-th level is defined. First define: $S_j = \{A \in H \mid l(A) \leq j\}$. Next, for $A \in H \setminus S_j$, $l(A) = i + 1$ if $\mathfrak{M}, root \models \langle \downarrow^* \rangle (\hat{A} \wedge [\downarrow][\downarrow^*] \bigvee_{B \in S_j} \hat{B})$. It is not hard to show that (H, l) is ordered, saturated, rooted, nominalized and A_{root} contains *root* and χ.

Now assume (H, l) is a rooted nominalized saturated ordered set of Hintikka Sets and there is an $A_{root} \in H$ with $\{root, \chi\} \subseteq A_{root}$. For each nominal i, let A_i be the Hintikka Set containing i. Let $\mathcal{T} = (T, R_{\downarrow})$ be a tree with root t_0 of depth $l(A_{root})$ and branching width $|H|$. The function $\mathsf{depth}(\cdot)$ measures the depth of nodes in the tree (with $\mathsf{depth}(t_0) = 0$). Let $h : T \longrightarrow H$ be a partial function satisfying

root $h(t_0) = A_{root}$.
max if $tR_{\downarrow}t'$ and h is defined on t and t', then $h(t)$ child $h(t')$.
min if $h(t)$ child B then either there exists a $t' \in T$ such that $tR_{\downarrow}t'$ and $h(t') = B$, or B contains a nominal.
nom for each nominal $i \in Cl(\{\gamma, \chi\})$ there exists exactly one t such that $h(t) = A_i$ and for any t', $h(t')$ child A_i implies that $\mathsf{depth}(t) > \mathsf{depth}(t')$.

It is straightforward to show that such h can be defined (by a step-by-step construction for instance). Now we turn $\mathcal{T}$ into an NDAG. First let $\mathcal{T}'$ be the largest subtree of $\mathcal{T}$ on which h is total. Second, let $\mathcal{T}''$ be $\mathcal{T}'$ with the following arrows added:

if $h(t)$ child $h(t')$ and $\mathsf{depth}(t) < \mathsf{depth}(t')$ and $id \in h(t')$, then add $tR_{\downarrow}t'$.

We claim that $(\mathcal{T}'', NOM)$ with $NOM = \{t \mid id \in h(t)\}$ is a rooted NDAG satisfying

up-min if B child $h(t')$ and $id \in h(t')$, then there exists a t such that $tR_{\downarrow}t'$ and $h(t) = B$.
nom-min if $h(t)$ child B and B contains a nominal, then there exists a $t' \in T$ such that $tR_{\downarrow}t'$ and $h(t') = B$.

By construction $(\mathcal{T}'', NOM)$ is a rooted NDAG. To show **up-min**, assume that B child $h(t')$ holds. Then $l(B) > l(h(t'))$. By Definition 3.2 there exists a path A_{root} child $\ldots$ child B, say of length k. Whence, by **min**, there exists a node t with $h(t) = B$ and $\mathsf{depth}(t) = k$. By **max**, the depth of t' must be strictly larger than k because $l(B) > l(h(t'))$. Thus an arrow from t to t' has been added. The proof for **nom-min** is similar.

```
begin
  L    :=  {A ∈ Choice(γ,χ) | leaf ∈ A};
  Pool :=  Choice(γ,χ)\ L;
  S    :=  L;
  k    :=  0;
  l    :=  {(A,k) | A ∈ L};
  do  L ≠ ∅  →
      L    := {A ∈ Pool | (S∪{A},l∪(A,k+1))
              is a down--saturated ordered set of
              Hintikka Sets };
      Pool := Pool \ L;
      S    := S ∪ L;
      k    := k+1;
      l    := l ∪ {(A,k) | A ∈ L}
  od
end
```

Fig. 4. The algorithm *elimination of Choice*(γ, χ)

Let $V(p) = \{t \mid p \in h(t)\}$, for p a nominal or a propositional variable. Then $\mathfrak{M} = (T'', NOM, V)$ is a model, because every nominal is true at exactly one node in NOM.

We claim that $\mathfrak{M} \models \gamma$ and $\mathfrak{M}, t_0 \models \chi$. By assumption $\chi \in h(t_0)$ and γ is in every Hintikka set, thus it is sufficient to prove the truth lemma

for all $\psi \in CL(\gamma, \chi)$, for all nodes t, $\mathfrak{M}, t \models \psi$ if and only if $\psi \in h(t)$.

The base case is by definition of V. The case for *id* is by definition of NOM. The boolean cases are by the conditions on Hintikka sets. The left to right direction for both modalities follows from max. The other direction for $\langle\downarrow\rangle\psi$ and $\langle\uparrow\rangle\psi$ follows from min, nom-min and up-min and saturation.

The algorithm. We now describe the algorithm for finding a saturated, ordered, rooted and nominalized set of Hintikka Sets. It consists of five different stages. Let γ, χ be the formulas for which we decide the existence of a model in which γ holds everywhere and χ at the root.

(1) Create Hintikka Sets The algorithm creates $HS(\gamma, \chi)$. $HS(\gamma, \chi)$ contains $2^{O(|\gamma \wedge \chi|)}$ sets of size $O(|\gamma \wedge \chi|^2)$.

(2) Choose Named Elements Choose a set $NAMED \subseteq HS(\gamma, \chi)$ having a Hintikka set containing χ and the root symbol *root* and exactly one Hintikka set containing the nominal i for each $i \in CL(\gamma, \chi)$.
There are at most $|HS(\gamma, \chi)| \cdot \ldots \cdot |HS(\gamma, \chi)|$ (as many as there are nominals in $\gamma \wedge \chi$ plus one) many choices, that is at most $2^{O(|\gamma \wedge \chi|^2)}$. Let $Choice(\gamma, \chi)$ be $NAMED \cup HS(\gamma, \chi) \setminus \{A \in HS(\gamma, \chi) \mid A$ contains a nominal or *root*$\}$.

(3) Create Down–Saturated Ordered Set Run the algorithm from Figure 4.

Lemma 3. *1. Elimination of Choice(γ, χ) terminates after at most $|HS|$ rounds of the do loop.*

 2. The statement "$\langle S, 1 \rangle$ is a down–saturated ordered set of Hintikka sets" holds after the do loop.

Proof. (1) The bound function of the do loop is the size of Pool which is being reduced in every round, or the loop terminates because L= $\emptyset$. The initial size of Pool is bounded by $|HS|$.

(2) Because the statement "$Choice(\gamma, \chi) =$ Pool $\uplus$ S and $\langle S, 1 \rangle$ is a down–saturated ordered set of Hintikka sets" holds before the do loop and is an invariant of the do loop.

(4) Make (H, l) Rooted Let (H, l) be the output $\langle S, 1 \rangle$ of the previous stage. Delete all elements from H for which condition 2 in Definition 3 does not hold.

(5) Test Let (H, l) be the output of the previous stage. Check whether (H, l) contains a Hintikka set A with $root \in A$ and $\chi \in A$. Check whether (H, l) is up–saturated. And check whether (H, l) contains for each nominal $i \in Cl(\gamma, \chi)$ a Hintikka set containing i.

Lemma 4. *If all these checks succeed, (H, l) is an up and down–saturated rooted and nominalized ordered set of Hintikka sets.*

The algorithm succeeds iff there is a choice in stage 2 for which the checks in stage 5 succeed.

The algorithm is correct by Lemma 2. Let us check that the algorithm runs in time exponential in the length of the input. The first stage is clear. For the second stage it has to perform the rest of the algorithm for at most $2^{O(|\gamma, \chi|^2)}$ many choices. So it is sufficient to show that stages 3–5 can be performed in exponential time. The algorithm of stage 3 terminates after at most $|HS(\gamma, \chi)| \leq 2^{O(|\gamma, \chi|)}$ rounds of the do loop. As in [21], the tests inside the do loop take time polynomially bounded by $|HS(\gamma, \chi)|$. Thus stage 3 takes time exponentially bounded by $|\gamma, \chi|$. It is clear that stages 4 and 5 can all be performed in time polynomially bounded by $|HS(\gamma, \chi)|$. Thus the algorithm is in EXPTIME.

6   Conclusions

We have given an exponential time decision algorithm for a modal language with nominals interpreted on finite rooted nominalized DAGs. This is –as far as we know– the first result which yields a decision algorithm for XPath query containment in the presence of XML's ID/IDREF referencing mechanism.

Obviously the algorithm is not that easy to implement, so that's a next research question. Another question is whether we can get the same exponential upper bound if we interpret the language with both sibling axis on ordered NDAG's.

The function h defined in the proof of Lemma 2 is almost a surjective bounded morphism (the zag direction might break for $R_\uparrow$ from Hintikka sets not containing id). This leads us to conjecture that a slight improvement of that Lemma can be used to prove a completeness theorem for this logic. At present no axiomatization is known.

References

[1] S. Abiteboul, P. Buneman, and D. Suciu. *Data on the web*. Morgan Kaufman, 2000.

[2] N. Alechina, S. Demri, and M. de Rijke. A modal perspective on path constraints. *Journal of Logic and Computation*, 13:1–18, 2003.

[3] M. Benedikt, W. Fan, and G. Kuper. Structural properties of XPath fragments, 2003. In Proc. ICDT'03, 2003. To appear.

[4] P. Blackburn. Representation, reasoning, and relational structures: a hybrid logic manifesto. *Logic Journal of the IGPL*, 8(3):339–365, 2000.

[5] P. Blackburn, M. de Rijke, and Y. Venema. *Modal Logic*. Cambridge University Press, 2001.

[6] P. Blackburn, B. Gaiffe, and M. Marx. Variable free reasoning on finite trees. In *Proceedings of Mathematics of Language (MOL-8), Bloomington*, 2003.

[7] J. Clark. XML Path Language (XPath). http://www.w3.org/TR/xpath.

[8] World-Wide Web Consortium. Xml path language (xpath): Version 2.0. http://www.w3.org/TR/xpath20/.

[9] Alin Deutsch and Val Tannen. Containment of regular path expressions under integrity constraints. In *Knowledge Representation Meets Databases*, 2001.

[10] J. Doner. Tree acceptors and some of their applications. *J. Comput. Syst. Sci.*, 4:405–451, 1970.

[11] E. Emerson and J. Halpern. "Sometimes and Not Never"; revisited: on branching versus linear time temporal logic. *Journal of the ACM (JACM)*, 33(1):151–178, 1986.

[12] M. Fisher and R. Ladner. Propositional dynamic logic of regular programs. *J. Comput. Syst. Sci.*, 18(2):194–211, 1979.

[13] D. Harel, D. Kozen, and J. Tiuryn. *Dynamic Logic*. MIT Press, 2000.

[14] M. Kracht. Syntactic codes and grammar refinement. *Journal of Logic, Language and Information*, 4:41–60, 1995.

[15] M. Kracht. Inessential features. In Christian Retore, editor, *Logical Aspects of Computational Linguistics*, number 1328 in LNAI, pages 43–62. Springer, 1997.

[16] M. Marx. XPath and modal logic of finite trees. In *Proceedings of M4M 2003, Nancy*, 2003.

[17] G. Miklau and D. Suciu. Containment and equivalence for an XPath fragment. In *Proc. PODS'02*, pages 65–76, 2002.

[18] F. Neven and T. Schwentick. XPath containment in the presence of disjunction, DTDs, and variables. In *ICDT 2003*, 2003.

[19] A. Palm. *Transforming tree constraints into formal grammars*. PhD thesis, Universität Passau, 1997.

[20] A. Palm. Propositional tense logic for trees. In *Sixth Meeting on Mathematics of Language*. University of Central Florida, Orlando, Florida, 1999.

[21] V. Pratt. Models of program logics. In *Proceedings of the 20th IEEE symposium on Foundations of Computer Science*, pages 115–122, 1979.

[22] M. Rabin. Decidability of second order theories and automata on infinite trees. *Transactions of the American Mathematical Society*, 141:1–35, 1969.

[23] K. Reinhardt. The complexity of translating logic to finite automata. In E. Grädel et al., editor, *Automata, Logics, and Infinite Games*, volume 2500 of *LNCS*, pages 231–238. Springer, 2002.

[24] B-H. Schlingloff. Expressive completeness of temporal logic of trees. *Journal of Applied Non–Classical Logics*, 2(2):157–180, 1992.

[25] P. Wood. On the equivalence of XML patterns. In *Proc. 1st Int. Conf. on Computational Logic*, volume 1861 of *LNCS*, pages 1152–1166, 2000.

Tableaux, Path Dissolution, and Decomposable Negation Normal Form for Knowledge Compilation*

Neil V. Murray[1] and Erik Rosenthal[2]

[1] Department of Computer Science, State University of New York
Albany, NY 12222, USA
nvm@cs.albany.edu
[2] Department of Mathematics, University of New Haven
West Haven, CT 06516, USA
erosenthal@newhaven.edu

Abstract. *Decomposable negation normal form* (DNNF) was developed primarily for knowledge compilation. Formulas in DNNF are linkless, in negation normal form (NNF), and have the property that atoms are not shared across conjunctions. Full dissolvents are linkless NNF formulas that do not in general have the latter property. However, many of the applications of DNNF can be obtained with full dissolvents. Two additional methods — regular tableaux and *semantic factoring* — are shown to produce equivalent DNNF. A class of formulae is presented on which earlier DNNF conversion techniques are necessarily exponential; path dissolution and semantic factoring handle these formulae in linear time.

1   Introduction

The last decade has seen a virtual explosion of applications of propositional logic. One emerging application is *knowledge compilation*: preprocessing the underlying propositional theory. While knowledge compilation is intractable, it is done once, in an off-line phase, with the goal of making frequent on-line queries efficient. Both the off-line and on-line phases are considered in this paper.

Horn clauses, ordered binary decision diagrams, tries, and sets of *prime implicates/implicants* have all been proposed as targets of such compilation — see, for example, [1, 2, 10, 14, 23, 24]. Some of these target languages employ negation normal form (NNF), although most research has restricted attention to CNF. This may be because the structure of NNF formulae can be surprisingly complex. A comprehensive analysis of that structure can be found in [16] and in [18]. That analysis includes operations on NNF formulae that facilitate the use of NNF in systems.

Decomposable negation normal form (DNNF), the subject of this paper, is a class of formulae studied by Darwiche [4, 6]. They are linkless, in negation

---

* This research was supported in part by the National Science Foundation under grant CCR-0229339.

M. Cialdea Mayer and F. Pirri (Eds.): TABLEAUX 2003, LNAI 2796, pp. 165–180, 2003.

normal form, and have the property that atoms are not shared across conjunctions. Every DNNF formula is automatically a *full dissolvent* — the end result of applying path dissolution to a formula until it is linkless. Full dissolvents are linkless NNF formulas, but in general they may share atoms across conjunctions. Nonetheless, many of the applications of DNNF depend primarily on the linkless property and are both available and equally efficient with full dissolvents. Moreover, full dissolvents can be advantageous both in time and in the size of the resulting formula. In particular, a class of formulae is presented on which the methods of [4, 6] require exponential time and space to obtain DNNF. The full dissolvent can be obtained in linear time, is in DNNF, and is actually smaller than the original formula. Two additional methods are shown to yield a DNNF equivalent of their input formulas. Regular tableaux can handle any set of clauses, and *semantic factoring* works on any formula whatsoever. Moreover, semantic factoring is shown to produce DNNF equivalents of the class of formulae alluded to above in linear time.

Knowledge compilation is in some sense a harder problem than SAT. A SAT solver need only produce a yes-no answer, possibly with a satisfying assignment. On the other hand, to compile a propositional theory, one often requires the result to represent *all* satisfying assignments (or all consequences) of the theory. Thus, computing the full dissolvent or the DNNF equivalent of a formula will answer the question of satisfiability but may also incur more overhead than a SAT solver. On the other hand, progress with SAT solvers might not contribute to progress in knowledge compilation.

A brief summary of the basics of NNF formulae, their two-dimensional representation, and path dissolution is presented in Section 2; greater detail can be found in [18]. Operations that are useful with DNNF formulae are discussed in Section 3, along with efficient counterparts for full dissolvents. In Section 4, regular tableaux and semantic factoring are described as methods for conversion to DNNF. Some concluding remarks and suggestions for future work are made in Section 5.

2 Background

2.1 Path Dissolution and Negation Normal Form

Path dissolution [18] is an inference mechanism that works naturally with formulae in negation normal form. It is strongly complete in the sense that any sequence of link activations will eventually terminate, producing a linkless formula called the *full dissolvent*. The paths that remain are models of the original formula. Full dissolvents have been used effectively for computing the prime implicants and implicates of a formula [22, 23]. Path dissolution has advantages over clause-based inference mechanisms, even when the input is in CNF, since CNF can be *factored*, i.e., put into more compact NNF with applications of the distributive laws. The time savings is often significant, and the space savings can be dramatic.

Formally, a logical formula is said to be in *negation normal form* if $\wedge$ and $\vee$ are the only (binary) connectives, and if all negations are at the atomic level. The structure of NNF formulae can be surprisingly complex, and it is convenient to represent formulae in NNF in a two-dimensional format, sometimes called a *semantic graph*[1].

$$((\overline{C} \wedge A) \vee D) \wedge (\overline{A} \vee (B \wedge C)) \quad \equiv \quad$$

$$
\begin{array}{c}
\overline{C} \\
\wedge \quad \vee \quad D \\
A \\
\wedge \\
\quad\quad B \\
\overline{A} \quad \vee \quad \wedge \\
\quad\quad C
\end{array}
$$

Formula   Semantic
                                          Graph

Two literal occurrences are said to be *c-connected* if they are conjoined within the formula (and *d-connected* if they are disjoined within the formula). For example, the literals D and C are c-connected. A *c-path* (*d-path*) is a maximal set of c-connected (d-connected) literal occurrences; the semantic graph in the figure contains four c-paths: $\{\overline{C}, A, \overline{A}\}$, $\{\overline{C}, A, B, C\}$, $\{D, \overline{A}\}$, and $\{D, B, C\}$. Of these, $\{\overline{C}, A, \overline{A}\}$, and $\{\overline{C}, A, B, C\}$ are unsatisfiable. A *link* is a complementary pair of c-connected literals; in the figure, $\{A, \overline{A}\}$ and $\{C, \overline{C}\}$ are links.

There are several inference rules that use paths and links. Path dissolution is one that works with formulae in NNF and is especially well suited for the propositional case. Dissolution operates on links in a formula by restructuring the formula so that all c-paths through the link are eliminated. Since the number of links is finite, the (linkless) full dissolvent must eventually be produced. Dissolution can be defined intuitively as follows (see [18] for a precise definition): Suppose the formula $\mathcal{F}$ contains the conjunction $\mathcal{G} \wedge \mathcal{H}$ with A in $\mathcal{G}$ and $\overline{A}$ in $\mathcal{H}$. Let $CPE(A, \mathcal{G})$ (the *c-path extension of A in $\mathcal{G}$*) be the part of $\mathcal{G}$ that contains all c-paths through $\mathcal{G}$ that contain A, and let $CC(A, \mathcal{G})$ (the *c-path complement of A in $\mathcal{G}$*) be the part of $\mathcal{G}$ that contains all c-paths through $\mathcal{G}$ that do not contain A. Similarly define $CPE(\overline{A}, \mathcal{H})$ and $CC(\overline{A}, \mathcal{H})$ in $\mathcal{H}$. Then dissolution on the $\{A, \overline{A}\}$ link replaces $\mathcal{G} \wedge \mathcal{H}$ in $\mathcal{F}$ with

$$
\begin{array}{ccccc}
CPE(A, \mathcal{G}) & & CC(A, \mathcal{G}) & & CC(A, \mathcal{G}) \\
\wedge & \vee & \wedge & \vee & \wedge \\
CC(\overline{A}, \mathcal{H}) & & CC(\overline{A}, \mathcal{H}) & & CPE(\overline{A}, \mathcal{H})
\end{array}
$$

It is easily seen that the formula above is equivalent to (and has the identical set of c-paths as) each of the more succinct constructions:

---

[1] Good sources for descriptions of that structure — and to get an idea of how complex that structure can be — are [16] and [18].

$$
\begin{array}{cc}
\mathcal{G} & CC(A,\mathcal{G}) \\
\wedge \quad \vee \quad \wedge \\
CC(\overline{A},\mathcal{H})) \quad CPE(\overline{A},\mathcal{H}))
\end{array}
\quad and \quad
\begin{array}{cc}
CPE(A,\mathcal{G}) & CC(A,\mathcal{G}) \\
\wedge \quad \vee \quad \wedge \\
CC(\overline{A},\mathcal{H}) & \mathcal{H}
\end{array}
$$

Consider the link $\{A,\overline{A}\}$ in the formula $\mathcal{F}$ in the figure below: *All* c-paths through it are unsatisfiable. Dissolving on that link restructures the graph as indicated in the figure. Note that the dissolvent contains precisely the c-paths that miss either A or $\overline{A}$ (or both). Repeated applications of dissolution eventually produce the full dissolvent, FD($\mathcal{F}$), which has no unsatisfiable c-paths.

$$
\begin{array}{ccc}
\overline{C} & \overline{C} & \\
\wedge \vee D & \wedge \vee D & \\
A & A \qquad D & \qquad D \qquad\qquad D \\
\wedge & \wedge \vee \wedge & \wedge \quad \vee \quad \wedge \\
B & B \quad \overline{A} & B \qquad\qquad \overline{A} \\
\overline{A} \vee \wedge & \wedge & \wedge \\
C & C & C
\end{array}
$$

$$
\qquad\quad \mathcal{F} \qquad\qquad\qquad \text{Dissolvent w.r.t } \{A,\overline{A}\} \qquad\qquad \text{FD}(\mathcal{F})
$$

2.2 Factoring and the Prawitz Rule

There is growing evidence that restriction to clause form can be a serious limitation. The NNF representation of a formula often has a considerable space advantage. There are efficient satisfiability preserving translations to clause form from NNF, but they are not equivalence preserving. On the other hand, many CNF formulae can be *factored* into an NNF equivalent that is exponentially smaller. Factoring a formula means applying the distributive laws so as to combine multiple occurrences of subformulae. With logical formulae, since two distributive laws hold, both conjunctive and disjunctive factoring can be done; conjunctive factoring is the more useful by far with inference operations that are based on (conjunctive) links [19].

An automated deduction system that employs clause form cannot factor formulae since the factored formula will in general not be in clause form. On the other hand, any technique that uses NNF can factor, and, if the technique is path based, factoring may improve performance since it (often substantially) reduces the number of c-paths as well as the formula's size. When factoring was added to *Dissolver* [18], a dissolution-based system, a dramatic speedup was achieved with *every* propositional formula that was input in clause form. (The time to factor the input was included in the running time.)

Recall that a unit resolution step (at the ground level) produces a resolvent that is smaller than and subsumes the larger parent. Resolving a pair of two-literal clauses will create a clause no larger than the parents. These particular cases of resolution are often given high priority because they tend to limit or even reverse the growth of the formula.

There are similar cases for ground dissolution, and the payoff is even more dramatic. One is unit dissolution, which arises when $\mathcal{G}$ consists of a single literal A. Then $CC(A, \mathcal{G})$ is empty, and the dissolvent of the link $\{A, \overline{A}\}$ in the subgraph $\mathcal{G} \wedge \mathcal{H}$ is $\mathcal{G} \wedge CC(\overline{A}, \mathcal{H})$. Observe that the effect of dissolving is to replace $\mathcal{H}$ by $CC(\overline{A}, \mathcal{H})$, and that $CC(\overline{A}, \mathcal{H})$ is formed by deleting $\overline{A}$ and anything directly conjoined to it. Hence no duplications whatsoever are required, and both the size of the formula and the number of c-paths are reduced.

Another special case of dissolution is the *Prawitz rule*. Let the smallest conjunction containing the link $\{A, \overline{A}\}$ be $\mathcal{G} \wedge \mathcal{H}$, with A in $\mathcal{G}$ and $\overline{A}$ in $\mathcal{H}$. If $\mathcal{G} \wedge \mathcal{H}$ has the form on the left, then it may be replaced by the formula on the right; this is called the *Prawitz Rule*.[2]

$$
\begin{array}{ccc}
A \vee \mathcal{G}' & A & \overline{A} \\[2mm]
\wedge & \wedge \quad \vee \quad \wedge \\[2mm]
\overline{A} \vee \mathcal{H}' & \mathcal{H}' & \mathcal{G}'
\end{array}
$$

The size of the formula is unchanged, and the c-path through the link $\{A, \overline{A}\}$ is removed. But also removed are *all* c-paths through both $\mathcal{G}'$ and $\mathcal{H}'$. When $\mathcal{G}'$ and $\mathcal{H}'$ are large, the reduction in c-paths is enormous, and the Prawitz step is far superior to an ordinary dissolution step. In addition, as with any dissolution step, the linked literals change from being c-connected to being d-connected. But note that the (potentially large) subformulae $\mathcal{G}'$ and $\mathcal{H}'$ have also become d-connected. This would not be the case in a routine dissolution step, and thus the Prawitz rule pushes the formula more aggressively towards DNNF.

It might appear that the requirements for applying the rule are so strong that there are likely to be few opportunities to employ it. But note that the case in which $\mathcal{G}'$ and $\mathcal{H}'$ are each a disjunction of literals is merely that of two linked clauses. Thus, for a CNF formula, there are many opportunities (initially) for application of the Prawitz rule. Moreover, the judicious use of factoring along with the Prawitz rule on CNF formulas will speed progress without increasing the size of the formula.

Combining factoring and the Prawitz rule can be particularly effective for constructing a DNNF equivalent of a set of clauses. Given any set S of clauses, let S_A be the subset consisting of those clauses containing either A or $\overline{A}$, and let $S' = S - S_A$. Factoring out A and $\overline{A}$ from their clauses in S_A, produces the NNF formula on the left in Figure 1.

Factoring has created a single link to which the Prawitz rule applies. The result (with S' included) is the formula on the right in Figure 1. Factoring plus the Prawitz rule may be applied to the CNF subformulas. But note that variables

---

[2] The descriptions here of both path dissolution and the Prawitz Rule focus on single links, but both rules have been defined to operate on multiple links under appropriate circumstances — see [18].

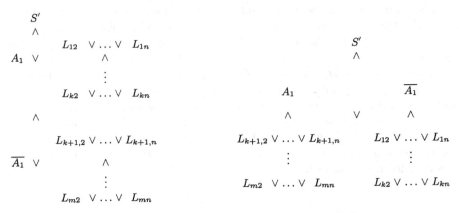

Fig. 1. Factoring on A_1 for the Prawitz Rule

(other than A) will in general be shared between S' and the other parts of the formula. Thus, unlike in the initial operation, not all occurrences of a variable can be factored into one occurrence. One way to solve this problem is to distribute the conjunction with S' across the disjunction, creating a disjunction of two CNF formulas. In fact, *any* NNF formula can be multiplied out to CNF (or to a disjunction of CNF formulas), but avoiding this is precisely the reason for focusing on NNF.

Observe that the initial application of factoring plus Prawitz is equivalent to applying the following identity, which holds for *any* formula $\mathcal{G}$, to S_A.

$$\mathcal{G} = (A \wedge \mathcal{G}[TRUE/A]) \vee (\overline{A} \wedge \mathcal{G}[FALSE/A])$$

where $\mathcal{G}[\mathcal{B}/A]$ denotes the replacement of all occurrences of atom A by $\mathcal{B}$ in $\mathcal{G}$. Observe that any formula $\mathcal{G}$ may be replaced by the formula on the right. In fact, this rule, called *semantic factoring*, can be applied to any subformula and, in particular, to the smallest part of the formula containing all occurrences of the variable being factored. The term semantic factoring reflects the fact that all occurrences of the atom A within the subformula under consideration have in effect been 'factored' into one positive and one negative occurrence. In addition to the *conditioning* operation discussed in Section 3.1, both the original Prawitz Rule [20] and Step 4 in Rule III of the Davis-Putnam procedure [8] are closely related to semantic factoring.

3   Decomposable Negation Normal Form

Let **atoms**$(\mathcal{F})$ denote the atom set of a formula $\mathcal{F}$. An NNF formula $\mathcal{F}$ (possibly containing boolean constants) is said to be in *decomposable negation normal form* (DNNF) if $\mathcal{F}$ satisfies the *decomposability property*: If $\alpha = \alpha_1 \wedge \alpha_2 \wedge ... \alpha_n$ is a conjunction in $\mathcal{F}$, then $i \neq j$ implies that **atoms**$(\alpha_i) \cap$ **atoms**$(\alpha_j) = \emptyset$; i.e., no two conjuncts of α share an atom. Observe that a DNNF formula is necessarily linkless since a literal and its complement cannot be conjoined — after all, they share the same atom. The structure of formulae in DNNF is much simpler than the more general NNF. As a result, many operations on DNNF formulae can be performed efficiently. Of course, obtaining DNNF can be expensive, but if this is done once as a preprocessing step, the "expense" can be spread over many queries.

3.1   Conjoining Literals and Conditioning

A useful operation presented in [6] is conjoining a literal L (or, iteratively, a set of literals) to a knowledge base $\mathcal{K}$ while preserving whatever normal form $\mathcal{K}$ may have. If $\mathcal{K}$ is in DNNF and contains occurrences of L or of $\overline{L}$, then $\mathcal{K} \wedge L$ is no longer in DNNF. The *conditioning* operation, denoted $(\mathcal{K}|L)$, produces an equivalent DNNF: Replace all occurrences of L in $\mathcal{K}$ by $TRUE$ and all occurrences of $\overline{L}$ by $FALSE$ and conjoin the result to L. This produces a formula that is equivalent to $\mathcal{K} \wedge L$ and is easily seen to be in DNNF. The operation is linear in the size of $\mathcal{K}$.

Suppose now that $\mathcal{K}$ is a full dissolvent. The conjunction $(\mathcal{K} \wedge L)$ may have links between $\mathcal{K}$ and L. For each such link a unit dissolvent can be generated that removes the c-extension of $\overline{L}$ in $\mathcal{K}$, in effect substituting false for $\overline{L}$ throughout $\mathcal{K}$. This leaves occurrences of L but removes all links. Thus, a full dissolvent results, and literal conjoining via unit dissolution preserves full dissolvents just as conditioning preserves DNNF. It is worth noting that if $\mathcal{K}$ happens to be in DNNF, there are semantic graph-based tools that not only enable literal conjoining but also result in a DNNF formula.[3]

3.2   Testing for Entailment

One of the most fundamental types of queries is whether a knowledge base $\mathcal{K}$ logically entails a clause C. If so, then of course $(\mathcal{K} \wedge \neg C)$ is unsatisfiable. Since $\neg C$ is a conjunction of literals, $\mathcal{K}$ may be conditioned on each literal in $\neg C$, and the entailment test amounts to a satisfiability test on the resulting formula.

A DNNF formula can be tested for satisfiability in linear time. The reason is that the test can always be performed independently on subformulae. For disjunctions, this is true for any NNF formula: the formula is satisfiable if and only if at least one of its disjuncts is. However, a conjunction may be unsatisfiable even if all of its conjuncts are (separately) satisfiable. But this is not true for DNNF because the conjuncts do not share variables. The satisfying interpretations of different conjuncts cannot conflict, and so can be combined into a single

---

[3] The occurrences of L in $\mathcal{K}$ left behind by unit dissolutions on $\overline{L}$ violate DNNF but form conjunctive *anti-links* [21] with L. Activating these with the appropriate anti-link operator removes the d-extension of occurrences of L in $\mathcal{K}$, which has the effect of replacing L by $TRUE$. So with unit dissolution and anti-links, $(\mathcal{K} \wedge L)$ can be converted into an equivalent DNNF formula.

satisfying interpretation for the entire conjunction. This independence for both disjunction and conjunction yields a linear satisfiability test for DNNF immediately. (Note that the truth constants $TRUE$ and $FALSE$ are allowed in the formula; otherwise, the test would be in constant time.)

Full dissolvents, on the other hand, do share variables across conjunctions. Nonetheless, entailment can be determined in linear time. For any formula $\mathcal{K}$, $\mathcal{K} \models \mathcal{C}$ iff $\mathcal{K} \wedge \neg \mathcal{C}$ is unsatisfiable. In particular, if $\mathcal{K}$ is a full dissolvent, $\mathcal{K}$ contains no links, and all links in $\mathcal{K} \wedge \neg \mathcal{C}$ go between $\mathcal{K}$ and $\neg \mathcal{C}$. Unit dissolutions on those links strictly decrease the size of the formula. Each operation is no worse than linear in the amount by which the formula shrinks. So the sum total of a series of unit dissolutions can require no worse than time linear in the size of $\mathcal{K}$. Activating all links results again in a linkless formula. If this formula is non-empty, the absence of links implies that it is satisfiable. (With dissolution, we usually assume that truth constants are simplified away; if not, then the additional analysis used for DNNF applies.)

For implementation purposes, there is a better approach. We need not actually perform the unit dissolutions on $\mathcal{K}$. The next theorem is proved in [22].

Theorem 1. In any non-empty formula $\mathcal{K}$ in which no c-path contains a link, every implicate of $\mathcal{K}$ is subsumed by some d-path of $\mathcal{K}$.   □

As a result, a clause $\mathcal{C}$ can be tested for entailment as follows: First, mark all complements of literals of $\mathcal{C}$ in $\mathcal{K}$ and call the resulting subgraph $\mathcal{K}_{\overline{\mathcal{C}}}$. Then determine whether $\mathcal{K}_{\overline{\mathcal{C}}}$ contains a full d-path through $\mathcal{K}$. This computation can be done in a recursive manner analogous to the satisfiability test for DNNF. A marked literal is a d-path from $\mathcal{K}_{\overline{\mathcal{C}}}$ through itself, and an unmarked literal contains no d-path from $\mathcal{K}_{\overline{\mathcal{C}}}$. $\mathcal{K}_{\overline{\mathcal{C}}}$ contains a d-path through a conjunction if there is one through one of the conjuncts, and it contains a d-path through a disjunction if there is one through each of its disjuncts. Although this is a somewhat informal description, the algorithm is easily seen to be linear in the size of $\mathcal{K}$.

3.3   Projection

If $\mathcal{A}$ is a set of atoms, then an $\mathcal{A}$-*sentence* is defined to be a formula, all of whose atoms come from $\mathcal{A}$. An $\mathcal{A}$-*literal* is a literal whose atom is in $\mathcal{A}$, and an $\overline{\mathcal{A}}$-*literal* is a literal whose atom is not in $\mathcal{A}$. In this paper, we assume all $\mathcal{A}$-sentences to be in NNF. If $\mathcal{F}$ is a formula, then the *projection of $\mathcal{F}$ onto $\mathcal{A}$*, denoted **Project**$(\mathcal{F}, \mathcal{A})$, is defined to be the formula produced by substituting *true* for each $\overline{\mathcal{A}}$-literal (whether it occurs positively or negatively) in $\mathcal{F}$.[4]

The next theorem is proved in [6] for DNNF formulae and generalized here to arbitrary linkless formulae. The path-based proof below is considerably simpler than the one in [6]. Observe that when a formula contains the constants *true* and *false*, the formula can easily be simplified to an equivalent formula that contains no constants, and for most implementation purposes, this is desirable. For the

---

[4] These definitions come from [6], although the definition of projection is stated differently there.

proof of the theorem, however, it is more convenient to assume the projection is not so simplified.

Theorem 2. If $\mathcal{F}$ is a linkless formula, if $\Gamma = \textbf{Project}(\mathcal{F}, \mathcal{A})$, and if β is an $\mathcal{A}$-sentence, then $\Gamma \models \beta$ iff $\mathcal{F} \models \beta$.

Proof. Suppose first that $\mathcal{F} \models \beta$; i.e., that $\mathcal{F} \wedge \neg\beta$ is unsatisfiable. We must show that $\Gamma \models \beta$; i.e., that $\Gamma \wedge \neg\beta$ is unsatisfiable. It suffices to show that every c-path p through $\Gamma \wedge \neg\beta$ contains a link or the constant $false$. Let $\tilde{p}$ be the path through $\mathcal{F} \wedge \neg\beta$ corresponding to p; i.e., $\tilde{p}$ contains the same $\mathcal{A}$-literals as p but has the $\overline{\mathcal{A}}$-literals that were replaced by $true$ in Γ. Since $\mathcal{F} \wedge \neg\beta$ is unsatisfiable, $\tilde{p}$ contains a link or the constant $false$. If $\tilde{p}$ contains the constant $false$, then p must also contain the constant since $false$ cannot be introduced into a formula by the projection operation. If $\tilde{p}$ contains a link, since $\mathcal{F}$ is linkless, that link must contain one literal from $\mathcal{F}$ and one from $\neg\beta$ or both from $\neg\beta$. But β and thus $\neg\beta$ is an $\mathcal{A}$-sentence, so that link must exist on p.

Conversely, suppose that $\Gamma \models \beta$; i.e., that $\Gamma \wedge \neg\beta$ is unsatisfiable. We must show that $\mathcal{F} \wedge \neg\beta$ is unsatisfiable. Let $\tilde{p}$ be a path through $\mathcal{F} \wedge \neg\beta$, and let p be the corresponding path through $\Gamma \wedge \neg\beta$. Then p contains a link or the constant $false$, and that link or constant must also be in $\tilde{p}$ since every literal in p is also in $\tilde{p}$.   □

Note that the second half of the proof made no use of the properties of $\mathcal{F}$, and so that half of the theorem is valid for any NNF formula. Hence,

Corollary. If $\mathcal{F}$ is any NNF formula, if $\Gamma = \textbf{Project}(\mathcal{F}, \mathcal{A})$, and if β is an $\mathcal{A}$-sentence such that $\Gamma \models \beta$, then $\mathcal{F} \models \beta$.   □

Projection can be quite useful for handling queries over a fixed subset $\mathcal{A}$ of the variables in a compiled knowledge base $\mathcal{K}$. Depending upon $\mathcal{K}$ and $\mathcal{A}$, $\textbf{Project}(\mathcal{K}, \mathcal{A})$ may be much smaller than $\mathcal{K}$ and yet entails exactly the same $\mathcal{A}$-sentences entailed by $\mathcal{K}$. So queries based on entailment that are confined to $\mathcal{A}$ can be processed correspondingly more efficiently. Furthermore, since projection preserves structure and introduces no new atoms, it also preserves the DNNF or the linkless status of formulas. Yet none of these observations make full use of the corollary above.

$\textbf{Project}(\mathcal{F}, \mathcal{A})$ is missing atoms from $\mathcal{F}$ and so may be linkless or in DNNF even if $\mathcal{F}$ is not. After such a case has been identified, queries confined to $\mathcal{A}$ for which the projection provides an affirmative answer are in effect answered without compilation at all, even if the knowledge base $\mathcal{K}$ is arbitrary NNF. Furthermore, if the projection does not have the desired normal form, it still entails the same subset of the $\mathcal{A}$-sentences entailed by $\mathcal{K}$. If it is much smaller than $\mathcal{K}$, then compiling the projection will likely be much more efficient than compiling $\mathcal{K}$, and the compiled version of the projection can then be used to provide answers for that subset of queries. Since only the projection need be compiled, if compilation is required at all, this provides a very efficient sound but not complete query answering method.[5]

Various queries from planning, diagnosis, and Boolean circuit analysis are naturally confined to subsets of the knowledge base variables [6]. With the theorem and corollary above, projection seems to be a promising technique for these situations.

4   Compilation Techniques

Full dissolvents provide efficiency equal to that of DNNF for the operations discussed so far. Furthermore, compiling to DNNF will sometimes create a much larger formula than will computing a full dissolvent. However, there are several operations that are efficient for DNNF, and that cannot obviously be handled efficiently with full dissolvents. In this section, after a brief discussion of several of these operations, alternative techniques for producing DNNF are introduced.

An NNF formula $\mathcal{F}$ is *smooth* if for every disjunction $\beta = \beta_1 \vee \ldots \vee \beta_n$ in $\mathcal{F}$, $\mathbf{atoms}(\beta) = \mathbf{atoms}(\beta_i), 1 \leq i \leq n$. A formula β can be smoothed as follows: Let $\overline{B_i} = \mathbf{atoms}(\beta) - \mathbf{atoms}(\beta_i)$, $1 \leq i \leq n$. For each i, form the conjunction $\wedge_{p_i \in \overline{B_i}}(p_i \vee \overline{p_i})$ and conjoin it with β_i to produce β_i', and call the resulting disjunction β'. This operation is polynomial (roughly quadratic), preserves both equivalence and DNNF, and produces a smooth β'. Smoothness is convenient for minimization, discussed below.

The *minimum cardinality* of a formula $\mathcal{F}$ is defined to be the minimum number of negated atoms in any satisfying model. If the formula is in DNNF, this can be computed in linear time because the computations on subformulas are independent. The algorithm is a straightforward recursion, but the disjointness of atom sets across conjunctions is crucial. The *minimization* of a formula $\mathcal{F}$ is a formula Γ, all of whose models are minimum cardinality models of $\mathcal{F}$. When $\mathcal{F}$ is a smooth DNNF, Γ can be computed in linear time by merely dropping from all disjunctions those disjuncts whose minimum cardinality is greater than the minimum cardinality of the disjunction. Minimum cardinality and minimization (and thus smoothness) have applications in model-based diagnosis and in planning [5].

Two techniques for compiling a formula into DNNF are *regular tableaux* and *semantic factoring*. The first is a restriction of the tableau method developed by Letz [12, 13]; the second was defined in Section 2.2. It is a variant of Darwiche's conditioning [6] and is closely related to earlier methods of Prawitz [20] and of Davis and Putnam [8]. Semantic factoring is polynomial on the class $\mathcal{C}_n^n$ of formulas defined in Section 4.2. The compilation techniques in [6] are exponential for $\mathcal{C}_n^n$; they do employ conditioning, but the intractability is not due to conditioning itself.

---

[5] This is an efficient approach that is similar to the *approximate compilation* defined in [6], in which some atoms are ignored. There, the restriction to a given atom set occurs during the compilation of $\mathcal{K}$.

4.1   Regular Tableaux

There are variations — all minor — in the way different authors define the tableau method. One issue is whether the initial formula is part of the tableau. Letz [12], for example, brings subformulas into the tableau only after the application of a beta rule. The definition below forms an initial tableau with the original formula, which is convenient for formulas in negation normal form. While the difference is minor, it does have some effect. For example, unit clauses are handled differently. It should also be noted that many authors do not restrict attention to clause form. Three of the many good references for the tableau method are [25, 9, 3].

There is a natural correspondence between tableau proof trees and semantic graphs, and, since our goal here is to produce DNNF formulas, it is convenient to cast tableaux in terms of semantic graphs. The generation of the tableau from a formula in NNF requires the usual α, β and closure rules. We restrict attention here to sets of clauses because it is far from obvious how to obtain DNNF from an arbitrary NNF formula using tableaux. With that restriction, α rules are unnecessary.

Definition. A *tableau* for a set (conjunction) of clauses $S = \{C_1, C_2, ..., C_n\}$ is a tree representing a semantic graph. The root is unlabeled, and the remaining nodes are labeled with clauses or literals.

1. The tree consisting of a single branch on which the nodes are labeled with the clauses of S is the *initial* tableau.
2. If T is a tableau, a new tableau may be obtained by conjoining to a leaf in T the disjunction of the literals of any clause in T. This is the *beta rule* or *beta extension*. Each of the extending literals thus becomes a leaf.
3. If a beta rule produces a leaf whose literal is the negative of a literal labeling a node already on that branch, the branch is marked closed. This is the *closure rule*.
4. A beta extension of path θ is *regular* if none of the resulting leaf literals also label a node of θ. A tableau is said to be regular if every extension used to produce it is regular.
5. A clause may be deleted from the tableau if the resulting tableau is logically equivalent.

Note that it is never useful to extend unit clauses, and we assume for the remainder of the discussion that such unit extensions are never done. Furthermore, we consider a node on the initial branch labeled with a unit clause to be in effect labeled with the literal of that clause.

A tree branch is the conjunction of its nodes but may also be viewed as a collection of c-paths. That view makes it easy to see that a closure amounts to dissolving on the link consisting of the leaf literal and its negative along the branch that enabled the closure. We call a tableau *closed* if every branch has been closed and note that a closed tableau may be interpreted as the empty clause. We call a tableau *complete* if every node is labeled with a literal. Thus, closed tableaux are complete (all clauses are in effect deleted) and represent unsatisfiable formulas. Observe that the branches in a complete tableau are single c-paths.

Complete tableaux are interesting when a formula is satisfiable. In particular, a complete regular tableaux is in DNNF. The reason is simple: Regularity ensures that no literal occurs more than once along a single branch, and all conjunctions in a tableau are represented by branches. Note that completeness is required for DNNF; otherwise, a literal and a clause containing the literal might appear on the same branch.

One way to obtain a complete tableau is to extend each clause along every branch and then delete it. In general, however, such tableaux are not regular. However, as the next theorem demonstrates, if we apply all possible regular extensions to every non-unit clause, then we do obtain a complete, regular tableau; i.e., we do obtain DNNF.

Theorem 3. Let $S = \{C_1, C_2, ..., C_n\}$ be a set of clauses, and let T be the tableau obtained by applying all possible regular extensions to each non-unit clause and then deleting the clause. Then T is a complete, regular tableau logically equivalent to S. In particular, T is a DNNF equivalent of S.

Proof. Proceed by induction on n, the number of clauses in S. If $n = 1$, the result is trivial, so suppose it holds for all clause sets with at most n clauses, and suppose S has $n + 1$, say $S = \{C_1, C_2, ..., C_{n+1}\}$.

If $S' = \{C_1, C_2, ..., C_n\}$, then the induction hypothesis applies to S'; i.e., if we apply all possible regular extensions to each non-unit clause in S' and delete the clause, the resulting tableau T' is complete, regular, and logically equivalent to S'. Applying the same operations to S produces a tableau T'_{n+1} identical to T' but with C_{n+1} on the initial branch. Let T be the result of applying all possible regular extensions of C_{n+1} in T'_{n+1} and then deleting it. Let T^* be the result of extending C_{n+1} on all branches in T'_{n+1} (whether or not the extension is regular) and then deleting it.

The proof will be complete if we show that T is logically equivalent to S. Since T' is logically equivalent to S', and since $T'_{n+1} = S' \wedge C_{n+1} = S$, T^* is logically equivalent to S. Thus, it suffices to show that every branch in T'_{n+1} on which an irregular extension is performed in producing T^* is subsumed by one of the extensions. Let $b = \{p_1, p_2, ..., p_k\}$ be (the node labels of) any such branch. Thus, for some i, $p_i \in C_{n+1}$. As a result, if I is an interpretation satisfying b, then I also satisfies the branch $\{p_1, p_2, ..., p_k, p_i\}$ in T^*.   □

4.2   Semantic Factoring

One disadvantage of using regular tableaux to produce DNNF, shared by the methods of [6], is the reliance on CNF for the initial formula set. In this section, repeated application of semantic factoring, as introduced in Section 4.2, is shown to produce DNNF from *any* formula; i.e., no normalization whatsoever is required. Recall that the method is based on the identity:

$$\mathcal{G} = (A \wedge \mathcal{G}[TRUE/A]) \vee (\overline{A} \wedge \mathcal{G}[FALSE/A])$$

where $\mathcal{G}[\mathcal{B}/A]$ denotes the replacement of all occurrences of atom A by $\mathcal{B}$ in $\mathcal{G}$.

It is immediate that the formula on the right has the decomposability property with respect to atom A. Applying this identity iteratively over the atom set of $\mathcal{G}$ will thus produce a DNNF equivalent of $\mathcal{G}$.

There will of course be formulas on which semantic factoring is exponential. (We consider the size of a formula to be the number of literal occurrences.) Since applying the rule seems to almost double the formula, one might suspect that in fact it is always exponential. This is not the case, however, because the rule may be applied to any subformula. In particular, subformulas that do not share atoms would clearly be processed more efficiently when processed independently. The reason is that when factoring on atom A in $\mathcal{G}$, any part of $\mathcal{G}$ that is sufficiently 'far away' from A will occur twice in the result. (For NNF, subformulas outside of both the c-extension and the d-extension of A-literals are sufficiently far away.) Note also that for conversion to DNNF, even formulas that do share variables may be processed independently if they are d-connected.

Consider $\mathcal{C}_n$, the *complete formula* on n variables: $\mathcal{C}_1$ is $A_1 \wedge \overline{A_1}$, and $\mathcal{C}_{n+1}$ is defined by taking two copies of $\mathcal{C}_n$, adding A_{n+1} to each clause in one copy and $\overline{A_{n+1}}$ to each clause in the other copy. The formulas that result are in CNF. Handling these formulas efficiently is somewhat uninteresting because they are as large as their truth tables. They are also unsatisfiable and compile to false. However, any proper subset of $\mathcal{C}_n$ is satisfiable.

Let $\mathcal{C}_n^m$ be any subset of $\mathcal{C}_n$ containing m clauses. Then $\mathcal{C}_n^n$ has n clauses of n literals each, and all clauses have identical atom sets. A typical member $\mathcal{G}$ of $\mathcal{C}_n^n$ is depicted on the left in Figure 2, where the atom set is $\{A_1, A_2, \ldots, A_n\}$ and $L_{ij} = A_j$ or $L_{ij} = \overline{A_j}$, $1 \le i, j \le n$. Furthermore, we assume that $L_{i1} = A_1, 1 \le i \le k$, and that $L_{i1} = \overline{A_1}, k+1 \le i \le n$. The situation is essentially that of Figure 1 except for the assumptions about the atom set. If we apply semantic factoring to $\mathcal{G}$ on A_1, the formula on the right in Figure 2 results. It is a disjunction whose arguments are each essentially in CNF (except for A_1 and for $\overline{A_1}$ which play no further role).

$$
\begin{array}{ccc}
A_1 \vee L_{12} \vee \ldots \vee L_{1n} & A_1 & \overline{A_1} \\
\vdots & & \\
A_1 \vee L_{k2} \vee \ldots \vee L_{kn} & \wedge \qquad \vee \qquad \wedge & \\
\overline{A_1} \vee L_{k+1,2} \vee \ldots \vee L_{k+1,n} & L_{k+1,2} \vee \ldots \vee L_{k+1,n} \qquad L_{12} \vee \ldots \vee L_{1n} & \\
\vdots & \vdots \qquad\qquad\qquad \vdots & \\
\overline{A_1} \vee L_{n2} \vee \ldots \vee L_{nn} & L_{n2} \vee \ldots \vee L_{nn} \qquad L_{k2} \vee \ldots \vee L_{kn} &
\end{array}
$$

Fig. 2. Semantic Factoring on A_1

Notice that the size of the formula after semantic factoring has not increased. In fact, the number of literal occurrences has decreased by $n-2$. (If $n=1$, this is an increase but is irrelevant since the formula is already in DNNF.) Furthermore, two disjoined clause sets remain, one a member of $\mathcal{C}_{n-1}^k$ and the other a member of $\mathcal{C}_{n-1}^{n-k}$. Each set can be processed independently, and the above analysis applies. Hence, this procedure produces DNNF in linear time, and the resulting DNNF is smaller than the original formula. This proves

Theorem 4. The class $\mathcal{C}_n^n$ of CNF formulas can be compiled into DNNF in linear time and space with semantic factoring.   □

The methods introduced in [6] rely on a recursive decomposition of the initial clause set, in effect inducing partitions of increasing rank. When single clauses are encountered, they are returned (perhaps partially evaluated) as already in DNNF. But in all other cases, we have a conjunction of DNNF formulas, and this triggers an iterative conditioning operation over all instances of certain atom sets. These atom sets are pruned in recursive invocations for efficiency. But in the initial calls, they amount to the intersection of the atom sets of the initial partion's blocks. For the examples above, any two clauses share the same n atoms, and so an $O(2^n)$ process results.[6] The output is a disjunction in which each of 2^n disjuncts is either TRUE or FALSE conjoined with one of the 2^n instantiations of the variable set. Unfortunately, the constant is TRUE in all but n cases, creating exponentially large output regardless of whether the constants are removed through simplification.

We note here that in combination with (ordinary algebraic) factoring and the Prawitz Rule, dissolution is also linear on $\mathcal{C}_n^n$ and produces DNNF. This is easy to see from Figure 2. We simply factor out both P_1 and $\overline{P_1}$ from the clauses where they occur; the resulting single link admits the Prawitz rule, and the result is exactly that of Figure 2 on the right.

Corollary. Path dissolution methods are sufficient to compile the class $\mathcal{C}_n^n$ of CNF formulas into DNNF in linear time and space.   □

5   Future Work

Tableaux inference systems are plentiful, and many can be set to obey regularity. Therefore, a closer look at the efficiency of regular tableaux for compiling to DNNF is merited. We conjecture that this method will turn out to be at least as efficient as those of [6].

Another question that remains is whether path dissolution can be guided so as to produce full dissolvents that are in DNNF. In some sense, the answer is yes, since dissolution can be viewed as a generalization of tableaux in which use of

---

[6] In [6], a class of formulas is discussed for which the *treewidth*, a measure of the degree to which clauses share atoms, is unbounded, yet the methods there are quadratic. It turns out that both semantic factoring and dissolution are also quadratic on this class, and the full dissolvent produced by dissolution is in DNNF.

the distributive laws is avoided fully or partially. A more precise, deterministic answer is likely to provide insights useful for knowledge compilation.

Acknowledgement

The authors are indebted to the reviewers for their careful reading of the paper; a number of corrections and improvements resulted.

References

[1] Bryant, R. E., Symbolic Boolean manipulation with ordered binary decision diagrams, *ACM Computing Surveys* **24**, 3 (1992), 293–318.

[2] Cadoli, M., and Donini, F. M., A survey on knowledge compilation, *AI Communications* **10** (1997), 137–150.

[3] D'Agostino, M., Gabbay, D. M., Hähnle, R., and Posegga, J., *Handbook of Tableau Methods*, Kluwer Academic Publishers, 1999.

[4] Darwiche, A., Compiling devices: A structure-based approach, Proc. *International Conference on Principles of Knowledge Representation and Reasoning (KR98)*, Morgan-Kaufmann, San Francisco (1998), 156–166.

[5] Darwiche, A., Model based diagnosis using structured system descriptions, *Journal of A. I. Research*, **8**, 165-222.

[6] Darwiche, A., Decomposable negation normal form, *J.ACM* **48**,4 (2001), 608–647.

[7] Darwiche, A. and Marquis, P., A knowledge compilation map, *J. of AI Research* **17** (2002), 229–264.

[8] Davis, M. and Putnam, H. A computing procedure for quantification theory. *J.ACM* **7**, 1960, 201–215.

[9] Fitting, M., *First-Order Logic and Automated Theorem Proving (2nd ed.)*, Springer-Verlag, New York, (1996).

[10] Forbus, K. D. and de Kleer, J., *Building Problem Solvers*, MIT Press, Cambridge, Mass. (1993).

[11] Goubault, J. and Posegga, J. BDD's and automated deduction. *Proceedings of the 8th International Symposium on Methodologies for Intelligent Systems*, Charlotte, NC, Oct. 1994. In *Lecture Notes in Artificial Intelligence*, Springer-Verlag, Vol. 869, 541–550.

[12] Letz, R., *First-order calculi and proof procedures for automated deduction*, Ph.D. dissertation, TU Darmstadt, 1993.

[13] Letz, R., Mayr, K. and Goller, C. Controlled integration of the cut rule into connection tableau calculi. *Journal of Automated Reasoning* **13**,3 (December 1994), 297–338.

[14] Marquis, P., Knowledge compilation using theory prime implicates, Proc. *International Joint Conference on Artificial Intelligence (IJCAI)* (1995), Morgan-Kaufmann, San Mateo, California, 837-843.

[15] Murray, N. V. and Ramesh, A. An application of non-clausal deduction in diagnosis. *Proceedings of the Eighth International Symposium on Artificial Intelligence*, Monterrey, Mexico, October 17-20, 1995, 378–385.

[16] Murray, N. V. and Rosenthal, E. Inference with path resolution and semantic graphs. *J. ACM* **34**,2 (1987), 225–254.

[17] Murray, N. V. and Rosenthal, E., Reexamining tractability of analytic tableaux, *Proc. of the 1990 Symposium on Symbolic and Algebraic Computation*, 1990, 52–59.

[18] Murray, N. V., and Rosenthal, E. Dissolution: Making paths vanish. *J.ACM* **40,3** (July 1993), 504–535.

[19] Murray, N. V., and Rosenthal, E. On the relative merits of path dissolution and the method of analytic tableaux, *Theoretical Computer Science* **131** (1994), 1–28.

[20] Prawitz, D. A proof procedure with matrix reduction. *Lecture Notes in Mathematics* **125**, Springer-Verlag, 1970, 207–213.

[21] Ramesh, A., Beckert, B., Hähnle, R. and Murray, N. V. Fast subsumption checks using anti-links. *Journal of Automated Reasoning* **18,1**, Kluwer, 47–83 (1997).

[22] Ramesh, A., Becker, G. and Murray, N. V. CNF and DNF considered harmful for computing prime implicants/implicates. *Journal of Automated Reasoning* **18,3** (1997), Kluwer, 337–356.

[23] Ramesh, A. and Murray, N. V. An application of non-clausal deduction in diagnosis. *Expert Systems with Applications* **12,1** (1997), 119-126.

[24] Selman, B., and Kautz, H., Knowledge compilation and theory approximation, *J.ACM* **43,2** (1996), 193-224.

[25] Smullyan, R. M., *First-Order Logic*, second corrected edition, Dover Press (1995).

A More Efficient Tableaux Procedure for Simultaneous Search for Refutations and Finite Models

Nicolas Peltier

Leibniz-IMAG
46, Avenue Felix Viallet, 38031 Grenoble, France
Nicolas.Peltier@imag.fr

Abstract. We describe a (many-sorted) tableaux procedure that has the following properties: it is sound, refutationally complete **and** complete for finite satisfiability (i.e. the procedure terminates if the formula has a finite model). As for standard tableaux methods, models can be extracted from finite open branches. As similar existing procedures, our method relies on a modified δ-rule allowing to reuse existing variables occurring in the same branch. An original notion of complexity measure is introduced in order to control the application of this rule (which is potentially time consuming). The procedure is semantically guided: an interpretation (provided by the user) is used for pruning the search space. This interpretation is refined dynamically during proof search until a model or a contradiction is found. The method has been implemented and some preliminary experimental results are presented.

1 Introduction and Motivations

Tableaux calculi (see for example [7]) are based on the following principle: they try to enumerate the (possibly infinite) interpretations potentially satisfying the considered formula. To this purpose, a tree labeled by first-order formulae is constructed, using expansion rules reflecting the semantics of the logical symbols. Each branch in the tree can be associated to a representation set of a potential model of the formula. Branches are closed when a contradiction is found, which indicates that the interpretation corresponding to the branch does not satisfy the formula at hand. Branches that cannot be closed correspond to models. These procedures are very natural and rather close to human reasoning, which makes them suitable for interactive theorem proving. They can be easily adapted to specific syntaxes. Moreover, if a tableaux procedure terminates without detecting a contradiction, then it provides as a by-product a model of the formula at hand. This is an advantage over resolution calculi in which no model is explicitly constructed. Additional post-processing algorithms have to be designed for extracting models from clause sets that are saturated under resolution, which is possible only in some particular cases (see for example [6]).

Unfortunately, tableaux procedures seldom terminate, due to the fact that infinite open branches (i.e. infinite models) can be generated. This is true even

M. Cialdea Mayer and F. Pirri (Eds.): TABLEAUX 2003, LNAI 2796, pp. 181–195, 2003.

if the formula does have finite models, because no effort is made to minimize the interpretation constructed during the search. Consider for example the formula

$$\phi = (p(a_0, a_1) \wedge (\forall x, y)(p(x, y) \Rightarrow (\exists z)p(y, z))).$$

ϕ has an obvious finite model $\{p(a_0, a_1), p(a_1, a_0)\}$. Standard tableaux calculi will generate the following branch:

$$\{p(a_0, a_1), p(a_1, a_2), p(a_2, a_3), \ldots, p(a_i, a_{i+1}), \ldots\}.$$

At each step $i \in \mathbb{N}$, the γ rule instantiates the formula $(\forall x, y)p(x, y) \Rightarrow (\exists z)p(y, z)$ with the substitution $x \to a_i, y \to a_{i+1}$ which produces the formulae $p(a_i, a_{i+1}) \Rightarrow (\exists z)p(a_{i+1}, z)$ and (by modus ponens, after closing the branch corresponding to $\neg p(a_i, a_{i+1})$) $(\exists z)p(a_{i+1}, z)$. To handle the existential quantifier, a new constant symbol a_{i+2} satisfying the desired property is introduced, which produces the atom $p(a_{i+1}, a_{i+2})$. This last step is an application of the so-called δ-rule, that can be formulated as follows (c denotes a new constant symbol, not occurring in the branch).

$$\frac{(\exists x)\phi(x)}{\phi\{x \to c\}}$$

If no mechanism is added to detect potential loops, the process will continue for ever, yielding the following (infinite) model: $\{p(a_i, a_{i+1}) \mid i \in \mathbb{N}\}$. Note that using skolemization would produce exactly the same result: the existential quantification $(\exists z)p(y, z)$ would be replaced by the formula $p(y, f(x, y))$ (where f denotes a new function symbol) and an infinite branch of the form $\{p(a_0, a_1), p(a_1, f(a_0, a_1)), p(f(a_0, a_1), f(a_1, f(a_0, a_1))), \ldots\}$ will be generated. The use of unification delays the application of the substitution but does not affect the termination behaviour of the proof process.

Of course this is not a problem for refutational completeness since such branches will eventually be closed in case the formula is unsatisfiable. However, if one want to use the tableaux procedure for satisfiability detection and model building, the generation of infinite branches should be avoided if possible.

The solution is to adapt the δ-rule in order to make it more flexible. The standard rule systematically introduces new constant symbols in the branch. Allowing to "reuse" existing symbols may reduce the size of the model and can avoid divergence. This rule, usually called the δ^*-rule [13], can be formalized as follows (c denotes a new constant symbol, not occurring in the branch and $\{c_1, \ldots, c_n\}$ are symbols occurring in the branch).

$$\frac{(\exists x)\phi(x)}{\phi\{x \to c_1\} \quad | \quad \ldots \quad | \quad \phi\{x \to c_n\} \quad | \quad \phi\{x \to c\}}$$

The branches corresponding to the constants c_i are called "ghost" in [10] and the branch corresponding to c is said to be "primary".

For refutational completeness it is sufficient to assert that the formula ϕ can be fulfilled by a variable c, possibly distinct from all the constants already occurring in the branch. But for model building and satisfiability detection, it may be useful to check whether ϕ could be fulfilled by an existing constant symbol. Note that the ghost branches should be explored before (or in parallel to) the primary branch. It is clear that if the primary branch can be closed then the ghosts can be closed too.

Using the modified δ-rule instead of the standard one preserves soundness and refutational completeness. Moreover, it also guarantees that the procedure is complete w.r.t. finite satisfiability, i.e. if a formula has a model on a domain of finite cardinality then the tableau has a finite branch. To the best of our knowledge, this idea was first introduced in [10] (see also [13]) for providing an alternative to Mc Carthy's circumscription [15]. The goal was to minimize the models constructed by the tableau procedure in order to define a notion of consequence closer to natural reasoning (the notion of mini-consequence). In [20], a similar idea to used for designing a calculus for "proving unprovability" i.e. for checking that a given formula has a finite model. It is shown that this procedure is complete for finite satisfiability but the possibility of combining it with the search for refutation was not investigated. A procedure combining the two features (search for refutations and finite models) is presented in [3]. It is designed for a subclass of first-order logic, the so-called **P**ositive Formulae with **R**estricted **Q**uantification (or PRQ-formulae for short) that has exactly the same expressive power as first-order logic. Note, however, that these procedures have been investigated *only* for first-order logic *without* function symbols. The extension to the functional case has – to the best of our knowledge – never been considered. Clearly the δ^*-rule is of no use if the formula is in skolem normal form. From a theoretical point of view this is not a problem since the adding of function symbols does not increase the expressive power of first-order logic. But in case the formula already contains function symbols, a pre-processing normalization step is required to "flatten" the terms and replace all function symbols by relations. Clearly, this normalization is not very natural. Moreover, it can reduce the efficiency, since one has to specify the properties of the "functional" predicate symbols (e.g. introduce axioms such as $(\forall y_1, \ldots, y_n)(\exists x)p(y_1, \ldots, y_n, x)$ ensuring that the interpretation of p is a function) instead of encoding them into the proof procedure itself. Moreover, it is clear that the systematic construction of ghost branches will often increase the size of the tableaux hence will reduce efficiency. This is due to fact that – if we want to insure that the procedure is complete w.r.t. finite satisfiability – all the ghost branches have to be closed before the primary branch is considered (otherwise some finite models may be missed). On the other hand, the use of the δ^* rule is not always useful. In many cases, the primary branch *itself* is finite (either because it is closed or because it corresponds to a finite model). In this case, the exploration of the ghost branches is simply useless. In order to overcome this problem it is worthwhile to investigate whether one can formulate tractable criteria allowing to decide – given a formula $(\exists x)\phi$ occurring in a branch – which of the two rules, the δ and δ^* rule is to be applied. Applying systematically the δ-rule insure refutational completeness, but some finite models may be missed hence completeness for finite satisfiability is lost. On the other hand, using only the δ^*-rule ensure that the procedure is complete both for refutation and finite satisfiability, but introduces redundant branches in the search space, which reduces efficiency. Is it possible to control the application of the δ^*-rule is such a way that completeness for finite satisfiability is preserved and that (some) redundant branches are removed ?

The present paper is an answer to these problems. We propose a tableaux procedure using both the δ and δ^*-rule in such a way that the use of δ^* is avoided when it is possible, which improves the performance. In contrast to the approaches mentioned before [20, 10, 13, 3] this procedure can handle any first-order formula with function symbols (possibly containing equational literals) without any normalization step, which avoids having to introduce artificially additional predicate symbols for denoting functions. A-priori normalization of the original formula (such as skolemization, transformation into clause normal form or PRQ-formulae) is avoided. Moreover, an original feature of our procedure is that it uses an interpretation (fixed arbitrarily before the beginning of the search) in order to prune the search space, by discarding some applications of the expansion rules. The idea of using semantics for pruning the search space of proof procedures is rather old and has been thoroughfully investigated. The Geometry Theorem Proving machine described in [8] used a figure in order to prune the search space. [18] formalized this idea in the context of the resolution calculus and defined the notion of "semantic strategy" in which an interpretation is used for discarding useless applications of the Resolution principle. This strategy has been extensively used in Automated Theorem Proving and often appears to be very efficient (note that hyperresolution is a particular case of semantic resolution). [17] described a proof calculus where an interpretation is used to cut some branches and choose the right instantiation of the variables. [16] (see also [5]) proposes a deduction procedure based on clever grounding of the clause sets in which the giving of an interpretation plays a central role (together with ordering restrictions). Model Generation Theorem Provers (such as SATCHMO [14, 4], MGTP [9] or Hyper-Tableaux [2]) combine the hyper-resolution rule with instantiation and splitting for building models of sets of clauses. Semantics are used to prune the search space, in the sense that only the instances that are falsified by the interpretation corresponding to the branch are considered for applying the γ-rule (see for example [1] for more details). The procedure FINIMO [3] can be seen as an extension of these techniques to first-order formulae containing existential quantifiers but without any function symbols. In some sense, our method extends these last approaches by considering arbitrary interpretations and arbitrary first-order formulae. The models constructed by our procedure are built by starting from an arbitrary interpretation (which is assumed to be given) and by "refining" it dynamically until a model is found (or a contradiction is detected).

Due to space restriction, the proofs are omitted (they can be found on http://www-leibniz.imag.fr/ATINF/Nicolas.Peltier/).

2   Preliminaries

In this section we briefly review the basic definitions and notations that are necessary for the understanding of our work. We assume familiarity with the usual notions in Logic and Automated Theorem Proving (see for example [7]).

We assume given a set of *sort symbols* S, a set of *function symbols* Σ and a set of variables $\mathcal{V}$. Each variable is associated to a unique sort s and each function symbol $f \in \Sigma$ is associated to a unique profile of the form $s_1 \times \ldots \times s_n \to s$, where $s_1, \ldots, s_n, s$ are sort symbols. The fact that f has profile P is denoted by $f : P$. The set of *terms of sort s* and the set of *first-order formulae* are defined as usual on the set of function symbols Σ, the set of variables $\mathcal{V}$ using the (unique) predicate symbol $\approx$ and the set of logical symbols $\vee, \wedge, \Rightarrow, \Leftrightarrow, \forall, \exists$. Note that we assume that the formulae contain no predicate symbol other than $\approx$[1]. Quantifications on a variable of sort s are denoted by $(\forall x : s)\phi$ and $(\exists x : s)\phi$ respectively. In order to simplify notations, $\phi(x)$ will frequently denote a first-order formula containing x as the only free variable. Then $\phi(t)$ denotes the formula obtained by replacing the variable in ϕ by t.

The notions of interpretation, model, validity, etc. are defined as usual. The truth value of a formula ϕ in an interpretation $\mathcal{I}$ over a variable assignment σ is denoted by $\phi^{(\mathcal{I}, \sigma)}$. If ϕ contains no free variable, then $\phi^{\mathcal{I}}$ denotes the truth value of ϕ in $\mathcal{I}$ (σ is not needed).

Positions are (possibly infinite) sequences of natural numbers that will be used to denote branches in a tableau or subterms in a term. ϵ denotes the empty position and $p.q$ denotes the concatenation of the positions p and q. We write $p \preceq q$ iff p is a prefix of q. The notion of *position in a term* is defined as usual. If s, t are terms and p is a position in t, then $t[s]_p$ denotes the term obtained by replacing the subterm at position p in t by s.

We assume that Σ contains an infinite (countable) number of constant symbols of each sort noted $\mathcal{D}_s$ and that c_s is an (arbitrarily chosen) element of $\mathcal{D}_s$. Elements of $\mathcal{D}_s$ are usually called the *skolem constants* (in order to distinguish them from the original constants occurring in the formula). They will be used to define the domains of the interpretation constructed by the proof procedure. We treat them as constant symbols in order to simplify notations. We assume given a precedence $\prec$ on skolem constants and an *injective* function α mapping each term t of sort s (resp. each closed formula of the form $(\exists x : s)\phi(x)$) to a constant symbol $c \in \mathcal{D}_s$ strictly greater than any skolem constant occurring in t (resp. in $\phi(x)$).

A substitution is a function mapping each variable to a term of the same sort. As usual substitutions can be extended to terms, formulae etc. The image of a term (resp. formula) t by a substitution σ is denoted by $t\sigma$.

An interpretation $\mathcal{I}$ of a set of functional symbols Σ' is said to be *canonic* iff the domain of the sort s is included into $\mathcal{D}_s$, i.e. if for all $s \in \mathcal{S}$, $s^{\mathcal{I}} \subseteq \mathcal{D}_s$ and

---

[1] This is not restrictive because non equational formulae such as $p(x)$ could be replaced by equations of the form $p(x) \approx true$, where *true* is a new constant symbol of sort *boolean*.

for all skolem constant $c \in \Sigma'$, $c^{\mathcal{I}} = c$ (note that Σ' may be a proper subset of Σ hence does not need to contain all the elements of $\mathcal{D}_s$).

A term t is called a *disconnected subterm* of s if t can be obtained from s by removing some of the positions in s. For example $f(a), f(b)$ are disconnected subterms of $f(g(a,b))$. More formally, the set of disconnected subterms of s is inductively defined as follows. (1) t is a disconnected subterm of t; (2) if t is of the form $f(t_1, \ldots, t_n)$ and s is a disconnected subterm of one of the t_i ($1 \leq i \leq n$) then s is a disconnected subterm of t; (3) if t is of the form $f(t_1, \ldots, t_n)$ and for all $i \in [1..n]$, s_i is a disconnected subterm of t_i then $f(s_1, \ldots, s_n)$ is a disconnected subterm of t. This notion can be extended straightforwardly to first-order formulae (for example $(\forall x)(\neg p(x))$ is a disconnected subformula of $(\forall x)(\exists y)\neg(p(x) \vee p(y))$).

A *complexity measure* $\mathcal{C}$ is a function mapping each term and formula to a natural number and satisfying the following properties:

- if t is a disconnected subterm (resp. subformula) of s and $t \neq s$, then $\mathcal{C}(t) < \mathcal{C}(s)$.
- for any natural number k, there exists a finite number of formulae and terms of complexity lower than k (i.e. the set $\{t \mid \mathcal{C}(t) \leq k\}$ is finite).
- for any $c \in \mathcal{D}$, if $c = \alpha(t)$ then $\mathcal{C}(c) = \mathcal{C}(t)$ (see above for the definition of the function α).

Our notion of complexity measure is closely related to the notion of *atomic complexity measure* as defined in [11]. However, it will be used for very different purposes: [11] introduces this notion for proving termination of the hyperresolution calculus on particular subclasses of first-order logic, whereas our notion of complexity measure is used in the core of the proof procedure, to restrict the application of the inference rules.

We need to introduce the following notation. Informally, $w^+(\phi)$ (w standing for "witness") is essentially identical to ϕ, excepted that existential quantifications of the form $(\exists x : s)\psi$ occurring in ϕ are replaced by the stronger formula $\psi\{x \to c_s\}$ (note that witness formulae are used only for detecting redundancies, not for proof search). Note that since we want to avoid normalization, we must also take into account the fact that subformulae may occur on the scope of a negation symbol (or more generally under an odd number of negation symbols).

Let ϕ be a formula. We denote by $w^+(\phi)$ (resp. $w^-(\phi)$) the formula defined as follows:

- $w^+(\phi) =_{def} w^-(\phi) =_{def} \phi$ if ϕ is atomic.
- $w^+(\phi \vee \psi) =_{def} w^+(\phi) \vee w^+(\psi)$, $w^+(\phi \wedge \psi) =_{def} w^+(\phi) \wedge w^+(\psi)$.
- $w^+(\neg\phi) =_{def} \neg w^-(\phi)$, $w^+(\phi \Rightarrow \psi) =_{def} w^-(\phi) \Rightarrow w^+(\psi)$.
- $w^+(\phi \Leftrightarrow \psi) =_{def} (w^+(\phi) \wedge w^+(\psi)) \vee (\neg w^-(\phi) \wedge \neg w^-(\psi))$.
- $w^+((\exists x : s)\phi) =_{def} \phi\{x \to c_s\}$, $w^-(\phi \Rightarrow \psi) =_{def} w^+(\phi) \Rightarrow w^-(\psi)$.
- $w^+((\forall x : s)\phi) =_{def} (\forall x : s)w^+(\phi)$, $w^-(\phi \vee \psi) =_{def} w^-(\phi) \vee w^-(\psi)$.
- $w^-(\phi \wedge \psi) =_{def} w^-(\phi) \wedge w^-(\psi)$, $w^-(\neg\phi) =_{def} \neg w^+(\phi)$.
- $w^-(\phi \Leftrightarrow \psi) =_{def} (w^-(\phi) \wedge w^-(\psi)) \vee (\neg w^+(\phi) \wedge \neg w^+(\psi))$.
- $w^-((\exists x : s)\phi) =_{def} (\exists x : s)w^-(\phi)$, $w^-((\forall x : s)\phi) =_{def} w^-(\phi\{x \to c_s\})$.

It follows from the definition than $w^+(\phi)$ expresses a *stronger* property than the original formula ϕ, e.g. that ϕ is a logical consequence of $w^+(\phi)$, i.e. we have, for any interpretation $\mathcal{I}$, if $\mathcal{I} \models w^+(\phi)$ then $\mathcal{I} \models \phi$ and if $\mathcal{I} \not\models w^-(\phi)$ then $\mathcal{I} \not\models \phi$.

Clearly, the definition of $w^+(\phi)$ does not depend on the terms occurring in ϕ (that are merely copied without any modification) but only on the boolean structure of the formula. Therefore, given a formula ϕ and a substitution σ, it is equivalent to apply the substitution σ to ϕ and then to compute the formula $w^+(\phi\sigma)$ or to apply the substitution σ to $w^+(\phi)$. This is expressed by the following:

Proposition 1. *Let ϕ be a formula and σ a substitution . $w^+(\phi\sigma) = w^+(\phi)\sigma$.*

3   Proof Procedure: The $S\text{-}TAB\text{-}RFM(\mathcal{I}, \mathcal{C})$ Calculus

In this section we present our tableaux calculus for simultaneous search for refutations and models. This calculus, called $S\text{-}TAB\text{-}RFM(\mathcal{I}, \mathcal{C})$, (for Semantically guided **T**ableau Procedure for **R**efutation and **F**inite **M**odel Construction) is parametrized by an interpretation $\mathcal{I}$ and by a complexity measure $\mathcal{C}$. $\mathcal{I}$ and $\mathcal{C}$ will be used to prune the search space by discarding some applications of the expansion rules. Both $\mathcal{I}$ and $\mathcal{C}$ may be chosen arbitrarily. For instance we can choose for $\mathcal{I}$ an interpretation mapping each term t of sort s not occurring in $\mathcal{D}_s$ to c_s. Alternatively, $\mathcal{I}$ can be provided by the user together with the problem to solve[2]. Similarly, several distinct complexity measures can be used resulting in different behaviours of the proof procedure. In our current implementation, we use the most simple complexity measure: the one based on the depth of the considered term (resp. formula). Precisely, this measure is defined as follows:

- for any $c \in \mathcal{D}_t$, if $c = \alpha(t)$ then $\mathcal{C}(c) = \mathcal{C}(t)$ (this follows from the definition).
- $\mathcal{C}(f(t_1, \ldots, t_n)) =_{def} \max\{\mathcal{C}(t_i) \mid i \in [1..n]\} + 1$.
- $\mathcal{C}(\phi \star \psi) =_{def} \max(\mathcal{C}(\phi), \mathcal{C}(\psi)) + 1$ (if $\star \in \{\vee, \wedge, \Rightarrow, \Leftrightarrow\}$).
- $\mathcal{C}(\star\phi) =_{def} \mathcal{C}(\phi) + 1$ if $\star \in \{\neg, (\forall x), (\exists x)\}$.

Other complexity measures could be considered instead (for example affecting different weights to each function and logical symbol, or taking into account the total number of symbols instead of the depth). Heuristics for designing interpretations and complexity measures well adapted to the particular problem at hand could deserve to be investigated in the future.

The notion of tableau is defined as usual. If $\mathcal{T}$ is a tableau and p a (possibly infinite) position in $\mathcal{T}$, then $\mathcal{T}(p)$ denotes the set of formulae occurring along the branch p in $\mathcal{T}$. A branch p in $\mathcal{T}$ is said to be *closed* if *false* $\in \mathcal{T}(p)$ and $\mathcal{T}$ is said to be closed if any branch in $\mathcal{T}$ is closed.

---

[2] Note that the fact that $\mathcal{I}$ is assumed to be canonic is not really restrictive here because it is easy to transform automatically a non canonic interpretation into a canonic one.

Tableaux will be constructed as usual from a root formula ϕ using a set of expansion rules. Each rule is described as follows:

$$\frac{S}{B_1 \,|\, \ldots \,|\, B_n}$$

meaning that a leaf p such that $T(p) = S$ can be extended by n branches $p.i$ with $T(p.i) = S \cup B_i$ (for the sake of readability we omit S from the conclusion of the rules). To ensure refutational completeness, we assume that the construction is fair, i.e. that for any infinite branch p in T, if a given expansion rule is applicable at all positions $r \prec p$ after a certain position $q \prec p$, then it must be applied at some point along p.

The following definition provides a criterion allowing to detect the formulae on which the expansion rules should be applied.

We first introduce a notation. Given a set of formulae S and a canonic interpretation $\mathcal{I}$, we denote by $\mathcal{I}_S$ the interpretation defined as follows.

- For any sort symbol s, $s^{\mathcal{I}_S}$ (the domain of the sort s) is the union of all the symbols $c \in \mathcal{D}_s$ occurring in S plus the symbol c_s.
- For any function symbol $f : s_1 \times \ldots \times s_n \to s$ and for any tuple $(c_1, \ldots, c_n) \in \mathcal{D}_{s_1} \times \ldots \mathcal{D}_{s_n}$, if there is no $c \in \mathcal{D}_s$ such that $f(c_1, \ldots, c_n) \approx c \in S$ then $f^{\mathcal{I}_S}(c_1, \ldots, c_n) =_{def} f^{\mathcal{I}}(c_1, \ldots, c_n)$.
- Otherwise, for any function symbol $f : s_1 \times \ldots \times s_n \to s$ and for any tuple $(c_1, \ldots, c_n) \in \mathcal{D}_{s_1} \times \ldots \mathcal{D}_{s_n}$, we have $f^{\mathcal{I}}(c_1, \ldots, c_n) =_{def} c$ iff c is the smallest (according to $\prec$) element in $\mathcal{D}_s$ such that $f(c_1, \ldots, c_n) \approx c \in S$.

Note that $\mathcal{I}_S$ is finite if S is finite (it may be infinite otherwise).

Informally, $\mathcal{I}_S$ is obtained by adapting the interpretation $\mathcal{I}$ in such a way that all the "minimal" equations in S are satisfied. In particular, if S is empty (or contains no equation) then $\mathcal{I}_S = \mathcal{I}$. If p is a branch in a tableau T, then $\mathcal{I}_{T(p)}$ is the interpretation corresponding to p: functions are interpreted as specified by the equations generated along the branch. Terms whose value is unknown are interpreted as in the interpretation $\mathcal{I}$ (this may be seen as a "default value").

Definition 1. *Let T be a tableau and p be a position in T.*

A formula ϕ is said to be redundant *(w.r.t. T and p) if there exists a set of formulae $\{\psi_1, \ldots, \psi_n\} \subseteq T(p)$ such that:*

- *For all $i \in [1..n]$, $\mathcal{C}(\psi_i) < \mathcal{C}(\phi)$;*
- *and $\bigwedge_{i=1}^{n} \psi_i \models \phi$.*

A formula ϕ is said to be eligible *(w.r.t. T and p) if ϕ is non redundant and $\mathcal{I}_S \not\models w^+(\phi)$.*

No expansion rule will be applied on a formula if it is not eligible, which reduces the number of generated formulae in the branch and may prune the search space.

Remark 1. For a practical point of view, checking whether $\bigwedge_{i=1}^{n} \psi_i$ models ϕ may be difficult (since this problem is semi-decidable). Fortunately, testing whether a formula is eligible or not is useful only for pruning the search space. Assuming that *any* formula is eligible would not affect the soundness and refutational completeness of our procedure, but would only increase the number of applicable rules, which obviously makes the procedure less efficient in most cases[3]. Therefore, we can either discard this condition (e.g. assume that it is never satisfied) or use instead some stronger criteria. In our current implementation, we only check whether ϕ can be deduced from the less complex formulae in the branch by propositional reasoning. Any other criteria can be used, provided that it is effectively decidable and still sufficient to ensure that the condition "$\bigwedge_{i=1}^{n} \psi_i \models \phi$" holds.

From a theoretical point of view, Definition 1 states – in some sense – the "weakest" conditions ensuring refutational completeness. From a practical point of view, it allows the elimination of redundant formulae in the branch, which may be crucial for performances.

A sort symbol s is said to be *equational* if the root formula contains an equation $t_1 \approx t_2$ where t_1 and t_2 are of sort s.

Propositional Rules

The following rules only apply on eligible formulae.

$$\frac{S \cup \{(\phi \vee \psi)\}}{\phi \mid \psi} \qquad \frac{S \cup \{(\phi \wedge \psi)\}}{\phi, \psi} \qquad \frac{S \cup \{(\phi \Rightarrow \psi)\}}{\neg\phi \mid \psi}$$

$$\frac{S \cup \{(\phi \Leftrightarrow \psi)\}}{\phi, \psi \mid \neg\psi, \neg\phi} \qquad \frac{S \cup \{\neg\neg\phi\}}{\phi} \qquad \frac{S \cup \{\neg(\phi \vee \psi)\}}{\neg\phi, \neg\psi}$$

$$\frac{S \cup \{\neg(\phi \wedge \psi)\}}{\neg\phi \mid \neg\psi} \qquad \frac{S \cup \{\neg(\phi \Rightarrow \psi)\}}{\phi, \neg\psi} \qquad \frac{S \cup \{\neg(\phi \Leftrightarrow \psi)\}}{\neg\phi, \psi \mid \psi, \neg\phi}$$

γ-Rule

$$\frac{S \cup \{(\forall x : s)\phi(x)\}}{\phi(e)}$$

If: $\phi(e)$ is eligible, $e \in \mathcal{D}_s$ and either e occurs in the branch or $e = c_s$.

Note that the condition "$\phi(e)$ is eligible" impose constraints on the *terms* on which the rule can be applied, which restrict the application of the rule, thus avoiding "blind" instantiation.

---

[3] Of course, in some cases this can actually make the procedure more efficient, just as in some very particular cases, unrestricted resolution may be more efficient than semantic resolution.

δ-Rule

$$\frac{S \cup \{(\exists x : s)\phi(x)\}}{\phi(e)}$$

If: $(\exists x : s)\phi(x)$ is eligible, s is non equational, $e = \alpha((\exists x)\phi)$ and if there exists a symbol $c \in \mathcal{D}$ occurring in the branch such that $\mathcal{C}(e) \leq \mathcal{C}(c)$.

δ^*-Rule

$$\frac{S \cup \{(\exists x : s)\phi\}}{\phi(e_1) \mid \cdots \mid \phi(e_n) \mid \phi(e)}$$

If: $(\exists x : s)\phi(x)$ is eligible, $e = \alpha((\exists x)\phi)$ $\{e_1, \ldots, e_n\}$ is the set of skolem constants of sort s occurring in the branch, and either s is equational or for all skolem constants c occurring in the branch, $\mathcal{C}(c) < \mathcal{C}(e)$.

Informally speaking, the principle of our approach is to restrict the δ^* rule to those symbols whose complexity is *greater* than the symbols already occurring in the branch. Otherwise, the δ-rule is applied instead, hence no "ghost" is created. A similar principle is used to restrict the application of the Term Decomposition rule (see below).

Paramodulation Rule

$$\frac{S \cup \{t \approx c, \phi(t)\}}{\phi(c)} \qquad \text{If } c \in \mathcal{D}.$$

Term Decomposition Rules

$$\frac{S \cup \{(s \approx t)\}}{s[e]_p \approx t, u \approx e}$$

If: $e = \alpha(u)$, $(s \approx t)$ is eligible, u is a term of a non equational sort occurring at position p in s, there exists a symbol $c \in \mathcal{D}$ occurring in the branch such that $\mathcal{C}(e) \leq \mathcal{C}(c)$ and either p is non empty or t is not a constant symbol in $\mathcal{D}$.

$$\frac{S \cup \{(s \approx t)\}}{s[e_1]_p \approx t, u \approx e_1 \mid \cdots \mid s[e_n]_p \approx t, u \approx e_n \mid s[e]_p \approx t, u \approx e}$$

If: $e = \alpha(u)$, $(s \approx t)$ is eligible, u occurs at position p in s, and either the sort of u is equational or if for all symbols $c \in \mathcal{D}$ occurring in the branch $\mathcal{C}(c) < \mathcal{C}(e)$, and either p is non empty or $t \notin \mathcal{D}$. $\{e_1, \ldots, e_n\}$ is the set of skolem constant of the same sort as u occurring in the branch.

$$\frac{S \cup \{(s \not\approx t)\}}{s[e]_p \not\approx t, u \approx e}$$

If: $e = \alpha(u)$, $(s \approx t)$ is eligible, u is a term of a non equational sort occurring at a position p in s and if there exists a symbol $c \in \mathcal{D}$ occurring in the branch such that $\mathcal{C}(e) \leq \mathcal{C}(c)$.

$$\frac{S \cup \{(s \not\approx t)\}}{s[e_1]_p \not\approx t, u \approx e_1 \quad | \quad \ldots \quad | \quad s[e_n]_p \not\approx t, u \approx e_n \quad | \quad s[e]_p \not\approx t, u \approx e}$$

If: $e = \alpha(u)$, $(s \approx t)$ is eligible, u occurs at position p in s, and either the sort of u is equational or for all symbols $c \in \mathcal{D}$ occurring in the branch $\mathcal{C}(c) < \mathcal{C}(e)$. $\{e_1, \ldots, e_n\}$ is the set of skolem constant of the same sort as u occurring in the branch.

In our current implementation, the Term Decomposition rules are restricted to subterms of the form $u = f(c_1, \ldots, c_n)$ where $c_1, \ldots, c_n$ are skolem constants. This is sufficient for completeness but more flexible strategies are of course possible.

Clash Rules

$$\frac{S \cup \{(s \not\approx s)\}}{\text{false}} \qquad \frac{S \cup \{(c \approx c')\}}{\text{false}} \qquad \text{If } c \neq c', \, c, c' \in \mathcal{D}.$$

The motivation of the second clash rule would become clearer from the soundness proof. Intuitively, the tableau is constructed in such a way that distinct constants will always denote distinct elements of the domain.

4 The Properties of the Calculus

In this section, we prove that $S\text{-}TAB\text{-}RFM(\mathcal{I}, \mathcal{C})$ is sound (i.e. that if a tableau for ϕ is closed, then ϕ is unsatisfiable) and refutationally complete (i.e. that if ϕ is unsatisfiable then any fair tableau for ϕ is closed). Both properties are very standard, but we need to prove them since our procedure cannot be simulated by existing calculi, due to the particular restrictions on the expansion rules. Moreover, we also show that the procedure is complete for finite model building, i.e. that if ϕ admits a finite model then any tableau for ϕ has a finite branch.

Theorem 1. *(soundness) Let ϕ be a formula and let $\mathcal{T}$ be a tableau for ϕ. If $\mathcal{T}$ is closed then ϕ is unsatisfiable.*

Theorem 2. *(refutational completeness) Let ϕ be a formula and let $\mathcal{T}$ be a fair tableau for ϕ. If ϕ is unsatisfiable then $\mathcal{T}$ is closed.*

The reader should note that $w^+(\phi)$ is needed to be carefully defined in order to ensure refutational completeness. In particular, replacement of existential quantification $(\exists x : s)\phi(x)$ by the stronger formulae $\phi(c_s)$ is essential. Assume for instance that $w^+(\phi)$ had been defined as identical to ϕ, and consider the following (unsorted) formula:

$$(\exists x)(p(x) \not\approx true) \wedge (\forall x)(p(x) \approx true) \wedge (\forall x)(\exists y)(x \not\approx y)$$

This formula is obviously unsatisfiable. Assume that $\mathcal{I}$ maps p to a function mapping each element of the domain to *false* (with *false* $\neq$ *true*). In this case, $(\exists x)(p(x) \not\approx true)$ is always true in $\mathcal{I}_S$, provided that there exists an element c such that $p(c) = true$ does not occur in S. In this case, one could obtain an infinite, unclosed branch, as follows:

1 $(\exists x)(p(x) \not\approx true)$	(by the $\wedge$-rule)	2 $(\forall x)p(x) \approx true$	
3 $(\forall x)(\exists y)(x \not\approx y)$		4 $(\exists y)(c_0 \not\approx y)$	(γ-rule on 3)
5 $(c_0 \not\approx c_1)$	(δ-rule on 4)	6 $p(c_0) \approx true$	(γ-rule on 2)
7 $(\exists y)(c_1 \not\approx y)$	(γ-rule on 3)	8 $(c_1 \not\approx c_2)$	(δ-rule on 7)
9 $p(c_1) \approx true$	(γ-rule on 2)		
etc.			

The point here is that the above infinite derivation is *fair*, despite the fact that the δ-rule is never applied on the formula $(\exists x)(p(x) \not\approx true)$. Indeed, it is easy to check that – at any position in the branch – the formula $(\exists x)(p(x) \not\approx true)$ is non eligible, since it is always true in the interpretation $\mathcal{I}_S$. Thus, the application of the expansion rule on this formula is simply prevented by the proof procedure, which leads to incompleteness ! On the other hand, if $w^+(\phi)$ is defined as the present paper, then $w^+((\exists x)(p(x) \not\approx true))$ becomes false as soon as the formula $p(c_0) \not\approx true$ is generated. Consequently, the application of the δ-rule is allowed which leads to the eventual closure of the branch. The key point here is that the "witness" element ensuring that an existential quantification holds in the underlying interpretation should *not* depend on the position in the branch. This explains why we need to replace existential quantification $(\exists x : s)\phi(x)$ by stronger formulae of the form $\phi(c_s)$. In this case, the element is constant and equal to c_s.

4.1   Completeness w.r.t. Finite Models

We now prove that $S\text{-}TAB\text{-}RFM(\mathcal{I}, \mathcal{C})$ is complete w.r.t. finite models. We need the following lemma, showing that our procedure satisfies (a form of) the subformula property:

Lemma 1. *Let ϕ be a formula and let T be a tableau for ϕ. The formulae occurring in T are either disconnected subformulae of ϕ or obtained from a disconnected subformula ψ of ϕ by replacing some of the terms occurring in ψ by skolem constants.*

We immediately deduce that a branch containing a finite number of distinct symbols must be finite:

Corollary 1. *Let ϕ be a formula and let T be a tableau for ϕ. If p is a branch in T such that the set of skolem constants contained in a formula in $T(p)$ is finite then p is finite.*

Theorem 3. *Let ϕ be a formula and let T be a fair tableau for ϕ. If ϕ admits a finite model, then T has a finite open branch.*

To summarize, given a formula ϕ, the procedure S-TAB-$RFM(\mathcal{I}, \mathcal{C})$ can behave in 3 different ways.

- If ϕ is unsatisfiable then the procedure terminates and returns a closed tableau for ϕ.
- If ϕ admits a finite model then there exists a finite open branch hence the procedure terminates (provided the branches are expanded in a fair way). From the constructed (still partial) tableau $\mathcal{T}$ and one finite open branch p, a finite model $\mathcal{I}_{\mathcal{T}(p)}$ of ϕ can be automatically extracted.
- If ϕ is satisfiable but has no finite model then the procedure does not terminate.

Of course several improvements are possible.

- Our procedure does not use unification. As well known, it is possible to significantly improve the performances of tableaux by introducing rigid variables (instead of concrete terms) when applying the γ-rule and use unification to compute the instances that are needed for closing the tableau. This avoids having to "guess" the value of the variables in advance. On the other hand, from the point of view of model building systematic instantiation of the variables appears to be essential. Is it possible to combine the use of rigid variables with the special δ^*-rule and with the search-space pruning strategies presented in the present paper ? In particular, the combination of our method with the *disconnection calculus* of [12] (a calculus that avoids instantiation but still permits branch saturation and extraction of models) should be considered in the future.
- In order to improve its efficiency, our procedure should be combined with the usual improvements of tableaux calculi. For instance, detection of lemmata from closed branches could help to discard redundant inferences. Additional techniques for restricting the application of the γ-rule and δ-rule could of course be considered.

5   Some Experimental Results

In this section, we provide some preliminary experimental results obtained with S-TAB-RFM_{Atinf}, our current implementation of S-TAB-RFM. S-TAB-RFM_{Atinf} is a prototype, implemented in GnuProlog with straightforward algorithms and data structures, and without any significant optimization. It is not intended to compete with the most powerful theorem provers available, but only to allow to roughly estimate the influence of the proposed strategies on the performances of the system. The program implements a very basic proof procedure, using a breath-first search strategy for choosing the instances of the universal formulae on which the γ-rule should be applied. The interpretation $\mathcal{I}$ used to prune the search space is a canonic interpretation mapping any atom distinct from $t \approx t$ to *false* and $\mathcal{C}$ is equal to the depth of the terms.

We use a small collection of problems from the TPTP library [19]. All the problems are formulae of first-order logic without equality. "-" means that the system did not terminate in a reasonable time (e.g. less than 5 mn). All the computation times are given in ms.

We provide for each problem the result obtained with 3 different strategies: The first one (**Basic**) corresponds to the basic procedure, i.e. no interpretation is used for pruning the search space and the δ-rule is not used (the δ^* rule is applied systematically on each existential quantifier). The second one (**Sem**) uses the empty interpretation for pruning the search space but the δ-rule is not used. The third one (δ) uses the δ-rule and restricted δ^* rule but not the semantic restrictions. The fourth one *S-TAB-RFM* corresponds to the procedure described in the present paper, i.e. combination of **Sem** and δ.

The fourth procedures share exactly the same theoretical properties (i.e. soundness, refutational completeness and completeness for finite models) which makes the comparison relevant. For some problems (e.g. SYN060+1, SYN061+1, SYN062+1, etc.) the computation times are too short for being significant hence we removed them from the list. Similarly we removed problems that are too hard for our current implementation.

Problems	Basic	Sem	δ	*S-TAB-RFM*
SYN036+1	101010	250	26570	170
SYN036+2	92310	670	21240	410
SYN049+2	360	490	70	200
SYN054+1	40	20	40	20
SYN055+1	30	20	30	20
SYN056+1	260	60	240	60
SYN057+1	50	30	50	20
SYN058+1	20	20	20	20
SYN059+1	130	50	70	40
SYN068+1	80	30	60	20
SYN069+1	7650	60	7140	50
SYN070+1	2460	160	2420	120
SYN084+1	160	60	90	50
SYN083+1	30	30	20	10
SYN319+1	-	-	-	20550
SYN320+1	-	-	-	50710

Generally speaking, it seems that the semantic strategy improves the performances of the system in a significant way. By comparison the use of the δ-rule is less effective. However, when the difficulty of the problem increases, the importance of the δ-rule becomes significant (see for instance SYN036+2, SYN036+1, or SYN319+1). The combination of the two strategies gives the best result. Note that for one particular problem (namely SYN049+2) disabling the semantic strategy actually *decreases* the computation time. On the other hand the δ rules appears to be quite effective in this case. Of course, further experimentations (with more powerful provers) and theoretical studies are needed before definite conclusions can be drawn.

References

[1] P. Baumgartner, P. Fröhlich, U. Furbach, and W. Nejdl. Semantically Guided Theorem Proving for Diagnosis Applications. In *Proc. 15th. Joint Conference on Artificial Intelligence (IJCAI)*, pages 460–465. Springer LNAI, 1997.

[2] P. Baumgartner, U. Furbach, and I. Niemelä. Hyper-tableaux. In *Logics in AI, JELIA '96*. Springer, 1996.

[3] F. Bry and S. Torge. A deduction method complete for refutation and finite satisfiability. In *Logics in AI, JELIA '98*. Springer LNAI 1489, 1998.

[4] F. Bry and A. Yahya. Positive unit hyperresolution. *Journal of Automated Reasoning*, 25(1):35–82, 2000.

[5] H. Chu and D. A. Plaisted. Semantically guided first-order theorem-proving using hyper-linking. In *Proc. of CADE-12*, pages 192–206. Springer, 1994. LNAI 814.

[6] C. Fermüller and A. Leitsch. Hyperresolution and automated model building. *Journal of Logic and Computation*, 6(2):173–203, 1996.

[7] M. Fitting. *First-Order Logic and Automated Theorem Proving*. Texts and Monographs in Computer Science. Springer-Verlag, 1990.

[8] H. Gelernter, J. Hansen, and D. Loveland. Empirical explorations of the geometry theorem-proving machine. In J. Siekmann and G. Wrightson, editors, *Automation of Reasoning, vol. 1*, pages 140–150. Springer, 1983. Originally published in 1960.

[9] R. Hasegawa, M. Hoshimura, and H. Fujita. MGTP: a parallel theorem prover based on lazy model generation. In *Proc. of CADE-11*, pages 776–780. Springer, 1992. LNAI 607.

[10] J. Hintikka. Model Minimization - An Alternative to Circumscription. *Journal of Automated Reasoning*, 4:1–13, 1988.

[11] A. Leitsch. Deciding clause classes by semantic clash resolution. *Fundamenta Informaticae*, 18:163–182, 1993.

[12] R. Letz and G. Stenz. Proof and model generation with disconnection tableaux. In *Logic for Programmaing, Artificial Intelligence and Reasoning (LPAR)*, pages 142–156. Springer, 2001.

[13] S. Lorenz. A Tableau Prover for Domain Minimization. *Journal of Automated Reasoning*, 4:375–390, 1994.

[14] R. Manthey and F. Bry. SATCHMO: A theorem prover implemented in Prolog. In *Proc. of CADE-9*, pages 415–434. Springer, LNCS 310, 1988.

[15] J. Mc Carthy. Circumscription – a form of non-monotonic reasoning. *Artificial Intelligence*, 13:27–39, 1980.

[16] D. A. Plaisted and Y. Zhu. Ordered semantic hyperlinking. *Journal of Automated Reasoning*, 25(3):167–217, October 2000.

[17] R. Reiter. A semantically guided deductive system for automatic theorem proving. *IEEE Transactions on Computers*, C-25(4):328–334, April 1976.

[18] J. R. Slagle. Automatic theorem proving with renamable and semantic resolution. *Journal of the ACM*, 14(4):687–697, October 1967.

[19] C. Suttner and G. Sutcliffe. The TPTP problem library. Technical report, http://www.cs.miami.edu/ tptp/. V-2.5.0.

[20] M. Tiomkin. Proving unprovability. In *Logic In Computer Science'88*, pages 22–26, 1988.

Automatic Abstraction of Equations in a Logic of Equality

Miroslav N. Velev

Department of Electrical and Computer Engineering
Carnegie Mellon University, Pittsburgh, PA 15213, USA
mvelev@ece.cmu.edu

Abstract. The paper presents a method to automatically abstract equations when translating formulas with equality to equivalent Boolean formulas, allowing the use of a SAT-checker to determine the validity of the original formula. The equations are abstracted with a special interpreted predicate that satisfies the properties of symmetry, reflexivity, transitivity, and functional consistency. This abstraction is both sound and complete. In contrast to previous methods that encode only low-level equations between term variables, the presented abstraction directly encodes top-level equations where the arguments can be nested-*ITE* expressions that select term variables. The automatic abstraction was used to formally verifying the safety of pipelined, superscalar, and VLIW processors, and reduced the CNF clauses by up to 50%, while speeding up the formal verification by up to an order of magnitude relative to the e_{ij} method where a new Boolean variable is used to encode each unique low-level equation between term variables. A heuristic for partial transitivity resulted in additional speedup for correct benchmarks that require transitivity.

1 Introduction

In formal verification of microprocessors, equations (equality comparisons) are used: 1) in the control logic, to express forwarding and stalling conditions, based on equality between a source and a destination register; 2) in mechanisms for correcting wrong speculations, when a predicted data value is not equal to the actual one; and 3) in the correctness formula, to compare the final architectural states of the implementation and the specification. The logic of Equality with Uninterpreted Functions and Memories (EUFM) [7] allows us to abstract functional units and memories, while completely modeling the control path of a processor. In EUFM, word-level values are abstracted with expressions called terms (see Sect. 2), whose only property is that of equality with other terms. In our previous work on using EUFM to formally verify pipelined, superscalar, and VLIW microprocessors [21], we imposed some simple restrictions on the style for defining high-level processors. The result was a significant reduction in the number of terms that appear in both positive and negated equations—and are so called *g-terms* (for general terms)—while increasing the number of

M. Cialdea Mayer and F. Pirri (Eds.): TABLEAUX 2003, LNAI 2796, pp. 196–213, 2003.

terms that appear only in positive (not negated) equations—and are so called *p-terms* (for positive terms). We will refer to equations that appear in both positive and negated polarity as *g-equations*, and to those that appear only in positive polarity as *p-equations*. The property of Positive Equality [21] allowed us to treat syntactically different p-terms as not equal when checking the validity of an EUFM formula, thus achieving significant simplifications, and orders of magnitude speedup—see [5] for a correctness proof.

In the current paper, the implementation and specification are described in the high-level hardware description language AbsHDL [25][27], based on the logic of EUFM. The formal verification is done with an automatic tool flow, consisting of: 1) the term-level symbolic simulator TLSim [25], used to symbolically simulate the implementation and specification, and to produce an EUFM correctness formula; 2) the decision procedure EVC [25] that exploits Positive Equality and other optimizations to translate the EUFM correctness formula to an equivalent Boolean formula, which has to be a tautology for the implementation to be correct; and 3) an efficient SAT-checker. This tool flow was used at Motorola [13] to formally verify a model of the M•CORE processor, and detected bugs. The tool flow was also used in an advanced computer architecture course [27][28], where undergraduate and graduate students designed and formally verified pipelined DLX [10] processors, including variants with exceptions and branch prediction, as well as dual-issue superscalar implementations.

While SAT-checkers are very quick to find a counterexample for a bug [26], they can be orders of magnitude slower when proving unsatisfiability of CNF formulas from correct designs. This paper proposes an approach to speed up the formal verification of correct models by abstracting the g-equations in a sound and complete way that results in a conceptually simpler solution space, fewer CNF clauses, and up to an order of magnitude reduction in the SAT-checking decisions and conflicts, relative to previous methods for encoding g-equations with Boolean variables [9][16].

2   Background

The formal verification is done by correspondence checking—comparison of a pipelined implementation against a non-pipelined specification, using flushing [7][8] to automatically compute an *abstraction function* that maps an implementation state to an equivalent specification state. The safety property (see Figure 1) is expressed as a formula in the logic of EUFM, and checks that one step of the implementation corresponds to between 0 and k steps of the specification, where k is the issue width of the implementation. F_{Impl} is the transition function of the implementation, and F_{Spec} is the transition function of the specification. We will refer to the sequence of first applying the abstraction function and then exercising the specification as the *specification side* of the commutative diagram in Figure 1, and to the sequence of first exercising the implementation for one step and then applying the abstraction function as the *implementation side* of the commutative diagram.

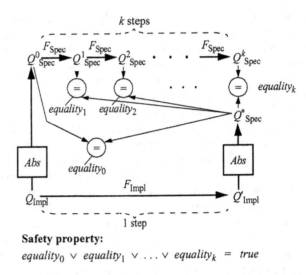

Safety property:

$$equality_0 \lor equality_1 \lor \ldots \lor equality_k = true$$

Fig. 1. The safety correctness property for an implementation with issue width k: one step of the implementation should correspond to between 0 and k steps of the specification, when the implementation starts from arbitrary initial state Q_{Impl} that may be restricted by invariant constraints

The safety property is a proof by induction, since the initial implementation state, Q_{Impl}, is completely arbitrary. If the implementation is correct for all transitions that can be made for one step from an arbitrary initial state, then the implementation will be correct for one step from the next implementation state, Q'_{Impl}, since that state will be a special case of an arbitrary state as used for the initial state, and so on for any number of steps. For some processors, e.g., where the control logic is optimized by using unreachable states as don't-care conditions, we may have to impose a set of *invariant constraints* for the initial implementation state in order to exclude unreachable states. Then, we need to prove that those constraints will be satisfied in the implementation state after one step, Q'_{Impl}, so that the correctness will hold by induction for that state, and so on for all subsequent states. See [1][2] for a discussion of correctness criteria.

To illustrate the safety property in Figure 1, let the implementation and specification have three architectural state elements—program counter (PC), register file, and data memory. Let PC^i_{Spec}, $RegFile^i_{Spec}$, and $DMem^i_{Spec}$ be the state of the PC, register file, and data memory, respectively, in specification state Q^i_{Spec} ($i = 0, ..., k$) along the specification side. Let PC^*_{Spec}, $RegFile^*_{Spec}$, and $DMem^*_{Spec}$ be the state of the PC, register file, and data memory in specification state Q^*_{Spec}, reached after the implementation side of the diagram. Then, each disjunct $equality_i$ ($i = 0, ..., k$) is defined as:

$$equality_i \leftarrow pc_i \land rf_i \land dm_i,$$

where

$$pc_i \leftarrow (PC^i_{Spec} = PC^*_{Spec}),$$
$$rf_i \leftarrow (RegFile^i_{Spec} = RegFile^*_{Spec}),$$
$$dm_i \leftarrow (DMem^i_{Spec} = DMem^*_{Spec}).$$

That is, $equality_i$ is the conjunction of the pair-wise equality comparisons for all architectural state elements, thus ensuring that the architectural state elements are updated in synchrony by the same number of instructions. In processors with more architectural state elements, an equality comparison is conjuncted similarly for each additional state element. Hence, for this implementation, the safety property is:

$$pc_0 \wedge rf_0 \wedge dm_0 \ \vee \ pc_1 \wedge rf_1 \wedge dm_1 \ \vee \ \ldots \ \vee \ pc_k \wedge rf_k \wedge dm_k = true. \quad (1)$$

The syntax of EUFM [7] includes *terms* and *formulas*. Terms are used to abstract word-level values of data, register identifiers, memory addresses, as well as the entire states of memory arrays. A term can be an Uninterpreted Function (UF) applied to a list of argument terms, a term variable, or an *ITE* operator selecting between two argument terms based on a controlling formula, such that $ITE(formula, term_1, term_2)$ will evaluate to $term_1$ if $formula = true$, and to $term_2$ if $formula = false$. The syntax for terms can be extended to model memories by means of the interpreted functions *read* and *write* [7][24]. Formulas are used to model the control path of a processor, as well as to express the correctness condition. A formula can be an Uninterpreted Predicate (UP) applied to a list of argument terms, a propositional variable, an *ITE* operator selecting between two argument formulas based on a controlling formula, or an equation between two terms. Formulas can be negated and combined with Boolean connectives. We will refer to both terms and formulas as *expressions*.

UFs and UPs are used to abstract functional units by replacing them with "black boxes" that satisfy no particular properties other than that of *functional consistency*—that the same combinations of values to the inputs of the UF (or UP) produce the same output value. Then, it no longer matters whether the original functional unit is an adder, or a multiplier, etc., as long as the same UF (or UP) is used to replace it in both the implementation and the specification. Thus, we will prove a more general problem—that the processor is correct for any functionally consistent implementation of its functional units. However, this more general problem is easier to prove.

Two possible ways to impose the property of functional consistency of UFs and UPs are Ackermann constraints [3], and nested *ITEs* [21]. The Ackermann scheme replaces each UF (UP) application in the EUFM formula F with a new term variable (Boolean variable) and then adds external consistency constraints. For example, the UF application $f(a_1, b_1)$ will be replaced by a new term variable c_1, another application of the same UF, $f(a_2, b_2)$, will be replaced by a new term variable c_2. Then, the resulting EUFM formula F' will be extended as $[(a_1 = a_2) \wedge (b_1 = b_2) \Rightarrow (c_1 = c_2)] \Rightarrow F'$. In the nested-*ITE* scheme, the first application of the above UF is still replaced by a new term variable c_1. However, the second is replaced by $ITE((a_2 = a_1) \wedge (b_2 = b_1), c_1, c_2)$, where c_2 is a new term variable. A third one, $f(a_3, b_3)$, is replaced by $ITE((a_3 = a_1) \wedge (b_3 = b_1), c_1, ITE((a_3 = a_2) \wedge (b_3 = b_2), c_2, c_3))$, where c_3 is a new term variable, and so on. UPs are eliminated similarly, but with new Boolean variables.

To compare the sequence of *write* operations that form the final states of memories after the implementation and specification sides of the diagram, the decision procedure EVC [25] automatically introduces a new term variable to serve as *comparison address* for each memory. Let cmp_addr be the new term variable introduced for the register file. Then, each equation $(RegFile^i_{Spec} = RegFile^*_{Spec})$ is replaced with:

$$(read(RegFile^i_{Spec}, cmp\_addr) = read(RegFile^*_{Spec}, cmp\_addr)),$$

thus checking whether an arbitrary address in the register file is modified in the same way by both sides of the diagram. EVC replaces a *read* from a sequence of *writes* with a sequence of nested *ITEs*, according to the forwarding property of the memory semantics, such that each *ITE* is controlled by an equation between the new term variable and the destination address of the eliminated *write*. These equations appear in dual polarity—positive when selecting the then-expression of the *ITE*, and negative when selecting the else-expression—i.e., are g-equations that need to be encoded with Boolean variables.

We will call *complete equality* the usual equality, where two (syntactically) different term variables a and b can be either equal or not equal to each other, and will use $=$ to denote it. Reasoning about complete equality requires a case split, in order to account for both cases, and so the need to encode it with Boolean variables when translating an EUFM formula to an equivalent Boolean formula. We will call *syntactic equality* the subset of complete equality where a term variable is equal only to itself, and will use $=_{SYN}$ to denote it. We will call *delta equality* the difference between complete equality and syntactic equality, and will use $=_\Delta$ to denote it. That is, if t_1 and t_2 are two terms consisting of *ITE* operators, term variables, and formulas controlling the *ITE* operators, then $(t_1 =_\Delta t_2)$ is defined as $(t_1 = t_2) \wedge \neg(t_1 =_{SYN} t_2)$, or equivalently, complete equality $(t_1 = t_2)$ is defined as $(t_1 =_{SYN} t_2) \vee (t_1 =_\Delta t_2)$. We will call *hybrid equality* the extension of syntactic equality with a proper subset of the delta equality between two terms, and will denote it with $=_{HYB}$.

The property of Positive Equality is due to the observation that the correctness formula (1) consist of top-level p-equations that are combined with the monotonically positive connectives of conjunction and disjunction, but are not negated. Then, if the formula is valid (true) when the complete equality in the top-level p-equations is replaced with syntactic equality, the formula will also be valid with the original complete equality in those equations, since then the formula can only get bigger due to the omitted delta equality that will be added through monotonically positive connectives. However, the benefit from using only syntactic equality for the top-level p-equations is the significant reduction of the solution space, resulting in orders of magnitude speedup. Similarly, we exploit syntactic functional consistency when eliminating UFs and UPs in that the property of functional consistency is enforced only for the cases of syntactic equality between corresponding arguments in applications of the same UF/ UP, unless both arguments are g-terms. Syntactic functional consistency is a conservative approximation, since functional consistency is enforced only for a subset of the conditions for complete functional consistency (based on complete equality). If F is a formula obtained after eliminating all UFs/UPs by accounting for only syntactic functional consistency, and F is valid, then so will be the formula obtained from F by accounting for complete functional consistency, e.g., by extending F with Ackermann constraints for complete functional consistency. However, g-equations could be either *true* or *false*, and need to be encoded with Boolean variables [9][16].

A *low-level g-equation* is one where both arguments are term variables. A *top-level g-equation* is one where the arguments can be either term variables or nested-*ITE* expressions selecting term variables. Previous methods for encoding g-equations with Boolean variables [9][16] eliminate top-level g-equations by pushing them to the argument term variables, and then encode the resulting low-level g-equations. The e_{ij} encoding [9] replaces each unique low-level equation between different term variables v_i and v_j with a new Boolean variable, called e_{ij}. The property of symmetry of equality is accounted for by sorting v_i and v_j according to their indices, e.g., so that $i < j$, before introducing a Boolean variable; and the property of transitivity of equality—if $v_i = v_j$ and $v_j = v_k$ then $v_i = v_k$—is enforced with transitivity constraints of the form $e_{ij} \wedge e_{jk} \Rightarrow e_{ik}$. In the small-domain encoding [16], each g-term variable is assigned a set of constants that it can take on in a way that allows it to be either equal to or different from any other g-term variable that it can be transitively compared for equality with. If a g-term variable is assigned a set of N constants, then those can be indexed with $\lceil log_2(N) \rceil$ Boolean variables. Two g-term variables are equal if their indexing Boolean variables select simultaneously a common constant. The property of transitivity is automatically enforced in this encoding. Depending on the structure of the g-term equality-comparison graphs, the small-domain encoding may introduce fewer Boolean variables than the e_{ij} encoding. That could mean a smaller search space. However, now a low-level g-equation is replaced with a Boolean formula—enumerating all cases when the argument g-term variables evaluate to a common constant—instead of a single Boolean variable. In our previous work [26], we found the e_{ij} encoding to outperform the small-domain encoding when formally verifying microprocessors. For other benchmarks, Seshia et al. [18] proposed a hybrid encoding, such that the e_{ij} and small-domain encodings are each used on a different connected component of low-level g-equations in the same correctness formula. The decision procedure EVC adds all transitivity constraints for the e_{ij} variables to the CNF correctness formula, while the decision procedure CVC [4] iteratively analyzes counterexamples, and includes transitivity constraints incrementally—just as many as to prevent the recurrence of a counterexample. However, Seshia et al. [18] found the incremental approach to result in significant overhead when checking validity of complex formulas.

3   Automatic Abstraction of Equations

In this section, we will assume that the interpreted functions *read* and *write*, as well as all UFs and UPs, have been eliminated from the EUFM correctness formula. That is, each term in the formula is either a term variable, or a nested-*ITE* expression selecting term variables. In this formula, instead of encoding low-level g-equations, we can automatically abstract the top-level g-equations with the special interpreted predicate *abs_equality* that satisfies the properties of transitivity, syntactic functional consistency, syntactic symmetry, and (syntactic) reflexivity. Note that the abstracted complete equality is: 1) transitive—if $a = b$ and $b = c$ then $a = c$; 2) symmetric—if $a = b$ then $b = a$; 3) reflexive, i.e., $a = a$ is *true*; and 4) functionally consistent—given two equations $a = b$ and $c = d$, the equality between arguments in the same positions, i.e., $a = c$ and $b = d$, implies that the two equations have equal values, as follows from the property of transitivity, since the four equations form a cycle, and if three of them are *true* then the fourth should also be *true*, while if one is *false* and two are *true* then the fourth should be *false* or otherwise there will be a cycle of three equations that are *true*, implying that the first should be *true* and so contradicting its value. Also note that the property of reflexivity is equivalent to syntactic equality, since the property holds when exactly the same term variable appears on both sides of an equation.

We can extend either the nested-*ITE* or the Ackermann-constraint scheme for elimination of uninterpreted predicates in order to eliminate applications of *abs_equality* by accounting for its properties of syntactic functional consistency, syntactic symmetry, and reflexivity. Transitivity can be imposed as in the case of low-level g-equations [26]—by triangulating the equality-comparison graph (where each vertex is a term used in a top-level g-equation, and each edge corresponds to a g-equation between two terms) with extra edges, added in a greedy manner to turn every two edges with a common vertex into a triangle (cycle of length 3), and then imposing three transitivity constraints for the CNF variables representing the values of g-equations in a triangle.

In order to adopt the nested-*ITE* scheme, each *ITE*-controlling formula is extended to account for syntactic symmetry, while the top *ITE* expression is disjuncted with the condition for syntactic equality between the two arguments, thus ensuring reflexivity. That is, the first application of $abs\_equality(t_1, t_2)$, where t_1 and t_2 are terms, is eliminated with $(t_1 =_{SYN} t_2) \vee E_1$, where E_1 is a new Boolean variable, and the disjunction of $(t_1 =_{SYN} t_2)$ ensures reflexivity. A second application $abs\_equality(t_3, t_4)$ is eliminated with $(t_3 =_{SYN} t_4) \vee ITE((t_3 =_{SYN} t_1) \wedge (t_4 =_{SYN} t_2) \vee (t_4 =_{SYN} t_1) \wedge (t_3 =_{SYN} t_2), E_1, E_2)$, where E_2 is a new Boolean variable, and the disjunction of $(t_3 =_{SYN} t_4)$ ensures reflexivity. In the controlling formula, the expression $(t_3 =_{SYN} t_1) \wedge (t_4 =_{SYN} t_2)$ ensures syntactic functional consistency, as in the original nested-*ITE* scheme for elimination of UFs and UPs, and the disjunction of $(t_4 =_{SYN} t_1) \wedge (t_3 =_{SYN} t_2)$ ensures syntactic symmetry. A third application $abs\_equality(t_5, t_6)$ is eliminated with $(t_5 =_{SYN} t_6) \vee ITE((t_5 =_{SYN} t_1) \wedge (t_6 =_{SYN} t_2) \vee (t_6 =_{SYN} t_1) \wedge (t_5 =_{SYN}$

$t_2), E_1, ITE((t_5 =_{SYN} t_3) \wedge (t_6 =_{SYN} t_4) \vee (t_6 =_{SYN} t_3) \wedge (t_5 =_{SYN} t_4), E_2, E_3))$, where E_3 is a new Boolean variable.

The Ackermann-constraint scheme can be customized similarly. Each of the three applications, $abs\_equality(t_1, t_2)$, $abs\_equality(t_3, t_4)$, and $abs\_equality(t_5, t_6)$, will be eliminated with a new Boolean variable—E_1, E_2, and E_3, respectively. Let F' be the resulting EUFM formula. To account for the properties of reflexivity, syntactic symmetry, and syntactic functional consistency of $abs\_equality$, we define separate constraints, conjunct them in formula $constraints$, and use it to restrict F', i.e., prove the validity of $constraints \Rightarrow F'$. In particular, to enforce reflexivity of the first, second, and third applications of $abs\_equality$, we use the constraints $(t_1 =_{SYN} t_2) \Rightarrow E_1$, $(t_3 =_{SYN} t_4) \Rightarrow E_2$, and $(t_5 =_{SYN} t_6) \Rightarrow E_3$, respectively. To account for syntactic functional consistency and syntactic symmetry of the second application with respect to the first, we use $[(t_3 =_{SYN} t_1) \wedge (t_4 =_{SYN} t_2) \vee (t_4 =_{SYN} t_1) \wedge (t_3 =_{SYN} t_2)] \Rightarrow (E_1 \Leftrightarrow E_2)$. To account for syntactic functional consistency and syntactic symmetry of the third application with respect to the first, we use $[(t_5 =_{SYN} t_1) \wedge (t_6 =_{SYN} t_2) \vee (t_6 =_{SYN} t_1) \wedge (t_5 =_{SYN} t_2)] \Rightarrow (E_1 \Leftrightarrow E_3)$. Finally, to account for syntactic functional consistency and syntactic symmetry of the third application with respect to the second, we add the constraint $[(t_5 =_{SYN} t_3) \wedge (t_6 =_{SYN} t_4) \vee (t_6 =_{SYN} t_3) \wedge (t_5 =_{SYN} t_4)] \Rightarrow (E_2 \Leftrightarrow E_3)$.

Example: Let t_1, t_2, t_3, t_4, t_5, and t_6 be six terms defined as follows:

$$t_1 \leftarrow a \qquad\qquad t_2 \leftarrow b$$
$$t_3 \leftarrow ITE(f_1, c, a) \qquad t_4 \leftarrow c$$
$$t_5 \leftarrow ITE(f_2, d, a) \qquad t_6 \leftarrow ITE(f_3, a, b)$$

where a, b, c, and d are term variables, and f_1, f_2, and f_3 are formulas. Let $(t_1 = t_2), (t_3 = t_4)$, and $(t_5 = t_6)$ be top-level g-equations in an EUFM formula.

To apply the e_{ij} encoding, the top-level g-equations will be pushed to their argument term variables: the first equation will remain unchanged, $(a = b)$, since both arguments are term variables, and will be replaced with the new Boolean variable e_{ab}; the second will become $ITE(f_1, c = c, a = c)$, i.e., $ITE(f_1, true, a = c)$, and will be replaced with $f_1 \vee e_{ac}$, after $a = c$ is encoded with the new Boolean variable e_{ac}; the third will become $ITE(f_2, ITE(f_3, a = d, b = d), ITE(f_3, a = a, a = b))$, and will be replaced with $f_2 \wedge f_3 \wedge e_{ad} \vee f_2 \wedge \neg f_3 \wedge e_{bd} \vee \neg f_2 \wedge f_3 \vee \neg f_2 \wedge \neg f_3 \wedge e_{ab}$, after $a = d$ is encoded with e_{ad} and $b = d$ is encoded with e_{bd}.

Using the special interpreted predicate $abs\_equality$, the top-level g-equations will be abstracted as $abs\_equality(t_1, t_2)$, $abs\_equality(t_3, t_4)$, and $abs\_equality(t_5, t_6)$. Then, using the nested-ITE scheme, the first application of $abs\_equality$ will be eliminated with $(a =_{SYN} b) \vee E_1$, which evaluates to E_1, since a and b are two (syntactically) different term variables, so that their syntactic equality evaluates to $false$. The second application will be eliminated with $(ITE(f_1, c, a) =_{SYN} c) \vee ITE((ITE(f_1, c, a) =_{SYN} a) \wedge (c =_{SYN} b) \vee (c =_{SYN} a) \wedge (ITE(f_1, c, a) =_{SYN} b), E_1, E_2)$, which evaluates to $f_1 \vee ITE(\neg f_1 \wedge false \vee false \wedge false, E_1, E_2)$, i.e., to $f_1 \vee E_2$, where f_1 expresses the condition for syntactic equality between the two arguments, while E_2 encodes the two possible values of the delta equality between the two argument terms. Eliminating the third application of *abs_equality*, and simplifying the resulting expression, we get $\neg f_2 \wedge f_3 \vee ITE(\neg f_2 \wedge \neg f_3, E_1, E_3)$, where formula $\neg f_2 \wedge f_3$ expresses the conditions for syntactic equality between the two arguments, while the *ITE* operator will select Boolean variable E_1 if formula $\neg f_2 \wedge \neg f_3$ is *true*, i.e., in the cases of syntactic functional consistency with the arguments of the first application of *abs_equality*, while the new Boolean variable E_3 encodes the delta equality between the two arguments in the cases when the arguments do not satisfy conditions for syntactic functional consistency or syntactic symmetry with respect to previous pairs of arguments.

Using Ackermann constraints to enforce reflexivity, syntactic functional consistency, and syntactic symmetry for *abs_equality*, the first, second, and third applications will be replaced with the new Boolean variables E_1, E_2, and E_3, respectively. Then, the resulting formula will be evaluated under the constraints: $f_1 \Rightarrow E_2$, enforcing reflexivity for *abs_equality*(t_3, t_4); $\neg f_2 \wedge f_3 \Rightarrow E_3$, enforcing reflexivity for *abs_equality*(t_5, t_6); and $\neg f_2 \wedge \neg f_3 \Rightarrow (E_1 \Leftrightarrow E_3)$, enforcing syntactic functional consistency between *abs_equality*(t_3, t_4) and *abs_equality*(t_5, t_6).

THEOREM 1. Let F be an EUFM formula that contains term variables, Boolean variables, logic connectives, *ITE* operators, and equations. Then, abstracting the top-level g-equations in F with the interpreted predicate *abs_equality* is sound and complete.

Proof: Let formula F' be obtained from F after abstracting the top-level g-equations with the interpreted predicate *abs_equality*.

Soundness—the validity of F' implies the validity of F. If F' is valid, then so will be any formula obtained from F' after replacing *abs_equality* with any predicate that has two arguments, and satisfies the properties of transitivity, reflexivity, syntactic symmetry, and syntactic functional consistency, including the original complete equality. Note that by its definition, *abs_equality* satisfies the property of syntactic equality, i.e., reflexivity. What is missing from complete equality are two constraints: 1) if the delta equality between terms a and b is *true*, then *abs_equality*(a, b) should be *true*; and 2) if the complete equality between terms a and b is *false*, then *abs_equality*(a, b) should be *false*. However, if F' is valid without such constraints, it will be valid with them:

$$[((a =_\Delta b) \Rightarrow abs\_equality(a, b)) \wedge (\neg(a = b) \Rightarrow \neg abs\_equality(a, b))] \Rightarrow F',$$

where the resulting formula is trivially valid, since F' is already valid.

Completeness—a counterexample in F' can be mapped to a counterexample in F. A counterexample in F' consists of assignments to variables E_i, used when eliminating *abs_equality*, and assignments to the other Boolean variables that also appear in F. The arguments of each abstracted g-equation are either term variables or nested-*ITE* expressions that select term variables, where the *ITE* operators are controlled by formulas that depend on applications of *abs_equality* and on the other Boolean variables. Hence, each counterexample results in an assignment to the *ITE*-controlling formulas, thus selecting some term variable a as the first argument of an application of *abs_equality*, and another term variable b as the second argument. Then, we can assign the value of that particular application of *abs_equality* to the low-level equation $a = b$ and can replace that application of *abs_equality* with the original complete equality. The correctness formula is a Directed Acyclic Graph (DAG), so that the value of the introduced top-level g-equation does not affect the arguments of that equation. Hence, the *ITE*-controlling formulas in the arguments will keep their values, and so a and b will still appear on the two sides of that g-equation, which will have the same value as the replaced application of *abs_equality*. We can similarly map the value of each remaining applications of *abs_equality* to a value of a low-level g-equation between the term variables selected by the nested-*ITE* arguments, and then replace that application of *abs_equality* with a top-level g-equation, which will get the same value as the one assigned to the low-level g-equation, i.e., as the one of the eliminated application of *abs_equality*. Thus, all abstractions of top-level g-equations will be undone, and we will get the original formula F. What remains to be proved is that these assignments to low-level g-equations will not violate the properties of equality. First, reflexivity is always preserved, since syntactic equality between the two arguments is always accounted for, and an application of *abs_equality* is forced to be *true* when exactly the same term variable is selected to appear on both sides of the abstracted g-equation, i.e., it is impossible for the same term variable to appear as both arguments of an application of *abs_equality* that evaluates to *false*. Second, constraints for syntactic functional consistency ensure that if the same pair of term variables is selected as arguments of different applications of *abs_equality*, then those applications will have the same value, i.e., it is impossible for the same low-level g-equation between term variables to get assigned contradicting values from different applications of *abs_equality*. Third, because of constraints for syntactic symmetry, it is similarly impossible for two symmetric low-level equations, $a = b$ and $b = a$, to get assigned different values. Fourth, transitivity will never be violated, since constraints for transitivity of equality ensure that applications of *abs_equality* do not violate transitivity, and, as described above, a counterexample determines a 1-to-1 mapping of every cycle of abstracted top-level g-equations to an isomorphic cycle of low-level g-equations, each having value identical to that of the corresponding abstracted top-level g-equation.   □

Note that each counterexample maps the value of an abstracted top-level g-equation to exactly one low-level g-equation between term variables in the support of the top-level g-equation. The low-level g-equations that are left unassigned are don't-care conditions. They do not affect the counterexample, and can be left unassigned or given any value that does not violate transitivity, when interpreting the counterexample.

As an optimization, we can choose not to enforce transitivity, or reflexivity, or both; these properties are not needed for models with in-order execution, as shown in the experiments (see Sect. 5). Alternatively, we can enforce partial transitivity—a heuristic for that is presented in Sect. 4.3, and was found to speed up the formal verification of processors with out-of-order execution and completion.

4   Using the Automatic Abstraction

4.1   Identifying Connected Equality-Comparison Components

We will again assume that the formula contains term variables, Boolean variables, logic connectives, *ITE* operators, and equations. Each equation is classified as a p-equation or a g-equation, if it was reached under an even or odd number of negations, respectively, before the uninterpreted functions and uninterpreted predicates were eliminated. Syntactic equations introduced when eliminating uninterpreted functions and uninterpreted predicates are also classified as p-equations. The arguments of g-equations are either term variables or nested-*ITE* expressions selecting term variables. For each g-equation, all term variables that can be selected to appear as an argument are grouped into an *equivalence class*. Equivalence classes that have common term variables are merged and their g-equations are marked to belong to the same connected equality-comparison component. The properties of transitivity, functional consistency, and symmetry need to be enforced only within a connected component, since the values of equations from a connected component have no way of affecting equations from another connected component. That is, we can use a different version of interpreted predicate *abs_equality* for each connected component.

In processors with branch prediction, the equations for the PC states, pc_i, in correctness formula (1) will contain term variables that are arguments to g-equations introduced by the mechanism for correcting branch mispredictions—if the actual and predicted branch targets are equal, then the prediction is correct and any speculative instructions are allowed to complete; otherwise, the speculative instructions are squashed. To ensure that such p-equations have values that are consistent with those of abstracted g-equations that control the speculation and have common term variables as arguments, we need to promote p-equations to g-equations. That is, for each p-equation, determine the equivalence class of term variables that may appear as an argument; if this equivalence class has a common element with another equivalence class that identifies a connected component of g-equations, then merge the two equivalence classes, and promote the p-equation to a g-equation from that connected component.

4.2   Mapping Abstract Counterexamples to Concrete Ones

A counterexample for the abstract model, where top-level g-equations are abstracted with *abs_equality*, is expressed by an assignment to the E_i variables—used when eliminating applications of *abs_equality*—and an assignment to the other Boolean variables—representing initial state of control signals, or introduced when eliminating uninterpreted predicates. We can map an abstract counterexample to a concrete one for the original model by following:

Step 1. Compute the value of each application of *abs_equality*, based on the counterexample assignment to E_i and other Boolean variables in the abstract model.

Step 2. For each application of *abs_equality* (the arguments are either term variables, or nested-*ITE* expressions selecting term variables), compute the values of *ITE*-controlling formulas in the two arguments.

Step 3. For each application of *abs_equality*, find the two term variables that will be selected for equality comparison in the abstracted g-equation, given the values of *ITE*-controlling formulas computed in Step 2. If this application of *abs_equality* evaluates to *true*, then the two term variables should be equal in order to trigger a corresponding counterexample in the concrete model; otherwise, they are not equal.

According to Theorem 1, the above steps will result in consistent assignments to low-level g-equations, without violating the properties of equality.

4.3   Heuristic for Partial Transitivity

In processors with out-of-order completion, the specification side of the commutative diagram (Figure 1) completes the instructions in program order—assuming the abstraction function completes the instructions in program order—while the implementation side may reorder them. In a correct implementation, out-of-order execution and completion occur only if that would not introduce write-after-read or write-after-write hazards [10]. That is, destination registers of younger instructions are compared for equality with both source and destination registers of older instructions (appearing earlier in program order). A younger instruction is issued/completed only if each older instruction is issued/completed, or if the younger instruction will not introduce a hazard for an older instruction. The absence of a write-after-write hazard, when the destination registers of two instructions are not equal, implies that term variable *cmp_addr*, used as comparison address for the final states of the register file (see Sect. 2), may equal only one of these destination registers, but not both, or that will violate a transitivity constraint. That is, if $dest_1$ and $dest_2$ are destination registers compared for equality by logic for preventing write-after-write hazards, then the comparison of the final register file states will introduce equations $(dest_1 = cmp\_addr)$ and $(dest_2 = cmp\_addr)$. However, at most one of them can be *true*, since transitivity of equality implies that $\neg(dest_1 = dest_2) \wedge (cmp\_addr = dest_1) \Rightarrow \neg(cmp\_addr = dest_2)$ and $\neg(dest_1 = dest_2) \wedge (cmp\_addr = dest_2) \Rightarrow \neg(cmp\_addr = dest_1)$.

Similar cycles of 3 equations, comparing two destination registers and a source register, may be introduced by the control logic in processors with out-of-order execution or completion—the equation between the two destination registers by logic checking for write-after-write hazards, and the two equations between a source register and each of the destination registers by logic checking for read-after-write or write-after-read hazards. Constraints for transitivity of equality are needed to prevent simultaneous forwarding of data from two older destination registers that are not equal, and whose instructions are reordered, to a younger source register. Hence, we can use a *heuristic for enforcing partial transitivity*. Let the names of all destination and source register identifiers contain the substring "Dest" and "Src", respectively. Then, we can automatically detect pairs of destination registers compared for equality by the control logic. We can enforce partial transitivity only for such destination registers and any source registers occurring in equations with them, including term variable *cmp_addr*. As noted earlier, partial transitivity is a conservative approximation, since it results in discarding constraints. If the resulting formula F' is valid, it will also be valid when extended with the omitted transitivity constraints, *extra_transitivity*, to a formula *extra_transitivity* $\Rightarrow F'$.

5   Results

The benchmarks used in the experiments are: 1dlx_c_mc_ex_bp, a single-issue pipelined DLX [10] with multicycle ALU, Instruction Memory, and Data Memory, as well as with exceptions and branch prediction, modeled and formally verified as described in [22]; 2dlx_cc_mc_ex_bp, a dual-issue superscalar DLX with in-order execution, and two identical execution pipelines with all of the above features [22]; 9vliw_mc_ex_bp, a 9-wide VLIW processor that also has all of the above features, as well as the same number and types of functional units as the Intel Itanium [11][19], and imitates it in predicated execution, register remapping, and advanced loads—modeled and formally verified as described in [23]; xscale, a model of the Intel XScale processor [12] with specialized execution pipelines, scoreboarding [10], out-of-order completion, and imprecise exceptions—modeled and formally verified as described in [20]; 12pipe, a superscalar processor that can issue up to 12 instructions in program order on every clock cycle, and is capable of executing only ALU instructions [26]; and 8pipe_ooo, a superscalar model that can issue up to 8 instructions out of program order on every cycle, and is also capable of executing only ALU instructions [26].

The experiments were performed on a Dell OptiPlex GX260 with a 3.06-GHz Intel Pentium 4 processor that had a 512-KB on-chip level-2 cache, 2 GB of physical memory, and was running Red Hat Linux 9. The SAT-checker Siege [17], a top-performer in the SAT'03 competition [14], was found to have best performance on these benchmarks and was used for the experiments. The reader is referred to [26] for the translation to CNF format. All constraints for transitivity of equality were added to the CNF formulas from buggy implementations and correct models that require transitivity (xscale and 8pipe_ooo), but were manually switched off for correct models that do not require transitivity. Transitivity was enforced by triangulating the equality comparison graphs [6], and adding transitivity constraints for each resulting cycle of length 3. All models were formally verified by computing the abstraction function with controlled flushing [8], where the user provides a flushing schedule that avoids the triggering of stalling conditions, thus simplifying the correctness formula.

Table 1 presents the results with the e_{ij} encoding. "Trans" CNF clauses represent constraints for transitivity of equality. The e_{ij} Boolean variables ranged from 62 to 2,724; the total Boolean variables were between 142 and 2,844; the CNF variables between 1,148 and 115,915; the CNF clauses between 6,207 and 8,395,649; the decisions made by the SAT-checker Siege were between 5,000 and 167,000,000, while the conflicts that it resolved were between 2,000 and 15,000,000; and the total verification time was between 0.18 seconds and 41,886 seconds (i.e., 11.6 hours).

Table 2 summarizes the results when abstracting the top-level g-equations, and using the nested-ITE scheme to eliminate the applications of predicate $abs\_equality$. The E_i Boolean variables—introduced when eliminating predicate $abs\_equality$—ranged between 47 and 3,600, while the total number of Boolean variables increased accordingly. Three of the benchmarks—2dlx_cc_mc_ex_bp, 9vliw_mc_ex_bp, and xscale—required fewer CNF variables compared with the e_{ij} encoding, with a reduction of 38% for xscale. Five of the benchmarks had fewer CNF clauses relative to the e_{ij} encoding—with a reduction of approximately 50% in the case of xscale and 12pipe, and a 94% reduction of the transitivity clauses for xscale.

The conceptually simpler solution space, resulting from the special interpreted predicate $abs\_equality$, reduced the number of decisions for the last 3 benchmarks by up to an order of magnitude—in the case of 12pipe, the decisions went from 167 million down to 16 million, and the conflicts from 15 million down to 1 million, speeding up the verification 13.73×; in the case of 8pipe_ooo, the decisions were reduced from 31 million to 4 million, and the conflicts from 15 million to 1 million, with the speedup being 14.33 times. Note that the EVC time for translation to SAT increased significantly for the two most complex benchmarks—from 57 seconds to 835 seconds (14.6×) in the case of 12pipe, and from 6 seconds to 53 seconds (8.8×) in the case of 8pipe_ooo—but that was more than offset by the dramatic reduction in the SAT time.

Using Ackermann constraints instead of nested ITEs when enforcing reflexivity, functional consistency, and functional symmetry—see Table 3—required up to 74% more CNF variables, and up to 10% more clauses in the case of 8pipe_ooo, resulting in smaller speedups of 9.23× for 12pipe, and 9.74× for 8pipe_ooo.

Applying the heuristic for partial transitivity when formally verifying the two correct benchmarks that require transitivity—see Table 4—resulted in a 98% reduction in the number of transitivity clauses, and a 13% reduction in the total number of clauses for 8pipe_ooo, increasing the speedup to 20× for that model.

The above benchmarks do not require reflexivity of equality, since the g-equations are between source and destination register identifiers, which are separate instruction fields. However, in the M•CORE processor [15], a register identifier is used as both a source and destination register for the same instruction. Modifying both 12pipe and 8pipe_ooo, so that one of the source registers also served as destination register for the same instruction, resulted in automatically added reflexivity constraints, since the symbolic conditions for enforcing reflexivity did not simplify to $false$ in EVC. However, those benchmarks also passed the safety check without reflexivity, since it is impossible for a register to be compared with itself in a correct implementation.

The mechanism for enforcing reflexivity was tested by modifying 8pipe_ooo to require this property after the model was extended with:

$$t \leftarrow ITE(new\_var, a, b)$$
$$f_1 \leftarrow (t = a)$$
$$f_2 \leftarrow (t = b)$$
$$f_3 \leftarrow f_1 \lor f_2$$

where a and b are arbitrary terms, new_var is a new Boolean variable, and formula f_3 was used as additional enabling condition in the forwarding logic of the processor. Note that when $t = a$ and $t = b$ are abstracted with $abs\_equality$, and the property of reflexivity is enforced, then f_3 will evaluate to $true$, since f_1 will be $true$ when new_var is $true$, while f_2 will be $true$ when new_var is $false$. However, without reflexivity, f_3 will evaluate to a symbolic expression that will not be constrained to evaluate to $true$, and the modified forwarding logic will be incorrect. When reflexivity was not enforced, the SAT-checker Siege took 12 seconds to find a counterexample. However, with reflexivity constraints added automatically, only for the applications of $abs\_equality$ where the conditions for enforcing reflexivity (i.e., the syntactic equality between the two arguments) do not simplify to $false$, Siege took time comparable to that for the original 8pipe_ooo.

Similarly, the mechanism for enforcing transitivity was tested with a variant of 8pipe_ooo. One level of the forwarding logic—where a result is forwarded to the ALU if a source register src equals a destination register $dest$—was modified to:

$$regs\_equal\_original \leftarrow (src = dest)$$
$$f_1 \leftarrow (t = src) \land (t = dest)$$
$$regs\_equal \leftarrow f_1 \lor regs\_equal\_original$$

where t is an arbitrary term, such that the new formula $regs\_equal$ was used to control forwarding of data, as opposed to the original formula $regs\_equal\_original$. Note that if transitivity is enforced, then $((t = src) \land (t = dest)) \Leftrightarrow (src = dest)$, i.e., $f_1 \Leftrightarrow regs\_equal\_original$, so that $regs\_equal \Leftrightarrow regs\_equal\_original$, and the modified processor will function like the original, where formula $regs\_equal\_original$ is used to control forwarding. However, without transitivity, f_1 may evaluate to $true$ when $regs\_equal\_original$ evaluates to $false$, so that data may be forwarded incorrectly. With partial transitivity, a counterexample was found in 15 seconds, but with complete transitivity, validity was proved in time comparable to that for the original 8pipe_ooo.

To evaluate the efficiency of $abs\_equality$ when formally verifying incorrect models, 10 buggy variants of 12pipe were created. While $abs\_equality$ reduced the number of SAT decisions by up to 2.5×, the number of conflicts by up to 5×, and the SAT-checking time by up to 5× as well, the total time was always longer compared with the e_{ij} encoding (up to 7×), due to the much increased time for SAT translation.

Table 1. Results from the e_{ij} encoding

Processor	Boolean Variables		CNF Vars	CNF Clauses		SAT-Checker Siege		CPU Time [sec]			
	e_{ij}	Total		Trans	Total	decisions	conflicts	TLSim	EVC	SAT	Total
1dlx_c_mc_ex_bp	62	142	1,148	0	6,207	5×10^3	2×10^3	0.04	0.04	0.1	0.18
2dlx_cc_mc_ex_bp	256	414	4,482	0	41,071	36×10^3	10×10^3	0.06	0.27	1.18	1.51
9vliw_mc_ex_bp	2,968	3,326	24,373	0	232,209	936×10^3	101×10^3	0.1	1.6	37	38.7
xscale	2,387	2,669	43,574	102,480	656,381	72×10^3	27×10^3	0.2	3.8	21	25
12pipe	2,724	2,844	115,915	0	8,395,649	167×10^6	15×10^6	0.4	57	41,829	41,886
8pipe_ooo	2,129	2,209	35,510	117,462	1,191,215	31×10^6	15×10^6	0.2	6	19,981	19,987

Table 2. Abstracting the top-level g-equations, and using the nested-*ITE* scheme. The speedup is the total time with the e_{ij} encoding divided by the new total time

Processor	Boolean Variables		CNF Vars	CNF Clauses		SAT-Checker Siege		CPU Time [sec]				Speedup
	E_i	Total		Trans	Total	decisions	conflicts	TLSim	EVC	SAT	Total	
1dlx_c_mc_ex_bp	47	128	1,188	0	6,415	5×10^3	1×10^3	0.04	0.04	0.04	0.12	1.50
2dlx_cc_mc_ex_bp	159	317	4,251	0	32,716	35×10^3	10×10^3	0.06	0.27	1	1.33	1.14
9vliw_mc_ex_bp	1,894	2,252	17,481	0	167,567	936×10^3	122×10^3	0.1	1.9	41	43	0.90
xscale	333	643	26,857	5,907	326,041	52×10^3	19×10^3	0.2	2.5	12.6	15	1.67
12pipe	3,600	3,720	136,800	0	4,216,460	16×10^6	1×10^6	0.4	835	2,215	3,050	13.73
8pipe_ooo	2,157	2,638	42,365	134,670	1,021,721	4×10^6	1×10^6	0.2	53	1,342	1,395	14.33

Table 3. Abstracting the top-level g-equations, and using the Ackermann-constraint scheme. The speedup is the total time with the e_{ij} encoding divided by the new total time

Processor	Boolean Variables		CNF Vars	CNF Clauses		SAT-Checker Siege		CPU Time [sec]				Speedup
	E_i	Total		Trans	Total	decisions	conflicts	TLSim	EVC	SAT	Total	
1dlx_c_mc_ex_bp	47	128	1,254	0	6,592	5×10^3	1×10^3	0.04	0.06	0.05	0.15	1.20
2dlx_cc_mc_ex_bp	159	317	4,627	0	33,756	33×10^3	9×10^3	0.06	0.27	1.16	1.49	1.01
9vliw_mc_ex_bp	1,894	2,252	20,168	0	175,593	848×10^3	115×10^3	0.1	1.95	40	42	0.92
xscale	333	643	27,839	5,907	328,924	44×10^3	19×10^3	0.2	2.7	13.4	16	1.56
12pipe	3,600	3,720	192,105	0	4,382,554	17×10^6	1×10^6	0.4	839	3,699	4,538	9.23
8pipe_ooo	2,157	2,638	73,680	134,670	1,128,542	4×10^6	0.9×10^6	0.2	52	2,001	2,053	9.74

Table 4. Using the heuristic for partial transitivity, abstracting the top-level g-equations, and applying the nested-*ITE* scheme. The speedup is the total time with the e_{ij} encoding divided by the new total time

Processor	Boolean Variables		CNF Vars	CNF Clauses		SAT-Checker Siege		CPU Time [sec]				Speedup
	E_i	Total		Trans	Total	decisions	conflicts	TLSim	EVC	SAT	Total	
xscale	330	612	26,826	4,089	324,223	52×10^3	23×10^3	0.2	2.3	17.1	19.6	1.28
8pipe_ooo	1,684	1,764	41,491	2,772	889,823	4×10^6	0.9×10^6	0.2	25	973	998	20.03

6 Conclusions

The paper presented a method for automatic abstraction of equations in a logic of equality by using a special interpreted predicate that satisfies the properties of transitivity, reflexivity, syntactic functional consistency, and syntactic symmetry. This abstraction is both sound and complete. The abstraction reduced the number of CNF clauses by up to 50%, and sped up the formal verification by up to an order of magnitude relative to the e_{ij} method, where a Boolean variable is used to encode each unique low-level equation between term variables. A heuristic for partial transitivity resulted in additional speedup for correct benchmarks that need transitivity. Abstracting the top-level equations had better performance, due to the concise encoding of many low-level equations between term variables with a single Boolean variable, thus resulting in an order of magnitude reduction in the number of decisions and the number of conflicts, resolved by a SAT-checker when evaluating the Boolean correctness formula.

References

[1] M. D. Aagaard, N. A. Day, and M. Lou, "Relating Multi-Step and Single-Step Microprocessor Correctness Statements," Formal Methods in Computer-Aided Design (FMCAD '02), M. D. Aagaard, and J. W. O'Leary, eds., LNCS 2517, Springer-Verlag, November 2002.

[2] M. D. Aagaard, B. Cook, N. A. Day, and R. B. Jones, "A Framework for Superscalar Microprocessor Correctness Statements," Software Tools for Technology Transfer, 2002.

[3] W. Ackermann, Solvable Cases of the Decision Problem, North-Holland, 1954.

[4] C. Barrett, D. Dill, and A. Stump, "Checking Satisfiability of First-Order Formulas by Incremental Translation to SAT," Computer-Aided Verification (CAV '02), E. Brinksma, and K. G. Larsen, eds., LNCS 2404, Springer-Verlag, July 2002.

[5] R. E. Bryant, S. German, and M. N. Velev, ôProcessor Verification Using Efficient Reductions of the Logic of Uninterpreted Functions to Propositional Logic,ö ACM Transactions on Computational Logic (TOCL), Vol. 2, No. 1 (January 2001).

[6] R. E. Bryant, and M. N. Velev, "Boolean Satisfiability with Transitivity Constraints," ACM Transactions on Computational Logic (TOCL), Volume 3, Number 4 (October 2002).

[7] J. R. Burch, and D. L. Dill, "Automated Verification of Pipelined Microprocessor Control," Computer-Aided Verification (CAV '94), D. L. Dill, ed., LNCS 818, Springer-Verlag, 1994.

[8] J. R. Burch, "Techniques for Verifying Superscalar Microprocessors," 33rd Design Automation Conference (DAC '96), June 1996, pp. 552–557.

[9] A. Goel, K. Sajid, H. Zhou, A. Aziz, and V. Singhal, "BDD Based Procedures for a Theory of Equality with Uninterpreted Functions," Computer-Aided Verification (CAV '98), A. J. Hu, and M. Y. Vardi, eds., LNCS 1427, Springer-Verlag, June 1998.

[10] J. L. Hennessy, and D. A. Patterson, Computer Architecture: A Quantitative Approach, 3rd edition, Morgan Kaufmann Publishers, San Francisco, CA, 2002.

[11] Intel Corporation, IA-64 Application Developer's Architecture Guide, May 1999. http://developer.intel.com/design/ia-64/architecture.htm.

[12] Intel Corporation, Intel XScale Technology, http://www.intel.com/design/intelxscale/.

[13] S. Lahiri, C. Pixley, and K. Albin, "Experience with Term Level Modeling and Verification of the M•CORE^{TM} Microprocessor Core," 6th Annual IEEE International Workshop on High Level Design, Validation and Test (HLDVT '01), November 2001, pp. 109–114.

[14] D. Le Berre, and L. Simon, "Results from the SAT'03 SAT Solver Competition," 6th International Conference on Theory and Applications of Satisfiability Testing (SAT '03), May 2003. http://www.lri.fr/~simon/contest03/results/.

[15] Motorola, Inc., M•CORE^{TM}: microRISC Engine Programmer's Refernce Manual, 1997.

[16] A. Pnueli, Y. Rodeh, O. Strichman, and M. Siegel, "The Small Model Property: How Small Can It Be?", Journal of Information and Computation, Vol. 178, No. 1 (October 2002).

[17] L. Ryan, Siege SAT Solver v.3, http://www.cs.sfu.ca/~loryan/personal/.

[18] S. A. Seshia, S. K. Lahiri, and R. E. Bryant, "A Hybrid SAT-Based Decision Procedure for Separation Logic with Uninterpreted Functions," 40th Design Automation Conference (DAC '03), June 2003, pp. 425–430.

[19] H. Sharangpani, and K. Arora, "Itanium Processor Microarchitecture," IEEE Micro, Vol. 20, No. 5 (September–October 2000), pp. 24–43.

[20] S. K. Srinivasan, and M. N. Velev, "Formal Verification of an Intel XScale Processor Model with Scoreboarding, Specialized Execution Pipelines, and Imprecise Data-Memory Exceptions," Formal Methods and Models for Codesign (MEMOCODE '03), June 2003.

[21] M. N. Velev, and R. E. Bryant, "Superscalar Processor Verification Using Efficient Reductions of the Logic of Equality with Uninterpreted Functions to Propositional Logic," Correct Hardware Design and Verification Methods (CHARME '99), L. Pierre, and T. Kropf, eds., LNCS 1703, Springer-Verlag, September 1999.

[22] M. N. Velev, and R. E. Bryant, "Formal Verification of Superscalar Microprocessors with Multicycle Functional Units, Exceptions, and Branch Prediction," 37th Design Automation Conference (DAC '00), June 2000, pp. 112–117.

[23] M. N. Velev, "Formal Verification of VLIW Microprocessors with Speculative Execution," Computer-Aided Verification (CAV '00), E. A. Emerson and A. P. Sistla, eds., LNCS 1855, Springer-Verlag, July 2000, pp. 296–311.

[24] M. N. Velev, "Automatic Abstraction of Memories in the Formal Verification of Superscalar Microprocessors," Tools and Algorithms for the Construction and Analysis of Systems (TACAS '01), T. Margaria and W. Yi, eds., LNCS 2031, Springer-Verlag, April 2001.

[25] M. N. Velev, and R. E. Bryant, "EVC: A Validity Checker for the Logic of Equality with Uninterpreted Functions and Memories, Exploiting Positive Equality and Conservative Transformations," Computer-Aided Verification (CAV '01), G. Berry, H. Comon, and A. Finkel, eds., LNCS 2102, Springer-Verlag, July 2001.

[26] M. N. Velev, and R. E. Bryant, "Effective Use of Boolean Satisfiability Procedures in the Formal Verification of Superscalar and VLIW Microprocessors," Journal of Symbolic Computation (JSC), Vol. 35, No. 2 (February 2003), pp. 73–106.

[27] M. N. Velev, "Integrating Formal Verification into an Advanced Computer Architecture Course," ASEE Annual Conference & Exposition, June 2003.

[28] M. N. Velev, "Collection of High-Level Microprocessor Bugs from Formal Verification of Pipelined and Superscalar Designs," International Test Conference (ITC '03), October 2003.

A Free Variable Sequent Calculus with Uniform Variable Splitting

Arild Waaler[1,2] and Roger Antonsen[2]

[1] Finnmark College, Alta, Norway
[2] Department of Informatics, University of Oslo, Norway

Abstract. A system with variable splitting is introduced for a sequent calculus with free variables and run-time Skolemization. Derivations in the system are invariant under permutation, so that the order in which rules are applied has no effect on the leaves. Technically this is achieved by means of a simple indexing system for formulae, variables and Skolem functions. Moreover, the way in which variables are split enables us to restrict the term universe branchwise.

1 Introduction

In free variable sequent calculi variable binding is separated from the rules and implemented by means of explicit substitutions. In consequence there are no side conditions which regulate the order of rule application. However, since free variables are then copied into different branches, this may lead to dependencies between branches that one will not have if a specific rule ordering is observed. To avoid such dependencies, the same variable can in some cases be bound to different values in different branches. The task of identifying exactly when this can be done is known as the variable splitting problem.

The aim of this paper is to present a solution to this problem for a cut-free system of classical logic without equality. Attempting both to reveal the nature of this problem and to motivate our contribution, we shall start out with a brief discussion of the types of quantifier rules used in three standard proof systems for classical logic. To facilitate comparison we address their sequent calculi formulations. Following the notation of Smullyan [10] L∀ and R∃ are called γ-rules, L∃ and R∀ are called δ-rules, branching rules are β-type rules and the remaining logical rules are of α-type. A γ-formula is a formula occurrence which potentially can be principal in a γ-type inference.

In comparing the complexity of search spaces one relevant factor is the length of proofs (counted in number of inference steps). Of no less importance, although much harder to measure, is the uniformity of the search space and the possibility of avoiding irrelevant steps in the search process. Some central questions related to proof-search are:

1. Is it possible to apply the rules in any order?
2. Does the system admit free variables and variable binding by means of explicit substitutions?

M. Cialdea Mayer and F. Pirri (Eds.): TABLEAUX 2003, LNAI 2796, pp. 214–229, 2003.

3. Given that the system has free variables and does not constrain the intrinsic order of rules, is the number of explicit copies of γ-formulae on a branch independent of the intrinsic rule order?

4. Can the number of explicit copies of γ-formulae on a given branch be locally bound by the term universe of the branch?

A negative answer to the first question implies that there are limited possibilities for goal-directed search, i.e. search driven by connections or potential axioms. It is then hard to prevent expansion of irrelevant formulae. If the answer to the second question is negative, we must choose instantiation terms along with applications of γ-type rules. We then run the risk of instantiating quantifiers with irrelevant terms, and this may give rise to irrelevant inferences. A positive answer to the first two questions greatly improves the possibility of performing *least commitment* search. Nevertheless, it may still be the case that unification-based goal-directed search has a cost in terms of proof length (addressed by the third and fourth questions). Should this be the case, the least commitment strategy may give rise to redundancies in the proof objects, and the benefits of the strategy become harder to measure. The fourth question has to do with termination within decidable fragments of the language. An affirmative answer greatly facilitates the formulation of an efficient termination criterion. On the contrary, a negative answer is likely to have a negative impact both on the complexity of the search space and on the length of proofs (due to redundant inferences). We may also fail to detect that a given sequent is unprovable in cases where the term universe of one open branch would have been finitely bounded, given another set of quantifier rules.

For systems without free variables, and which adopt Gentzen's eigenparameter condition (to avoid confusion we shall use 'eigenparameter' instead of the more usual 'eigenvariable'), the answers to the first two questions are negative. However, the answer to the fourth is positive. If we, on the other hand, let the γ-rules introduce free variables and the δ-rules introduce Skolem functions (the type of which is irrelevant here), we can postpone the choice of instantiation terms to the level of axioms and select them on the basis of appropriate equations. In consequence, the rule dependencies expressed by the eigenparameter condition are replaced by term dependencies defined by unification problems. This gives systems with a positive answer to questions 1 and 2.

However, to say that γ-rules generate free variables is not an exhaustive description of the free variable system. We must also specify a mechanism for selecting free variables, and this has an impact on the other two questions. In one extreme, we may select a fresh variable for each γ-rule application. This strategy generates *variable-pure* skeletons and is illustrated by the leftmost skeleton (π_1) below. If, in that skeleton, the two variables u_1 and u_2 are distinct, the skeleton is variable-pure. Note that the skeleton can be extended to a proof without any new application of a $L\forall$ inference; the final skeleton can be closed by the substitution $\{u_1/a, u_2/b\}$. Also note that in each case the inference which introduces a variable occurs above the inference which introduces the Skolem function of the binding; this property corresponds to the fulfilment of the eigenparameter condition. In the rightmost skeleton (π_2) the order is reversed. Note that the free variable u is copied into the two branches, creating a dependency which is absent in π_1. In π_2 we must apply a $L\forall$ once more in one of the branches to close the skeleton (φx is $\exists y(Pxy \land Qx)$ and ξx is $\exists y Pxy$).

$$
\begin{array}{cc}
u_1 = a & u_2 = b \\
\vdots & \vdots
\end{array}
$$

$$
\cfrac{
\cfrac{
\cfrac{\forall x \varphi x, \varphi u_1 \vdash \xi a}{\forall x \varphi x \vdash \xi a}\gamma_{u_1}
}{\forall x \varphi x \vdash \forall x \xi x}\delta_a
\quad
\cfrac{
\cfrac{\forall x \varphi x, \varphi u_2 \vdash Qb}{\forall x \varphi x \vdash Qb}\gamma_{u_2}
}{\forall x \varphi x \vdash \forall x Q x}\delta_b
}{\forall x \varphi x \vdash \forall x \xi x \land \forall x Q x}\beta
$$

$$
u = a \qquad\qquad u = b
$$
$$
\vdots \qquad\qquad\qquad \vdots
$$

$$
\cfrac{
\cfrac{
\cfrac{\forall x \varphi x, \varphi u \vdash \xi a}{\forall x \varphi x, \varphi u \vdash \forall x \xi x}\delta_a
\quad
\cfrac{\forall x \varphi x, \varphi u \vdash Qb}{\forall x \varphi x, \varphi u \vdash \forall x Q x}\delta_b
}{\forall x \varphi x, \varphi u \vdash \forall x \xi x \land \forall x Q x}\beta
}{\forall x \varphi x \vdash \forall x \xi x \land \forall x Q x}\gamma_u
$$

Variable-pure skeletons correspond to free variable tableaux. As the example illustrates, the answer to question 3 is negative for these systems. The answer to question 4 is in general also negative, unless a rule ordering is selected which is guaranteed to fulfil the eigenparameter condition. However, question 1 then receives a negative answer. To remedy this situation a strategy for identifying universal variables was proposed in [3]. Applied to π_2, the strategy identifies the occurrences of u as universal in the branches; u can then be bound to different terms in different branches. However, even if this idea works in this particular example, it has limited range. It does e.g. not work for the sequent $\forall x(Px \rightarrow (Qx \land Rx)) \vdash \forall x(Px \rightarrow Qx) \land \forall x(Px \rightarrow Rx)$ given that the left implication inference occurs below the one for right conjunction.

If the variables u_1 and u_2 in π_1 are identical, the skeleton is *variable-sharing*. This class of skeletons was identified in [11], where it is shown that leaf sequents of skeletons which are balanced (defined below) correspond to paths through matrices [6]. As this strategy for selecting variables generates freely permuting skeletons, the answer to question 3 is positive. However, since the cost of the nice permutation properties is strong dependencies among variable occurrences, question 4 receives a negative answer.

Attempting to solve the redundancy problem for matrices Bibel sketched an idea for variable splitting [5]. We believe that the system introduced in this paper can be taken as a sharp formulation of his idea, fully generalized to non-clausal formulae and not restricted to balanced skeletons. As skeletons of our system are freely permuting, the answers to the first three questions are the same as for variable-sharing systems. And since we can fully simulate proofs constructed in a calculus with eigenparameters, question 4 receives the same answer as for this calculus. We hence combine the best of the three quantifier treatments discussed in this section and can respond 'yes' to all four questions.

2   The Free Variable System

The core of the object language consists of *basic formulae*, inductively defined from disjoint, countable sets of predicate symbols, function symbols and *quantification variables* by means of the logical connectives $\land$, $\lor$, $\rightarrow$, $\lnot$, $\forall$ and $\exists$. In addition to this the object language defines additional sets of *instantiation vari-*

ables of the form u_m^n and *Skolem functions* of the form f_m^n; $m, n \in \mathbb{N}$ (the role of the pair $\frac{n}{m}$ is explained below). The set of *instantiation terms* is inductively defined from the instantiation variables, Skolem functions and function symbols of the language. Note that an instantiation term does not contain quantification variables. The set of *formulae* is defined from the set of basic formulae by substitution: (1) a basic formula is a formula; (2) if a quantification variable x occurs free in a formula and t is an instantiation term, then the result of replacing every occurrence of x with t is a formula. Note that an instantiation variable is never bound by a quantifier.

In a *closed* formula all quantification variables are bound. Formulae with instantiation terms will be generated by the rules of the calculus and do not exist outside such a context: their purpose is to provide a syntax for free variables and run-time skolemization. The arity of each Skolem function will always be clear from the context.

Formula occurrences are assumed to be representations of underlying formula trees (cnf. e.g. [13] or [9]). Each node in a formula tree contains a *label* (a connective or an atomic formula) together with a unique *index pair* $\frac{n}{m}$; the subscript m is called an *occurrence number* and the superscript n is called a *copy number*. In a given formula tree all copy numbers must be identical and all occurrence numbers must be unique. Each node also has a *polarity* (L or R, denoting their side in sequents) and a *principal type* (α, β, γ, δ - or none for atomic formulae). A node will be referred to by its index pair. Formula trees with copy numbers greater than 1 will always be generated from formula trees with lower copy numbers. The *dominance* relation is defined over uniquely indexed sets of formula trees as the least transitive relation such that: (1) $\frac{n}{m}$ dominates $\frac{n}{k}$ if the $\frac{n}{m}$-node is above the $\frac{n}{k}$-node in its formula tree; (2) if $\frac{n+1}{m}$ is a γ-node, it dominates $\frac{n}{m}$.

Two different nodes in a formula tree are *β-related* if their greatest common descendant in the formula tree is of principal type β and they are not in the same branch of the formula tree. If $\frac{n}{m}$ is a β-node, then the index pairs of the two immediate ancestors are called *β-options* for $\frac{n}{m}$ and the two index pairs are *dual*. Let S be a set of index pairs such that each of them is a β-option for some β-type index pair and each of them is a dual to another index pair in S. A *β-path through S* is defined as a maximal subset of S such that no two index pairs in it are β-related.

Example 1. $(\exists x ((Px)_3^1 \wedge (Qx)_4^1)_2^1)_1^1$ represents the rightmost formula tree. There are four β-paths through $\{\frac{1}{3}, \frac{1}{4}, \frac{1}{7}, \frac{1}{8}\}$; one of them is $\{\frac{1}{3}, \frac{1}{8}\}$. A shorthand notation for the formula occurrence above is $\exists x (Px \wedge Qx)^1$.

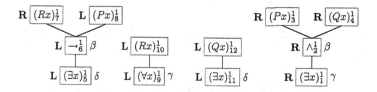

A set of β-options with no β-related index pairs is called a *splitting set*. A formula occurrence is *decorated* when it is labeled with a splitting set. By convention empty splitting sets are not displayed. As we shall see the β-rules of the system (which split branches of a skeleton) dynamically increase the splitting sets of their premises so that instantiation variables can be split accordingly later on. *Sequents* are ordered pairs of the form $\Gamma \vdash \Delta$, where Γ and Δ are sets of decorated, closed formula occurrences.

A *skeleton* is a finitely branching tree regulated by the rules of Fig. 1 with an endsequent in which all occurrence numbers are distinct, all copy numbers are identical to 1 and all splitting sets are empty. All skeletons addressed in this paper are finite (infinite skeletons naturally arise as limit objects generated by proof search operations).

Henceforth, we will use the term *formula* for the more pedantic 'decorated formula occurrence'. In Fig. 1 the formulae in Γ and Δ are called *extra formulae*, while the other formula in the conclusion is called the *principal formula* and the other formula(e) in the premiss(es) are called *active* formulae. Meta-language

α-rules

$$\frac{\Gamma, \varphi_k^n S, \psi_l^n S \vdash \Delta}{\Gamma, (\varphi_k^n \wedge \psi_l^n)_m^n S \vdash \Delta} \ \mathrm{L}\wedge$$

$$\frac{\Gamma \vdash \varphi_k^n S, \psi_l^n S, \Delta}{\Gamma \vdash (\varphi_k^n \vee \psi_l^n)_m^n S, \Delta} \ \mathrm{R}\vee$$

$$\frac{\Gamma, \varphi_k^n S \vdash \psi_l^n S, \Delta}{\Gamma \vdash (\varphi_k^n \rightarrow \psi_l^n)_m^n S, \Delta} \ \mathrm{R}\rightarrow$$

$$\frac{\Gamma, \varphi_k^n S \vdash \Delta}{\Gamma \vdash (\neg\varphi_k^n)_m^n S, \Delta} \ \mathrm{R}\neg$$

$$\frac{\Gamma \vdash \varphi_k^n S, \Delta}{\Gamma, (\neg\varphi_k^n)_m^n S \vdash \Delta} \ \mathrm{L}\neg$$

β-rules

$$\frac{\Gamma \uplus_k^n \vdash \varphi_k^n S, \Delta \uplus_k^n \quad \Gamma \uplus_l^n \vdash \psi_l^n S, \Delta \uplus_l^n}{\Gamma \vdash (\varphi_k^n \wedge \psi_l^n)_m^n S, \Delta} \ \mathrm{R}\wedge$$

$$\frac{\Gamma \uplus_k^n, \varphi_k^n S \vdash \Delta \uplus_k^n \quad \Gamma \uplus_l^n, \psi_l^n S \vdash \Delta \uplus_l^n}{\Gamma, (\varphi_k^n \vee \psi_l^n)_m^n S \vdash \Delta} \ \mathrm{L}\vee$$

$$\frac{\Gamma \uplus_k^n \vdash \varphi_k^n S, \Delta \uplus_k^n \quad \Gamma \uplus_l^n, \psi_l^n S, \vdash \Delta \uplus_l^n}{\Gamma, (\varphi_k^n \rightarrow \psi_l^n)_m^n S \vdash \Delta} \ \mathrm{L}\rightarrow$$

Weakening

$$\frac{\Gamma \vdash \Delta}{\Gamma, \varphi_m^n S \vdash \Delta} \ \mathrm{LW} \qquad \frac{\Gamma \vdash \Delta}{\Gamma \vdash \varphi_m^n S, \Delta} \ \mathrm{RW}$$

δ-rules

$$\frac{\Gamma \vdash \varphi_k^n[x/f_m^n u]S, \Delta}{\Gamma \vdash (\forall x \varphi_k^n)_m^n S, \Delta} \ \mathrm{R}\forall$$

$$\frac{\Gamma, \varphi_k^n[x/f_m^n u]S \vdash \Delta}{\Gamma, (\exists x \varphi_k^n)_m^n S \vdash \Delta} \ \mathrm{L}\exists$$

γ-rules

$$\frac{\Gamma, (\forall x \varphi_k^{n+1})_m^{n+1} S, \varphi_k^n[x/u_m^n]S \vdash \Delta}{\Gamma, (\forall x \varphi_k^n)_m^n S \vdash \Delta} \ \mathrm{L}\forall$$

$$\frac{\Gamma \vdash (\exists x \varphi_k^{n+1})_m^{n+1} S, \varphi_k^n[x/u_m^n]S, \Delta}{\Gamma \vdash (\exists x \varphi_k^n)_m^n S, \Delta} \ \mathrm{R}\exists$$

Fig. 1. The rules of the free variable system

conventions: In $\varphi_m^n S$, $_m^n$ denotes the index pair of φ and S the splitting set. Remarks about the rules:

β-*rules*: $\Gamma \uplus _k^l$ denotes $\{\varphi_m^n S \cup \{_k^l\} \mid \varphi_m^n S \in \Gamma\}$, the set of formulae in Γ where the index pair $_k^l$ has been added to the splitting sets. In a premiss of a β-rule, the index pair of the active formula is added to all splitting sets in the sequent except the splitting set of the active formula itself. It is immediate that the property of being a splitting set is preserved in this operation.

γ-*rules*: These rules introduce instantiation variables u_m^n, where $_m^n$ corresponds to the index pair of the principal formula. In addition, the γ-rules have built-in an implicit contraction operation so that they also introduce new copies of the principal formula. The new occurrence is obtained from the principal formula by incrementing the copy number.

δ-*rules*: These rules skolemize the principal formulae in the following way: If $(\forall x \varphi_k^n)_m^n S$ is the principal formula in which exactly the instantiation variables $\boldsymbol{u} = u_{m_1}^{n_1}, \ldots, u_{m_i}^{n_i}$ occur, then the Skolem term $f_m^n \boldsymbol{u}$ is introduced and substituted for the variable x. This δ-rule lies somewhere between a δ^+-rule [8] and a δ^{++}-rule [4]. It is δ^+-like in the sense that only variables in the current formula matter, not all variables in the conclusion, like the original δ-rule [7], or all *relevant* variables, like the δ^*-rule [2]. Moreover, all formulae with the same index pair will introduce identical Skolem functions, which is δ^{++}-like with respect to different branches. (A closer approximation to the δ^{++}-rule could be obtained by skipping the copy numbers of the Skolem functions altogether.) Convention: a_m^n denotes the Skolem function f_m^n in the case that this has arity 0, i.e. when it is a Skolem constant.

Weakening: The principal formula $\varphi_m^n S$ is called a *weakening formula*.

It is easy to see that a branch in a skeleton can be identified with a set B of index pairs; each β-inference which belongs to this branch has exactly one of its β-options in B. A splitting set in the branch B is thus always a subset of B. We shall simply refer to a branch by the set of index pairs which identifies it.

The implicit contraction in the γ-rules gives rise to a notational redundancy which should be avoided. Using the rules in an unconstrained way it is possible to generate a skeleton branch with a leaf of the form $\Gamma, (\forall x \varphi)_m^n S_1, (\forall x \varphi)_m^{n+1} S_2 \vdash \Delta$. If we apply L$\forall$ to this sequent with $(\forall x \varphi)_m^n S_1$ principal, this inference will generate a contraction formula $(\forall x \varphi)_m^{n+1} S_1$. We will thus have two occurrences of the formula $(\forall x \varphi)_m^{n+1}$ in the same sequent, only differing in their splitting sets. In this case we will use weakening to the occurrence with splitting set S_1. More generally, a skeleton is *normal* if the following conditions hold for all its inferences:

- if $\varphi_m^k S_1$ is principal, there is no extra formula of the form $\varphi_m^n S_2$ with $k > n$,
- if there is an active formula of the form $\varphi_m^n S_1$ and there is an extra formula of the form $\varphi_m^k S_2$ with $k \geq n$, then the active formula is a weakening formula,
- no occurrence is a weakening formula unless it satisfies the condition in the previous clause.

The first condition states that we must expand a formula with a lower copy number before a corresponding one with a higher copy number. In the second condition the extra formula $\varphi_m^k S_2$ has been generated by an implicit contraction in an inference further down in the branch. This occurrence is preferred over the active formula.

The following lemma holds for normal skeletons. Since this lemma is important, we will in the following assume that all skeletons are normal.

Lemma 1. *Let $\varphi_m^n S$ occur in the leaf sequent of a branch B of a normal skeleton. Then S is the set of all index pairs in B which are not dominated by $\frac{n}{m}$.*

The intuition behind the splitting sets is to identify instantiation variables by the splitting sets of the surrounding formulae. The uniformity of the splitting sets provides a machinery to split variables maximally without losing sight of logical dependencies between these. Splitting sets are assigned to instantiation terms and atomic formulae by the *color assignment* operator $\oplus$:

$$- (Pt_1,\ldots,t_k)\oplus S = P(t_1)\oplus S,\ldots,(t_k)\oplus S,$$
$$- (f_m^n t_1,\ldots,t_k)\oplus S = f_m^n(t_1)\oplus S,\ldots,(t_k)\oplus S,$$
$$- (u_m^n)\oplus S = u_m^n S.$$

The variable $u_m^n S$ is called a *colored variable*; $(t)\oplus S$ is a *colored term*, $(Pt)\oplus S$ is a *colored formula*. The splitting set S is in this context a *color*.

We shall see in subsequent examples that the splitting mechanism in the rules in some cases is too liberal and in other cases insufficient. To compensate for this we shall assign colors in a careful way and introduce a set of so-called secondary equations explicitly identifying colored variables which are syntactically different but logically identical. Key definitions follow.

A *connection* for a given skeleton branch B is an ordered pair of the form

$$(Ps_1,\ldots,s_k) \oplus (S \setminus T) \vdash (Pt_1,\ldots,t_k) \oplus (T \setminus S)$$

such that $(Ps_1,\ldots,s_k)_{m_1}^{n_1} S \vdash (Pt_1,\ldots,t_k)_{m_2}^{n_2} T$ is a subsequent of the leaf sequent of the branch. $S \setminus T$ consists of the index pairs in B dominated by $\frac{n_2}{m_2}$ and not dominated by $\frac{n_1}{m_1}$ (for intuitions see Ex. 4). The connection generates k *primary equations* of the form

$$(s_i)\oplus(S \setminus T) = (t_i)\oplus(T \setminus S).$$

The connection for B also generates a set of *auxiliary equations* in the following way. Let I be the set of all β-options in the skeleton (i.e. the union of all splitting sets which identify a branch in the skeleton). Let M be any β-path through I such that $B \subseteq M$. Note that M describes a potential extension of B. Let B' be the set of all index pairs in M not dominated by $\frac{n_1}{m_1}$ and not dominated by $\frac{n_2}{m_2}$ (the index pairs of the two connection formulae). If uU is a colored variable in a primary equation given by the connection for B, then

$$uU = u(U \cup B')$$

is an auxiliary equation for the given connection.

A set of connections C defines a set of primary equations as the collection of primary equations generated from connections in C. It also defines a set of *secondary* equations as the set of identities between colored variables in C which follow from the primary equations and the auxiliary equations for C, but do not follow from the primary equations alone. Intuitions about secondary equations can be found in the discussion of Ex. 5.

Let C be a set of connections. A *substitution for* C is a partial function σ which maps colored variables of C into the term universe of C (i.e. the terms generated by function symbols, Skolem functions and colored variables occurring in C). If ξ is a colored term or a colored formula, $\xi\sigma$ is the result of replacing every occurrence of each vS in both ξ and the domain of σ with $\sigma(vS)$. σ is *idempotent* if for all colored variables vS, $(vS)\sigma = ((vS)\sigma)\sigma$. It is a *solution* to an equation of colored terms $t_1 = t_2$ if $t_1\sigma = t_2\sigma$. It is *closing* for C if it is a solution to all the primary and secondary equations for C.

Let π be a skeleton. A set C of connection is *spanning* for π if there is exactly one connection in C for each branch in π. If C is spanning and σ is idempotent and closing for C, then $\langle \pi, C, \sigma \rangle$ is a *proof* of the endsequent of π.

Example 2. Assume that R$\wedge$ is the lowermost inference in a skeleton of which the sequent discussed in Section 1 is the endsequent. The two leaf sequents (without extra occurrences of γ-formulae) are displayed below. The atomic formulae which give rise to connections are underlined. No auxiliary equations are generated.

$$\underline{(Pu_1^1 f_2^1 u_1^1)_4^1\{_7^1\}}, (Qu_1^1)_5^1\{_7^1\} \vdash \underline{(Pa_7^1 u_8^1)_9^1} \quad \underline{(Pu_1^1 f_2^1 u_1^1)_4^1\{_{10}^1\}}, \underline{(Qu_1^1)_5^1\{_{10}^1\}} \vdash \underline{(Qa_{10}^1)_{11}^1}$$

$$\vdots \qquad\qquad\qquad\qquad \vdots$$

$$\forall x \exists y (Pxy \wedge Qx)_1^1\{_7^1\} \vdash (\forall x \exists y Pxy)_7^1 \quad \forall x \exists y (Pxy \wedge Qx)_1^1\{_{10}^1\} \vdash (\forall x Qx)_{10}^1$$

$$\underset{1\quad 2\qquad 4\quad 3\ 5}{\forall x \exists y (Pxy \wedge Qx)_1^1} \vdash \underset{7\ 8\quad 9\quad 6\ 10\ 11}{(\forall x \exists y Pxy \wedge \forall x Qx)_6^1}$$

Connections: $Pu_1^1\{_7^1\}f_2^1(u_1^1\{_7^1\}) \vdash Pa_7^1 u_8^1$ and $Qu_1^1\{_{10}^1\} \vdash Qa_{10}^1$. The substitution $\sigma = \{u_1^1\{_7^1\}/a_7^1, u_8^1/f_2^1(a_7^1), u_1^1\{_{10}^1\}/a_{10}^1\}$ is closing for the two connections and provides a proof.

For each inference in a skeleton, there is exactly one principal formula. We can thus relate inferences in a skeleton by means of how their principal formulae are related. Moreover, since every inference in a branch of a normal skeleton is uniquely determined by the index pair of its principal formula, we can use the notation $r[_m^n, B]$ to designate the inference in a branch given by the splitting set B. If there is no such inference in the branch, then $r[_m^n, B]$ is undefined. If the inference is of type γ we can display this in the notation by using the instantiation variable; $r[u_m^n, B]$ refers to the inference $r[_m^n, B]$ and also shows that it is of type γ. Similarly, $r[f_m^n, B]$ shows that the inference is of type δ. If a formula with index pair $_m^n$ is principal in only one branch B, or the branch B is clear from the context, we can omit B and write $r[_m^n]$, $r[u_m^n]$ or $r[f_m^n]$.

There are three important relations between inferences in a skeleton that we will consider. First, the dominance relation between formula occurrences gives rise to a dominance relation between inferences: $r[^n_m, B] \gg r[^l_k, B]$ holds if ^n_m dominates ^l_k. Second, a substitution σ gives rise to a *substitution relation* between inferences in the following way. If $\sigma(u^n_m S) = f^l_k t$, B denotes a branch with $S \subseteq B$, and both $r[u^n_m, B]$ and $r[f^l_k, B]$ are defined, then $r[u^n_m, B] \sqsupset r[f^l_k, B]$ holds *with respect to* S. Third, for all branches B and B' in a skeleton, $r[^n_m, B]$ is *contextually equivalent* to $r[^n_m, B']$, i.e. inferences whose principal formulae have the same index pair are contextually equivalent.

Skeletons can be represented at a higher level of abstraction by skeleton diagrams in which the diagram labels denote inferences and edges denote relations between them. The following diagram labels are used:

α-inference:	$\circ$	γ-inference:	u^n_m (or a meta-symbol)
β-inference:	$\triangle$	δ-inference:	$f^n_m u$ (or a meta-symbol)

We use v, w as metasymbols for instantiation variables and a, b, c for Skolem constants. There are three types of links between the labels, corresponding to the three relations introduced above. Arrows with solid lines display the dominance relation. Arrows with dashed lines display the substitution relation and are labeled with splitting sets; the dashed line is labeled with S if $r \sqsupset s$ holds with respect to S. A dotted line means that the two inferences are contextually equivalent.

Example 3.

The skeleton diagram represents one possible way of filling out the missing details of Ex. 2. The variable u^1_1 is colored in two ways: with $\{^1_7\}$ and $\{^1_{10}\}$. The colored variables $u^1_1\{^1_7\}$ and $u^1_1\{^1_{10}\}$ are assigned different values. Incidentally, all substitution arrows point downwards. This property of the diagram corresponds to the eigen-parameter condition (in Section 3 we shall call skeletons with this property *conforming*).

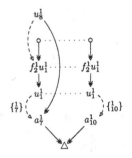

For the discussion of the following two examples we need some new concepts. To this end let $(Ps)^{n_1}_{m_1} S \vdash (Pt)^{n_2}_{m_2} T$ be the subsequent which gives rise to the connection $(Ps) \oplus (S \setminus T) \vdash (Pt) \oplus (T \setminus S)$. Let us say that it also gives rise to the *unpruned connection* $(Ps) \oplus S \vdash (Pt) \oplus T$ and that the real connection is a *pruning* of the unpruned one. Let us also say that a variable in the unpruned connection is *pruned* to a corresponding variable in the real connection (e.g. $u^n_m S$ is pruned to $u^n_m(S \setminus T)$).

In Ex. 2 the unpruned connections coincide with the real connections. The next example illustrates that this is not always the case and that it is in general incorrect to use unpruned connections as a basis for proofs.

Example 4. The sequent $\forall x(Px \lor Qx) \vdash \forall x Px, \forall x Qx$ is not valid. Let b,c abbrevi-
ate a_5^1, a_7^1, respectively. The leaf sequents obtained after expanding the formulae
in the sequent and one extra copy of the antecedent formula are given below
(the next extra copy is not displayed). The underlined formulae give rise to a
spanning set of connections; the corresponding primary equations are given next
to each leaf sequent. These primary equations cannot be solved simultaneously.
There are also four auxiliary equations, $u_1^1 = u_1^1\{^2_3\}$, $u_1^1 = u_1^1\{^2_4\}$, $u_1^2 = u_1^2\{^1_3\}$
and $u_1^2 = u_1^2\{^1_4\}$. No secondary equations result from these.

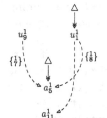

Leaf sequents:

$\underline{Pu_1^1\{^2_3\}}, Pu_1^2\{^1_3\} \vdash \underline{Pb\{^1_3{}^2_3\}}, Qc\{^1_3{}^2_3\}$

$Pu_1^1\{^2_4\}, \underline{Qu_1^2\{^1_3\}} \vdash Pb\{^1_4{}^2_3\}, \underline{Qc\{^1_3{}^2_4\}}$

$\underline{Qu_1^1\{^2_3\}}, Pu_1^2\{^1_4\} \vdash \underline{Pb\{^1_4{}^2_3\}}, Qc\{^1_4{}^2_3\}$

$Qu_1^1\{^2_4\}, \underline{Qu_1^2\{^1_4\}} \vdash Pb\{^1_4{}^2_4\}, \underline{Qc\{^1_4{}^2_4\}}$

Prim. eq.:

$u_1^1 = b$

$u_1^2 = c$

$u_1^2 = b$

$u_1^1 = c$

Observe that the corresponding unpruned connections can be closed by the
substitution $\{u_1^1\{^2_3\}/b, u_1^2\{^1_3\}/c, u_1^2\{^1_4\}/b, u_1^1\{^2_4\}/c\}$. It is this mapping which is
illustrated in the diagram above. The reason why the real connections cannot
be closed is that u_1^1 is a pruning of both $u_1^1\{^2_3\}$ and $u_1^1\{^2_4\}$. And the reason for
this is that the β-inference which causes the splitting of u_1^1 does not contribute
to any of the connections in which u_1^1 occurs.

Example 5. Secondary equations are in general necessary for consistency. Con-
sider a proof of the sequent: $\exists x(Rx \to Px), \forall x Rx, \exists x Qx \vdash \exists x(Px \land Qx)$.

Leaf sequents:

(1) $\underline{Ru_9^1\{^1_7\}}, Qa_{11}^1\{^1_7\} \vdash \exists x(Px \land Qx)\{^1_7\}, Ra_5^1$

(2) $\underline{Pa_5^1\{^1_3\}}, Qa_{11}^1\{^{11}_{38}\}, \forall x Rx\{^{11}_{38}\} \vdash \underline{Pu_1^1\{^1_8\}}$

(3) $Pa_5^1\{^1_4\}, Qa_{11}^1\{^{11}_{48}\}, \forall x Rx\{^{11}_{48}\} \vdash Qu_1^1\{^1_8\}$

Primary equations:

(1) $u_9^1\{^1_7\} = a_5^1$

(2) $a_5^1 = u_1^1\{^1_8\}$

(3) $a_{11}^1 = u_1^1$

Auxiliary equations:

(1a) $u_9^1\{^1_7\} = u_9^1\{^{11}_{37}\}$

(1b) $u_9^1\{^1_7\} = u_9^1\{^{11}_{47}\}$

(3a) $u_1^1 = u_1^1\{^1_8\}$

Secondary equations:

$u_1^1 = u_1^1\{^1_8\}$

Note that the formulae in the sequent correspond to the formula trees in Ex. 1.
The set of primary equations is solved by $\{u_9^1\{^1_7\}/a_5^1, u_1^1\{^1_8\}/a_5^1, u_1^1/a_{11}^1\}$, but there
is no substitution which *also* solves the set of secondary equations.

Also, note that $u_1^1\{^1_8\}$ is in connection (2) pruned to itself, but in connection (3) it is pruned to u_1^1. The function of the auxiliary equations is to reintroduce the logical identity between colored variables which are different prunings of the same unpruned variable.

3   Consistency

The system introduced in this paper is consistent and complete [12]. Both of these properties can be established by relating free variable proofs to proofs in a ground calculus with an eigenparameter condition (for which soundness and completeness is standard). Completeness is the easier direction; from a proof in the ground system it is straightforward to simultaneously construct a free variable skeleton and a closing substitution. The rest of this section gives an outline of the consistency argument. For any given proof in the free variable system the argument shows how one can transform it into a proof which is *cycle-free*, *conforming* and *projective* (all notions defined below). Given these properties, there is a proof of the endsequent in a ground calculus with the usual eigenparameter condition.

Take a proof $\langle \pi, C, \sigma \rangle$. The first step is to construct a *balanced* skeleton from π. A skeleton is balanced if for all contextually equivalent inferences r and s the following holds: if $r' \gg r$, there is an inference s' contextually equivalent to r' such that $s' \gg s$. Let π' be the skeleton obtained by balancing π.

Lemma 2. *The spanning set of connections C for π is spanning also for π'. Furthermore, the generated set of equations (primary, auxiliary, secondary) are identical. Thus, $\langle \pi', C, \sigma \rangle$ is a proof.*

Side comment. It follows from Lemma 2 that for balanced skeletons it is sufficient to define auxiliary equations of the form $uU = uV$, where uU is a pruning of the unpruned variable uV. The more complex definition in Section 2 accounts for unbalanced skeletons as well.

Next, we introduce a new relation $>$ between inferences: $r > s$ holds if there is an inference r' such that $r \gg r'$ and r' is contextually equivalent to s. The transitive closure of $> \cup \sqsupset$ defines the *reduction ordering* $\rhd$. A *cycle* is a finite sequence of inferences $r_1, \ldots, r_n$, for $n \geq 2$, such that $r_1 \rhd r_2, \ldots, r_n \rhd r_1$. We say that a proof *contains* the cycle $r_1, \ldots, r_n$; if a proof contains no cycle it is *cycle-free*. A proof is thus cycle-free if and only if $\rhd$ is irreflexive.

Due to the type of δ-rules used, proofs are in general not cycle-free (as they would have been if all instantiation variables in the conclusion had been arguments to the Skolem function). A simple example is the proof of $\vdash \exists x(Px \rightarrow \forall x Px)$ with only three rule applications. A more complex example is the following.

Example 6. Let φ_1^n be $(\forall x (\forall x Qx \rightarrow Px))^n$ and ψ_6^n be $\exists x(Qx \rightarrow \forall x Px)^n$. Below is a proof of $\varphi_1^1 \vdash \psi_6^1$.

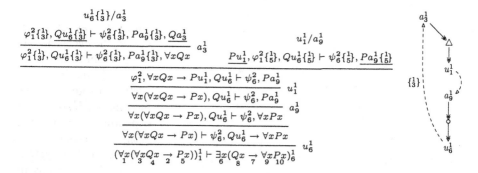

The set of primary equations is $\{u_6^1\{_3^1\} = a_3^1, u_1^1 = a_9^1\}$; there are no auxiliary equations. A closing substitution is given above the leaf sequents. The following is a cycle: $r[u_6^1] \sqsupset r[a_3^1]$, $r[a_3^1] > r[u_1^1]$, $r[u_1^1] \sqsupset r[a_9^1]$ and $r[a_9^1] > r[u_6^1]$.

Lemma 3. *A proof $\langle \pi', C, \sigma \rangle$ such that π is balanced can be extended to a proof which is cycle-free.*

The full details of this proof are fairly complex [12]. The idea is to systematically "break up" cycles by introducing fresh instantiation variables (from implicitly copied γ-formulae) and assigning substitution values to these instantiation variables such that cycles are eliminated one by one. By changing the underlying substitution in this way, the substitution relation, and consequently the reduction ordering, changes. A cycle provides enough information to pinpoint exactly the branches of the skeleton that should be extended and the γ-formulae in the leaf sequents that should be expanded to achieve this. The expanded γ-formulae always have a higher copy number than the γ-formulae principal in an inference in a cycle. All information about this is encapsulated in the diagrams.

Example 7. There are two ways of eliminating the cycle in Ex. 6. The first expands the rightmost leaf sequent, which is closed by the binding u_1^1/a_9^1. If this binding is removed, the cycle would be eliminated. In order to close the skeleton without this binding, and with the other bindings untouched, we can introduce a new instantiation variable u_1^2 and a new binding u_1^2/a_9^1. To get the variable u_1^2, we must expand φ_1^2 in the rightmost leaf sequent. The result of doing this gives a new skeleton in which the rightmost branch $\{_5^1 {}_5^2\}$ is closed by u_1^2/a_9^1, but the other new branch $\{_5^1 {}_3^2\}$ is not closed. One way to close this branch is to introduce the binding $u_6^1\{_3^1\}/a_3^1$, but this would give another cycle. Another way is to assign u_1^1 in that branch to the value a_9^1, but this is exactly the binding we want to avoid. In the same spirit as for the first expansion, we can expand ψ_6^2 in order to get a new variable u_6^2, which can be sent to a_3^2 by means of the binding $u_6^2\{_3^1\}/a_3^2$. Then a cycle-free proof is obtained. The leftmost diagram below is a representation of this cycle-free extension of the skeleton.

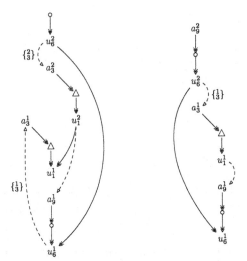

Another way of eliminating the cycle consists of removing the binding $u_6^1\{_3^1\}/a_3^1$ from the cycle, introducing the binding $u_6^2\{_3^1\}/a_3^1$ and expanding ψ_6^2 in the *leftmost* sequent of the original skeleton, i.e. the leaf sequent in which u_6^1 occurs. This cycle-free extension is represented by the rightmost diagram.

The next transformation step is a permutation operation. A *permutation variant* of a skeleton is a skeleton which differ only in the *order* of rule applications. Since the permutative operations are applied to balanced skeletons, it is sufficient to apply symmetrical permutation schemes [11]. By inspecting the patterns of these schemes, it is straightforward to verify the invariance property for balanced skeletons: the sets of leaf sequents are invariant under permutation. Hence any two permutation variants have identical leaf sequents.

Permutation operations are used to generate skeletons that are *conforming*. A skeleton *conforms to* the induced substitution ordering $\sqsupset$ if for all inferences r and s such that $r \sqsupset s$, r is above s in the skeleton. This property corresponds to the eigenparameter condition and depends only on the order of rule applications.

Lemma 4. *Every cycle-free proof has a conforming permutation variant.*

Proof. The proof is by induction on the sub-skeletons, using the $\sqsupset$-relation and the following fact. For any sub-skeleton with a non-atomic formula occurrence $\varphi_m^n S$ in the endsequent such that $\varphi_m^n S$ is expanded somewhere in the sub-skeleton, there is a permutation variant of the sub-skeleton which has $\varphi_m^n S$ as principal in the lowermost inference (see Lemma 2.14 in [11]). $\qquad\square$

The last part of the consistency argument deals with a particular feature of the new system. When an instantiation variable is assigned two different colors and the resulting colored variables are assigned to different terms, a direct translation into a ground calculus is blocked. We say that the induced substitution ordering $\sqsupset$ is *projective* if $r \sqsupset s_1$ and $r \sqsupset s_2$ implies that $s_1 = s_2$. If $\sqsupset$ is projective, we say that the proof is projective.

Lemma 5. *Every cycle-free and conforming proof has a projective extension.*

Like for cycle elimination, we can introduce fresh instantiation variables (from implicitly copied γ-formulae) and assign values to these instantiation variables in order to construct a projective proof. By repeatedly adding γ-inferences to the skeleton, removing the bindings from the substitution which makes the induced substitution ordering non-projective and introducing new bindings (keeping the substitution closing), it is possible to eliminate all "non-projective" parts of a closing substitution. Technically it is the primary and secondary equations which make this elimination go through. More precisely the argument rests on the following property. If $r[uS_1, B_1]$ and $r[uS_2, B_2]$ denote the same inference and $\sigma(uS_1) \neq \sigma(uS_2)$, then we can extend B_1 to B_1' such that B_1' has an inference $r[uS_1', B_1']$ and such that a substitution σ' is closing; σ' is undefined for uS_1, maps uS_1' to $\sigma(uS_1)$ and otherwise agrees with σ.

Example 8. Let a,b,c be abbreviations for appropriate Skolem functions. The extra copy of the γ-formula is not displayed in the skeleton. A proof of the sequent is given below.

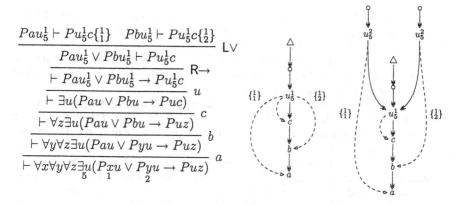

$$
\begin{array}{c}
\dfrac{Pau_5^1 \vdash Pu_5^1 c\{^1_1\} \quad Pbu_5^1 \vdash Pu_5^1 c\{^1_2\}}{Pau_5^1 \vee Pbu_5^1 \vdash Pu_5^1 c} \text{LV} \\[4pt]
\dfrac{}{\vdash Pau_5^1 \vee Pbu_5^1 \to Pu_5^1 c} \text{R}{\to} \\[4pt]
\dfrac{}{\vdash \exists u(Pau \vee Pbu \to Puc)} u \\[4pt]
\dfrac{}{\vdash \forall z \exists u(Pau \vee Pbu \to Puz)} c \\[4pt]
\dfrac{}{\vdash \forall y \forall z \exists u(Pau \vee Pyu \to Puz)} b \\[4pt]
\dfrac{}{\vdash \forall x \forall y \forall z \exists u(\underset{5}{P}xu \vee \underset{1}{P}yu \to \underset{2}{P}uz)} a
\end{array}
$$

The primary equations are $a = u_5^1\{^1_1\}, b = u_5^1\{^1_2\}, u_5^1 = c$. There are no secondary equations. A closing substitution is $\{u_5^1\{^1_1\}/a, u_5^1\{^1_2\}/b, u_5^1/c\}$. The substitution ordering is not projective since $r[u_5^1] \sqsupset r[a]$ and $r[u_5^1] \sqsupset r[b]$, and $r[a] \neq r[b]$. Observe that u_5^1 plays the role of *both* a rigid variable (u_5^1 *without* splitting set occurs in both branches) and a universal variable (u_5^1 occurs with *different* splitting sets in both branches). This skeleton can be made projective by expanding the γ-formula $\exists u(Pau \vee Pbu \to Puc)$ in both branches; thus removing both of the bindings $u_5^1\{^1_1\}/a$ and $u_5^1\{^1_2\}/b$. The diagrams suggest an interpretation of the operation in terms of detachment. Two of the arrows out from u_5^1 are detached from the node and attached to newly created nodes in the diagram to the right. It is not necessary to expand the β-subformula above these two new nodes, because a closing substitution is reached already after applying R$\to$.

4 Conclusion

Compared to standard sequent calculi or tableau systems with an eigenvariable condition our system is at least as good wrt. both proof length and the size of the search space, in addition to allowing full flexibility in the order of rule application. There is every reason to believe that the technique can be extended to freely permuting free variable systems for intuitionistic and modal logics surveyed in [11]. There are interesting questions that are not yet addressed, including complexity analyses and investigations of the cut rule.

Acknowledgements

We would like to thank two anonymous referees for very useful comments.

References

[1] R. Antonsen. Free variable sequent calculi. Master's thesis, University of Oslo, Language, Logic and Information, Department of Linguistics, May 2003.

[2] M. Baaz and C. G. Fermüller. Non-elementary speedups between different versions of tableaux. In R. H. Peter Baumgartner and J. Possega, editors, *4th Workshop on Theorem Proving with Analytic Tableaux and Related Methods*, volume 918 of *LNAI*, pages 217–230. Springer, 1995.

[3] B. Beckert and R. Hähnle. An improved method for adding equality to free variable semantic tableaux. In D. Kapur, editor, *Proceedings, 11th International Conference on Automated Deduction (CADE), Saratoga Springs, NY*, LNCS 607, pages 507–521. Springer, 1992.

[4] B. Beckert, R. Hähnle, and P. H. Schmitt. The *even more* liberalized δ-rule in free variable semantic tableaux. In G. Gottlob, A. Leitsch, and D. Mundici, editors, *Proceedings of the third Kurt Gödel Colloquium KGC'93, Brno, Czech Republic*, volume 713 of *LNCS*, pages 108–119. Springer-Verlag, Aug. 1993.

[5] W. Bibel. Computationally improved versions of Herbrand's theorem. In *Logic Colloquium '81*, pages 11–28. North-Holland, 1982.

[6] W. Bibel. *Automated Theorem Proving 2. Edition*. Vieweg Verlag, 1987.

[7] M. C. Fitting. *First-Order Logic and Automated Theorem Proving*. Graduate Texts in Computer Science. Springer-Verlag, Berlin, 2nd edition, 1996. 1st ed., 1990.

[8] R. Hähnle and P. H. Schmitt. The liberalized δ-rule in free variable semantic tableaux. *Journal of Automated Reasoning*, 13(2):211–222, Oct. 1994.

[9] C. Kreitz and J. Otten. Connection-based theorem proving in classical and non-classical logics. *Journal of Universal Computer Science*, 5(3):88–112, 1999.

[10] R. M. Smullyan. *First-Order Logic*, volume 43 of *Ergebnisse der Mathematik und ihrer Grenzgebiete*. Springer-Verlag, New York, 1968.

[11] A. Waaler. Connections in nonclassical logics. In A. Robinson and A. Voronkov, editors, *Handbook of Automated Reasoning*, volume II, chapter 22, pages 1487–1578. Elsevier Science, 2001.

[12] A. Waaler and R. Antonsen. A free variable sequent calculus with uniform variable splitting. Research Report 305, Department of Informatics, University of Oslo, June 2003.

[13] L. A. Wallen. *Automated deduction in nonclassical logics*. MIT Press, 1990.

The Tableaux Work Bench

Pietro Abate and Rajeev Goré

Australian National University, Canberra, Australia
[Pietro.Abate|Rajeev.Gore]@anu.edu.au
http://csl.anu.edu.au/~abate/twb

Abstract. The Tableaux Work Bench (TWB) is a meta tableau system designed for logicians with limited programming or automatic reasoning knowledge to experiment with new tableau calculi and new decision procedures. It has a simple interface, a history mechanism for controlling loops or pruning the search space, and modal simplification.

1 Introduction

Theorem provers for classical propositional modal logics have matured dramatically to the point where formulae with hundreds of symbols can be tested within a few seconds. Direct theorem provers like FaCT [12] and LWB [8] utilise many different optimisation techniques to speed up proof search in particular logics, while translational provers like MSPASS [13] utilise fast first-order theorem provers like SPASS. But these avenues are not always viable: FaCT cannot handle logics with an intuitionistic base; although the LWB can handle intuitionistic logic, it can handle only a fixed collection of logics; and although MSPASS gives a sound and complete prover for any first-order definable logic, *a priori*, it gives a decision procedure only for the ones that fall into decidable fragments of first-order logic like the two-variable fragment, or the guarded fragment. Indeed, MSPASS has many flags, and it is not at all obvious how to obtain a decision procedure for a particular first-order definable logic using MSPASS.

While generic tableau-based provers like `Blast_tac` [17] provide facilities for experimenting with new tableau calculi, as far as we are aware, the only system which allows a user to experiment with different optimisation techniques, different proof-search strategies and different tableau calculi together is `lotrec` [4], which we discuss in Section 6.

Existing proof editors like xpe [15], JAPE [2] and PESCA [16] provide only rudimentary proof search facilities (e.g. iterative deepening for PESCA), while `blobLogic` [11] contains a fixed collection of tableau rules.

The Tableaux Work Bench (TWB) is a generic meta-tableau system designed for expressing and combining new tableau rules into an underlying tableau proof (and disproof) engine. It provides a simple user-interface and facilities to incorporate or design optimisations and specific decision procedures. By dividing high level and low level optimisations, the user can concentrate on algorithmic aspects related to his or her tableau calculus, leaving to the developer more complex and generic performance issues about the underlying prover. The TWB

M. Cialdea Mayer and F. Pirri (Eds.): TABLEAUX 2003, LNAI 2796, pp. 230–236, 2003.

includes a generic history mechanism, facilities to perform simplification, to handle global assumptions and to implement more complex optimisation procedures. Currently, the TWB does not offer proof-editing capabilities. The TWB cannot possibly compete with FaCT, MSPASS or LWB in speed, but its versatility should be of use to the TABLEAUX community since there is no restriction to modal calculi, although currently, they must be propositional.

2   Generic Tableau Algorithm

The rules that can be specified with the TWB are those which respect the analytic super-formula property [7], allow history mechanisms [9, 10] and permit global assumptions [5]. At the moment it is not possible to specify calculi that require two pass tableau algorithms (a lá PLTL), infinitary calculi, or explicit geometric relational properties like weak-directedness. However using the history mechanism and clever calculi design, it is possible to handle most well-known modal calculi.

The TWB is based on a purely syntactic tableau algorithm. The user can specify the proof-search strategy governing the order of rule applications. Each step in the depth-first search corresponds with a rule application where a rule is selected if its pattern matches the formulae in the current node. Once a rule is selected and executed, the algorithm recursively continues the proof tree exploration until a closed tableau is found or it is not possible to apply any rules (in this case the branch is open). The only axiom embedded in the system is $\frac{X;A;\neg A}{\bot}$. However this axiom can be "turned off" if it is not necessary in the calculus (eg.: para-consistent logic).

3   User Interface

The TWB is designed to be easy to use, flexible and extensible. Negated Normal Form (NNF) is available but not mandatory, and the user can program his or her own rewriting system for normal forming; see the manual [1].

Connective Definitions The TWB has a number of hard-wired symbols to express connectives with one, two or three arguments. To add a new symbol to the language, at the moment, it is necessary to edit the source code of the lexer and symbols' parser, but in future releases, new connectives will be defined via the user interface itself.

Rule Definitions The first step toward the specification of a new calculus is to define a set of rules. A rule is (up to) a six-tuple `rule(pat, act, heuristic, branching, invertibility, name)` where: `pat` is a pattern for the numerator; `act` is a pattern for the denominator(s); `heuristic` is an ordering function that affects the principal formula selection strategy; `branching` is either `All` or `Exist` to indicate whether all or only one denominator must close;

invertibility is either Static or Trans to indicate whether the rule is invertible and name is a string in quotes to describe this rule. Heuristic, branching and invertibility are optional and their defaults are respectively: the null function (meaning no ordering), All and Static. The various components can be specified using a let statement inherited from OCaml in which an internal rule name can be specified for the defined rule. For example, the classical propositional tableau ($\vee$) rule is:

```
let or_r =
    let p =  pat { "{A v B}"; "X" }
    and a~=  act { "A" ; "X" | "B" ; "X" }
in rule (p, a, All, Static, "Or");;
```

$$\frac{A \vee B; X}{A\ ; X | B; X}$$

The numerator consists of a principal formula enclosed in braces and a list of sets. The denominator consists of a list of branches, defined by lists of sets. We prefer this terminology rather than "premiss" and "conclusion" since these latter terms can cause confusion when using sequent calculi. The rule pattern is matched against the formulae in the current node and the rule action is executed only if the pattern is satisfied. The heuristic function can be specified in each rule and affects the selection of the principal formula of that rule. It is basically a comparison function: for example, to select the formula with higher modal weight the heuristic function could be defined by the user as:

```
let weight f1 f2 =
    let rec w = function
        |term "A & B" |term "A v B" -> w (term "A") + w (term "B")
        |term "dia A" |term "box A" -> 1 + w (term "A")
        |term "~ A" -> w (term "A")
        |term "atom a" -> 0
        | _ -> failwith "error in weight"
in
    if (w f1) = (w f2) then 0
    else if (w f1) > (w f2) then 1
    else -1
```

The "or" rule definition above will consequently be modified to include the heuristic definition as rule (p, a, weight, All, Static, "Or"). It's also possible to define a set of additional side conditions that must be fulfilled in order to fire a rule. For example a basic modus ponens rule on atomic formulae for the contraction-free calculus for intuitionistic logic can be coded as below where member(x,Y) is a user-defined function that checks if x is in the set Y.

```
let mp =
    let p = pat { "{ atom p -> B }"; "X" ; member("atom p","X") }
    and a~= act { "atom p"; "B" ; "X"}
in rule (p, a, All, Static, "MP");;
```

Non-invertible rules invariably introduce non-determinism into a calculus and lead to choice-points in the search procedure which are explored by backtracking. In the TWB, we have chosen to specify such choice points explicitly via the "Existential" branching construct. For example the traditional (K)-rule is:

```
let k_r =
   let p =  pat { "{dia A}" ; "box X" ; "dia Y"; "Z" }
   and a~=
      matchset "dia Y" with
      []  ->  act { "A" ; "X" }
      | _ ->  act { "A" ; "X" | "box X" ; "dia Y" ; "Z"}
in rule (p, a, Exist, Trans,"K")
```

$$\frac{\Diamond A \;;\Box X;\Diamond Y;Z}{A;X \quad || \quad \Box X;\Diamond Y;Z}$$

where matchset is used to define two actions: one when the set $\Diamond Y$ is empty and the other when it is not. The Exist appellation declares to the prover that only one of the two denominators need close in the second case.

In the example above, the box X part matches *all* $\Box$-formulae in the current node, hence Z contains neither $\Box$- nor $\Diamond$-formulae.

Systematic Proof Search and Strategy Each set of rules is attached to a strategy defining the proof search procedure for the associated calculus. A strategy is defined in terms of two types of cycles (lists) containing rule names. A *-cycle executes the rules in list-order until none of them are applicable. A +-cycle executes the first applicable rule in its list only. The strategy stops if a closed tableau is detected or if no rule is applicable.

For example, the following strategy definition specifies the usual systematic procedure for modal logic **K** where the And and Or rules are executed until they are not applicable (saturation step), and the (K)-rule above is executed once (transitional step): let str = strategy [[and_r ; or_r]* ; k_r];;

History Mechanisms The TWB also has a history facility for efficient loop-checking as part of the rule definition. For example, the traditional rule for handling reflexivity in the logic **KT** requires an implicit contraction on the principal formula to make it invertible. But this rule can then be applied *ad infinitum*, and so a starring mechanism is usually employed to stop this behaviour as shown below left. Alternatively, explicitly specifying side-conditions and actions to be executed on a history Z suffices, as shown below at right where "–" is a separator:

$$\frac{\Box A \;;X}{A;(\Box A)*,X} \;\Box A \text{ not starred} \qquad\qquad \frac{\Box A \;;X - \Box A \notin Z}{A;\Box A \;;X - \Box A \cup Z}$$

This can be coded in the TWB using the construct – H{...} to express histories:

```
let t_r =
   let p  = pat { "{box A}"; "X" - H { isnotin("Z","box A") } } }
   and a~= act { "A"; "box A"; "X" - H { add("Z","box A") } }
in rule(p, a, All, Static, "T Rule History");;
```

The system also allows a light version of a history, called starring, where a formula is simply starred to avoid considering it more than once, rather than making an explicit copy of it into a history.

User-Defined Functions and Simplification Side conditions can be added to the pattern by specifying predefined functions like `isnotin` above or user-defined functions which accept a formula and return true or false. Every time a pattern is matched against a formula set, every such specified function is evaluated on every formula, and the rule is executed only if all conditions are satisfied. A user defined simplification procedure [14] is used as shown below:

$$\frac{A \vee B; X}{A; \ \ X[A:=\top] \ \ | \ \ B; \ \ X[B:=\top]}$$

```
let or_s_r =
 let p = pat { "{A v B}"; "X" }
 and a~= act { "A" ; "X"[ "A":= top ] | "B"; "X" [ "B":= top ] }
in rule (p, a, All, Static, "Or Simpl");;
```

Global assumptions can also be used in the TWB, and can be defined statically or dynamically: in the first case they must be specified on the command line and they are present for every input formula. In the latter, the user can specify a function that accepts an input formula and returns a set of global assumptions to be used against that formula in the proof.

4   Experiments and Performance

The flexibility of the TWB has been tested by implementing the traditional history-based calculi for the logics **K**, **KT**, and **S4** [9] in a modular fashion. We have tested the TWB using the LWB benchmarks: the TWB could solve only the first several formulae in the respective test formulae in under 50 seconds each. These times are an order of magnitude slower than the LWB and are hardly state-of-the-art. There are two basic reasons for such a difference. First, the LWB embeds many known optimisations and heuristics while the TWB currently allows only modal simplification. Second, the LWB uses more efficient data structures while the TWB currently uses naive lists.

Conversely, the TWB easily allows us to extend the calculus for **S4** into a calculus for **S4.3** and then into (the non-first-order-definable) **S4.3.1** while such an extension in the LWB is probably only possible for its authors.

5   Implementation

The TWB is implemented in OCaml [3], a strongly typed object oriented programming language available for many architectures. The TWB can be compiled either in native form, in byte code, or run via an OCaml shell, giving total flexibility. The rules and strategy definitions effectively become part of the system itself as they are compiled in byte code and dynamically linked to the prover engine. The basic data structures of the system are lists and hash tables leaving room for future performance improvements, but the modularity of the system allows easy customisation of the internal design.

The user interface inherits its syntax from OCaml and is in fact only syntactic sugar added to the real language itself. This adds flexibility and generality because it is always possible to write more complex rule definitions without using the user interface, but with the programming language itself. In this sense the prover can be seen just as a library for tableau oriented theorem proving.

6   Related Work

We now compare the TWB with related work in more detail.

Direct provers like LWB [8] and FaCT [12] are clearly superior if they can handle the sought-after logic, and the LWB, in particular, can handle intuitionistic logic and a long list of particular modal logics. But programming a new calculus into the LWB or FaCT is difficult for anyone except their authors.

Translational provers like MSPASS [13] are equally superior if the logic is first-order definable and falls into a decidable fragment like the two-variable fragment or the guarded fragment. SPASS can even handle "second-order" logics like **G** and **Grz** by using a non-standard translation into first-order logic that mimics the traditional tableau rules for these logics. But, a priori, MSPASS does not provide a decision procedure for a given decidable first-order-definable modal logic. The SCAN algorithm [6] for second-order quantifier elimination can often find first-order equivalents for many second-order relational conditions. But once again, this does not, a priori lead to a decision procedure unless the first-order equivalents fall into a decidable subset of first-order logic.

Blast_tac provides fairly basic facilities for designing new rules, and even allows certain rules to be marked as "undoable" (non-invertible), but it does not allow history mechanisms or further optimisation techniques like simplification. To be fair, Blast_tac deliberately trades completeness for versatility since it is designed to be used in an interactive setting like Isabelle and "completeness is hardly relevant to interactive proofs" [17].

The TWB is closest to lotrec [4] in that both are generic systems that allow a user to specify new rules and strategies for experimenting with proof-search. The main differences between them are the underlying execution models. lotrec works at a *global* level, keeping track of all tableau nodes, and the accessibility relations among them in an *explicit* manner. For example, the weak-directedness frame-conditions for the logic **S4.2** ($\forall x, y, z. \exists w. xRy \& xRz \Rightarrow xRw \& yRw$) can be coded explicitly in lotrec by referring explicitly to R in the rules. The TWB works on a local level and keeps only the information relevant to the current node, so a condition like weak-directedness must be captured implicitly using a particular form of cut on super-formulae; see [7].

Whereas the TWB makes histories explicit, these must be simulated in lotrec by the user using the various node and edge marking techniques provided by lotrec. Since lotrec uses labels to mark the nodes it creates, it can handle the difference operator, which the TWB cannot handle. Overall, lotrec is biased towards semantics while the TWB is biased towards proof theory. Which you use is probably best determined by the logic in question.

7   Conclusion and Further Work

The TWB allows users with little or no technical background in automatic reasoning to encode their own tableau prover in a simple yet flexible manner. The code and user manual is available at http://csl.anu.edu.au/~abate/twb.

The TWB is still a prototype and we envision much further work: we want to provide a sequent calculus front-end; provide facilities for calculi with "stoups"; provide facilities for hyper-sequents; allow nodes to contain multisets or lists rather than sets; allow rules which partition the side-formulae into two disjoint sets as needed in linear logic; and improve the speed of the underlying implementation.

References

[1] P Abate. Tableaux work bench (twb) – User Manual,
     http://arp.anu.edu.au/~abate/twb 2003.
[2] R Bornat and B Sufrin. Jape: A calculator for animating proof-on-paper. In
     W McCune (Ed), *CADE'97*, LNCS 1249:412–415, Springer.
[3] E. Chailloux, P. Manoury, and B. Pagano. *Dèveloppement d'applications avec
     Objective Caml*. O'Reilly, 2000.
[4] L Fariñas del Cerro, D Fauthoux, O Gasquet, A Herzig, D Longin, and F Massacci.
     Lotrec: the generic tableau prover for modal and description logics. In *IJCAR'01*,
     LNAI 2083:453-458. Springer Verlag, 2001.
[5] M. Fitting. *Proof Methods for Modal and Intuitionistic Logics*, volume 169 of
     Synthese Library. D. Reidel, Dordrecht, Holland, 1983.
[6] D Gabbay and H J Ohlbach. Quantifier elimination in second order predicate
     logic. In *Proc. KR-92*, 1992.
[7] R Goré. Chapter 6: Tableau methods for modal and temporal logics. In *Handbook
     of Tableau Methods*, pages 297–396. Kluwer, 1999.
[8] A Heuerding. LWBtheory: information about some propositional logics via the
     WWW. *Logic Journal of the IGPL*, 4(4):169–174, 1996.
[9] A Heuerding. *Sequent Calculi for Proof Search in some Modal Logics*. PhD thesis,
     Institute for Applied Mathematics and Computer Science, University of Berne,
     Switzerland, 1998.
[10] J.M. Howe. *Proof Search Issues in Some Non-Classical Logics*. PhD thesis,
     University of St Andrews, December 1998.
[11] C Howitt. bloblogic. http://users.ox.ac.uk/~univ0675/blob/, 2002.
[12] I Horrocks and P F Patel-Schneider. Optimising propositional modal satisfiability
     for description logic subsumption. In LNCS 1476, 1998.
[13] U. Hustadt and R.A. Schmidt. MSPASS: Modal reasoning by translation and
     first-order resolution. In TABLEAUX 2000, LNCS 1847:67–71, Springer, 2000.
[14] F Massacci. Simplification: a general constraint propagation technique for propositional and modal tableaux. In Proc. TABLEAUX 98, LNCS 1397:217-231.
     Springer, 1998.
[15] M Mouri. Theorem provers with counter-models and xpe. *Bulletin of the Section
     of Logic*, 30(2):79–86, 2001.
[16] S Negri and J von Plato *Structural Proof Theory*. CUP, 2001.
[17] L.C. Paulson. A generic tableau prover and its integration with Isabelle. *Journal
     of Universal Computer Science*, 5(3), 1999.

Decision Procedures for the Propositional Cases of Second Order Logic and Z Modal Logic Representations of a First Order L-Predicate Nonmonotonic Logic

Frank M. Brown

University of Kansas
Lawrence, Kansas 66045
brown@ku.edu

Abstract. Decision procedures for the propositional cases of two different logical representations for an L-Predicate Logic generalizing Autoepistemic Logic to handle quantified variables over modal scopes are described. The first representation is Second Order Logic. The second is Z Modal Logic which extends its S5 modal laws with laws stating what is logically possible. It is suggested that certain problems are more easily solved using one representation whereas other problems are more easily solved using the other.

1 Introduction

One interesting L-Predicate Logic is Rigid-worlds. Rigid-worlds of a theory Γ of First Order Logic (i.e. FOL) is the "infinite disjunction" of the worlds w which entail Γ and which for each sentence α_i with free variables ξ_i whose name occurs as an argument to an occurrence of the predicate L in Γ, for all ξ_i, w entails $L'\alpha_i$ if and only if $\Gamma \to \alpha_i$ holds in every world which gives L the same interpretation as did w. A world is a possible proposition that, for every other proposition, entails it or its negation. Entailment, written $[w]p$, is necessary implication. In Z Priorian Modal Second Order Logic [4] this is written as:

$$(Rigid\text{-}worlds\ \Gamma) =_{df} \exists w(w \land (world\ w) \land ([w]\Gamma) \land$$
$$\bigwedge_{i=1,n} \forall \xi_i(([w]L'\alpha_i) \leftrightarrow \forall u(((world\ u) \land$$
$$\bigwedge_{i=1,n} \forall \xi_i(([w]L'\alpha_i) \leftrightarrow ([u]L'\alpha_i))) \to ([u](\Gamma \to \alpha_i)))))$$

where the $'\alpha_i$ are arguments of L in Γ and L has a second unwritten argument binding the names of free variables in α_i to those variables. Rigid-worlds is interesting partly because it generalizes (propositional) Autoepistemic Logic [7] to a First Order Autoepistemic Logic such that quantifiers obey all the normal laws of FOL, the Barcan formula, and its converse, unlike [6]. Another reason, is that it is representable both as a sentence of Second Order Logic (i.e. SOL) and as a sentence of Z Modal Logic [3]. Herein, we examine the problem of representing Rigid-worlds in these two logics and deducing consequences in the propositional

M. Cialdea Mayer and F. Pirri (Eds.): TABLEAUX 2003, LNAI 2796, pp. 237–245, 2003.
© Springer-Verlag Berlin Heidelberg 2003

subcases. Section 2 describes how Rigid-world problems are represented in SOL. Section 3 describes their representation in Z. Section 4 compares automatic theorem provers for the propositional subcase of these logics on some propositional problems. Some conclusions are drawn in Section 5.

2 SOL Representation of Rigid-Worlds

Rigid-worlds of a theory Γ of FOL is equivalent to a sentence of SOL as given in theorem SOL1 below:

$$\text{SOL1: } (\textit{Rigid-worlds } \Gamma) \equiv (\Gamma \wedge \bigwedge_{i=1,n} \forall \xi_i ((L'\alpha_i)$$
$$\leftrightarrow \forall P_1...P_m ((\Gamma \rightarrow \alpha_i)\{\pi_j/P_j\}_{j=1,m})))$$

where $\pi_1...\pi_m$ are all the unmodalized predicates other than L in Γ and where the $'\alpha_i$ are arguments to occurrences of the L predicate in Γ. $\{\pi_j/P_j\}_{j=1,m}$ is the substitution for each j of P_j for the predicate π_j in the preceding sentence. However, fixed predicates can be handled with an asymptotically simpler subformula, which may change the asymptotic size of the overall formula as given in theorem SOL2 below:

$$\text{SOL2: } (\textit{Rigid-worlds } \Gamma \wedge \bigwedge_{i=1,f} \forall \xi_i ((L'(\rho_i\xi_i)) \leftrightarrow (\rho_i\xi_i)))$$
$$\equiv (\Gamma \wedge \bigwedge_{i=1,f} \forall \xi_i ((L'(\rho_i\xi_i)) \leftrightarrow (\rho_i\xi_i)) \wedge$$
$$\bigwedge_{i=1,n} \forall \xi_i ((L'\alpha_i) \leftrightarrow \forall P_1...P_m ((\Gamma \rightarrow \alpha_i)\{\pi_j/P_j\}_{j=1,m})))$$

where $\pi_1...\pi_m$ are all the predicates in Γ other than the ρ_i predicates and where the $'\alpha_i$ are arguments to occurrences of the L predicate in Γ. Restricting Γ and each α_i in the above formulas to have predicate symbols of only zero arity and eliminating any FOL object quantifiers gives propositional instances of the above theorems. The propositional instance of theorem SOL1 is:

$$\text{SOL1*: } (\textit{Rigid-worlds } \Gamma) \equiv$$
$$\Gamma \wedge \bigwedge_{i=1,n} ((L'\alpha_i) \leftrightarrow \forall P_1...P_m ((\Gamma \rightarrow \alpha_i)\{\pi_j/P_j\}_{j=1,m})))$$

which (with propositional constants replaced by unbound propositional variables) is equivalent to the formula previously given in [5] which was shown therein to represent Autoepistemic Logic. The propositional instance of theorem SOL2 is:

$$\text{SOL2*: } (\textit{Rigid-worlds}(\Gamma \wedge \bigwedge_{i=1,f} ((L'\rho_i) \leftrightarrow \rho_i)))$$
$$\equiv (\Gamma \wedge \bigwedge_{i=1,f} ((L'\rho_i) \leftrightarrow \rho_i) \wedge$$
$$\bigwedge_{i=1,n} ((L'\alpha_i) \leftrightarrow \forall P_1...P_m ((\Gamma \rightarrow \alpha_i)\{\pi_j/P_j\}_{j=1,m})))$$

A Propositional Logic example and an analogous FOL example of the SOL approach to deduction in Rigid-worlds are given below:

Propositional Example: $(\textit{Rigid-worlds}((L'p) \rightarrow p))$
By SOL1* we get: $((L'p) \rightarrow p) \wedge ((L'p) \leftrightarrow \forall P((((L'p) \rightarrow p) \rightarrow p)\{p/P\}))$
Applying the substitution gives: $((L'p) \rightarrow p) \wedge ((L'p) \leftrightarrow \forall P(((L'p) \rightarrow P) \rightarrow P))$

Which is equivalent to: $((L'p) \rightarrow p) \wedge ((L'p) \leftrightarrow \forall P(((L'p) \rightarrow \#f) \rightarrow P))$
Which is equivalent to: $((L'p) \rightarrow p) \wedge ((L'p) \leftrightarrow \forall P((\neg(L'p)) \rightarrow P))$
Pushing P to lowest scope gives: $((L'p) \rightarrow p) \wedge ((L'p) \leftrightarrow ((\neg(L'p)) \rightarrow \forall P \, P))$
which is equivalent to: $((L'p) \rightarrow p) \wedge ((L'p) \leftrightarrow ((\neg(L'p)) \rightarrow \#f))$
which is equivalent to: $((L'p) \rightarrow p) \wedge ((L'p) \leftrightarrow (L'p))$
which is equivalent to: $((L'p) \rightarrow p)$

FOL Example: $(Rigid\text{-}worlds(\forall x((L'(p\,x)) \rightarrow (p\,x))))$
where L has a second unwritten argument which is an association list binding
$'x$ to x.
By SOL1 or SOL2 we get:

$$(\forall x((L'(p\,x)) \rightarrow (p\,x)))$$
$$\wedge \, \forall x((L'(p\,x)) \leftrightarrow \forall P(((\forall x((L'(p\,x)) \rightarrow (p\,x))) \rightarrow (p\,x))\{p/P\}))$$

Applying the substitution gives:
$(\forall x((L'(p\,x)) \rightarrow (p\,x))) \wedge \forall x((L'(p\,x)) \leftrightarrow \forall P((\forall x((L'(p\,x)) \rightarrow (P\,x))) \rightarrow (P\,x)))$
The lemma given below shows that, the above sentence is just:
$\forall x((L'(p\,x)) \rightarrow (p\,x))$.

Lemma: $\forall x((L'(p\,x)) \leftrightarrow \forall P((\forall x((L'(p\,x)) \rightarrow (P\,x))) \rightarrow (P\,x)))$

Proof: It suffices to prove: $(L'(p\,x)) \leftrightarrow \forall P((\forall x((L'(p\,x)) \rightarrow (P\,x))) \rightarrow (P\,x))$.
We make explicit the unwritten second argument to the L predicate:
$(L'(p\,x)(('x.x))) \leftrightarrow \forall P((\forall x((L'(p\,x)(('x.x))) \rightarrow (P\,x))) \rightarrow (P\,x))$
We change the bound variable x to y:
$(L'(p\,x)(('x.x))) \leftrightarrow \forall P((\forall y((L'(p\,x)(('x.y))) \rightarrow (P\,y))) \rightarrow (P\,x))$
The proof divides into two parts:

1. $(L'(p\,x)(('x.x))) \rightarrow \forall P((\forall y((L'(p\,x)(('x.y))) \rightarrow (P\,y))) \rightarrow (P\,x))$
   It suffices to prove:
   $(L'(p\,x)(('x.x))) \rightarrow ((\forall y((L'(p\,x)(('x.y))) \rightarrow (P\,y))) \rightarrow (P\,x))$
   which holds by forward chaining.
2. $\forall P((\forall y((L'(p\,x)(('x.y))) \rightarrow (P\,y))) \rightarrow (P\,x)) \rightarrow (L'(p\,x)(('x.x)))$
   Letting P be $\lambda z(z \neq x)$, where x is the free x in the above sentence, gives
   $((\forall y((L'(p\,x)(('x.y))) \rightarrow ((\lambda z(z \neq x))y)))$
   $\quad \rightarrow ((\lambda z(z \neq x))x)) \rightarrow (L'(p\,x)(('x.x)))$
   By lambda conversion this is:
   $((\forall y((L'(p\,x)(('x.y))) \rightarrow (y \neq x))) \rightarrow (x \neq x)) \rightarrow (L'(p\,x)(('x.x)))$
   which is equivalent to:
   $((\forall y((y = x) \rightarrow \neg(L'(p\,x)(('x.y))))) \rightarrow \#f) \rightarrow (L'(p\,x)(('x.x)))$
   which is equivalent to: $((\neg(L'(p\,x)(('x.x)))) \rightarrow \#f) \rightarrow (L'(p\,x)(('x.x)))$
   which is a tautology.

[5] discusses the propositional example:

$$(Rigid\text{-}worlds(((\neg(L'\alpha)) \rightarrow \pi) \wedge \bigwedge_{i=1,f} ((L'\rho_i) \leftrightarrow \rho_i))) \rightarrow \pi$$

where α is a randomly produced expression of clausal form Propositional Logic constructed from the fixed zero arity predicates ρ_i and non-fixed zero arity predicates π_i. If the size of α is $\Theta(c)$ then the SOL1* representation:

$$(\Gamma \wedge ((L'\alpha) \leftrightarrow \forall P_1...P_m((\Gamma \to \alpha)\{\pi_j/Pj\}_{j=1,m}))\wedge$$
$$\bigwedge_{i=1,f}((L'\rho_i) \leftrightarrow \forall P_1...P_m((\Gamma \to \rho_i)\{\pi_j/P_j\}_{j=1,m}))) \to \pi$$

where Γ is: $(((\neg(L'\alpha)) \to \pi) \wedge \bigwedge_{i=1,f}((L'\rho_i) \leftrightarrow \rho_i))$
has size $\Theta(c + mf + f^2)$, but the SOL2* representation:

$$(\Gamma \wedge (\bigwedge_{i=1,f} ((L'\rho_i) \leftrightarrow \rho_i)) \wedge ((L'\alpha) \leftrightarrow \forall P_1...P_m((\Gamma \to \alpha)\{\pi_j/P_j\}_{j=1,m}))) \to \pi$$

where Γ is: $((\neg(L'\alpha)) \to \pi)$ has size $\Theta(c+m+f)$. This suggests that the SOL2* representation may be useful in proving theorems with many fixed predicates. This class of problems can be further simplified by renaming the P quantifiers and pushing them to lowest scope, giving:

$$((\neg(L'\alpha)) \to \pi) \wedge \bigwedge_{i=1,f}((L'\rho_i) \leftrightarrow \rho_i)\wedge$$
$$((L'\alpha) \leftrightarrow (((\neg(L'\alpha)) \to \exists P\,P) \to \forall P_1...P_m\alpha\{\pi_j/P_j\}_{j=1,m}))) \to \pi$$

which is equivalent to:

$$(((\neg(L'\alpha)) \to \pi)\wedge$$
$$\bigwedge_{i=1,f}((L'\rho_i) \leftrightarrow \rho_i) \wedge ((L'\alpha) \leftrightarrow \forall P_1...P_m\alpha\{\pi_j/P_j\}_{j=1,m})) \to \pi$$

which is equivalent to:

$$(((\neg(\forall P_1...P_m\alpha\{\pi_j/P_j\}_{j=1,m})) \to \pi)\wedge$$
$$\bigwedge_{i=1,f}((L'\rho_i) \leftrightarrow \rho_i) \wedge ((L'\alpha) \leftrightarrow \forall P_1...P_m\alpha\{\pi_j/P_j\}_{j=1,m})) \to \pi.$$

Since L does not occur in α, the equivalences defining L constitute a conservative extension, and therefore may be eliminated giving:

$$((\neg(\forall P_1...P_m\alpha\{\pi_j/P_j\}_{j=1,m})) \to \pi) \to \pi$$

which is equivalent to: $(\forall P_1...P_m\alpha\{\pi_j/P_j\}_{j=1,m}) \to \pi$
which is of size $\Theta(c + m)$. To prove or refute such a theorem we try to prove: $((\forall P_1...P_m\alpha\{\pi_j/P_j\}_{j=1,m}) \to \pi)$ using a tableaux sequent calculus for propositional logic with propositional quantifiers.

3   Z Modal Logic Representation of Rigid-Worlds

Rigid-worlds of a theory Γ of FOL is equivalent to a sentence of Z Modal Logic [2,3] as given in theorem ML1 below:

ML1: $(Rigid\text{-}worlds\ \Gamma) \equiv \exists k(k \wedge (k \equiv (\Gamma \wedge \bigwedge_{i=1,n} \forall \xi_i((L'\alpha_i) \leftrightarrow ([k]\alpha_i)))))$
where ξ_i is the sequence of free variables in α_i. Z includes FOL, propositional

quantifiers, S5 Modal Logic, and some axioms stating what is logically possible. However, a more interesting representation, also expressed in Z, is obtained by eliminating the L predicate as follows: First, using $\bigwedge_{i=1,n} \forall \xi_i((L'\alpha_i) \leftrightarrow ([k]\alpha_i))$, every occurrence of L' in Γ is replaced by $[k]$. As names of sentences $'\alpha_i$ are unquoted giving α_i, new occurrences of L may appear. Since these new occurrences of L appear under the scope of $[k]$ as do any occurrences of L in the α_i sentences and since the original necessary equivalence implies $[k] \bigwedge_{i=1,n} \forall \xi_i((L'\alpha_i) \leftrightarrow ([k]\alpha_i))$ those L' may also be replaced by $[k]$ ad infinitum. When all these replacements are made we derive the equivalent expression: $\exists k(k \wedge (k \equiv (\Gamma\{L'/[k]\} \wedge \bigwedge_{i=1,n} \forall \xi_i((L'\alpha_i) \leftrightarrow ([k]\alpha_i\{L'/[k]\})))))$ where $\Gamma\{L'/[k]\}$ represents the replacement of all L' by $[k]$. Since L does not occur in $\Gamma\{L'/[k]\}$ nor $\alpha_i\{L'/[k]\}$, $(\Gamma\{L'/[k]\} \wedge \bigwedge_{i=1,n} \forall \xi_i((L'\alpha_i) \leftrightarrow ([k]\alpha_i\{L'/[k]\})))$ is a conservative extension of $\Gamma\{L'/[k]\}$ and $\bigwedge_{i=1,n} \forall \xi_i((L'\alpha_i) \leftrightarrow ([k]\alpha_i\{L'/[k]\}))$ may thus be pulled out. The resulting representation, called the Kernel Representation, is:

ML2: *(Rigid-worlds Γ)* $\equiv$
$$\exists k(k \wedge (\bigwedge_{i=1,n} \forall \xi_i((L'\alpha_i) \leftrightarrow ([k]\alpha_i\{L'/[k]\}))) \wedge (k \equiv (\Gamma\{L'/[k]\})))$$

Restricting Γ and each α_i to have predicate symbols of only zero arity and eliminating any FOL object quantifiers gives propositional instances of the above theorems:

ML1*: *(Rigid-worlds Γ)* $\equiv \exists k(k \wedge (k \equiv (\Gamma \wedge \bigwedge_{i=1,n}((L'\alpha_i) \leftrightarrow ([k]\alpha_i)))))$

ML2*: *(Rigid-worlds Γ)* $\equiv$
$$\exists k(k \wedge (\bigwedge_{i=1,n}((L'\alpha_i) \leftrightarrow ([k]\alpha_i\{L'/[k]\}))) \wedge (k \equiv (\Gamma\{L'/[k]\})))$$

In this case the fixed-point solutions can be deduced by the following algorithm:

Procedure for Solving Modal Equivalences [1]:

Step 1: Each maximal subformula α which contains k and is equivalent to $([\,]\alpha)$ is pulled out of the equation causing it to be split into two cases using the following theorem schema to replace any instance of the left side by the corresponding instance of the right side: $(k \equiv (\phi\ \alpha)) \leftrightarrow ((\alpha \wedge (k \equiv (\phi\ \#t))) \vee ((\neg\alpha) \wedge (k \equiv (\phi\ \#f))))$.

Step 2: The resulting equivalence is simplified by the laws of Propositional Logic.

Step 3: On each disjunct the simplified value for k is back substituted into each such α or $(\neg\alpha)$ sentence thereby eliminating k from them.

Step 4: The α and $(\neg\alpha)$ sentences are then simplified using the laws of the Z Modal Logic. In the propositional case, a decision procedure for Propositional Logic such as a Tableaux Sequent Calculus may be used instead.

A Tableaux Sequent Calculus for Propositional Logic may then be used to deduce consequences from the disjunction of all the solutions. A Propositional Logic example and an analogous FOL example are given below:

Propositional Example: ($Rigid$-$worlds$(($L'p$) $\rightarrow$ p)).
By ML1* this is equivalent to: $\exists k(k \wedge (k \equiv (((L'p) \rightarrow p) \wedge ((L'p) \leftrightarrow ([k]p)))))$
By ML2* this is equivalent to: $\exists k(k \wedge ((L'p) \leftrightarrow ([k]p)) \wedge (k \equiv (([k]p) \rightarrow p)))$
Using the Procedure for Solving Modal Equivalences the necessary equivalence representing the kernel $k \equiv (([k]p) \rightarrow p)$ is solved as follows:
Step 1: $(([k]p) \wedge (k \equiv (\#t \rightarrow p))) \vee ((\neg([k]p)) \wedge (k \equiv (\#f \rightarrow p)))$
Step 2: $(([k]p) \wedge (k \equiv p)) \vee ((\neg([k]p)) \wedge (k \equiv \#t))$
Step 3: $(([p]p) \wedge (k \equiv p)) \vee ((\neg([\#t]p)) \wedge (k \equiv \#t))$
Step 4: $(k \equiv p) \vee (k \equiv \#t)$
Plugging the solutions to the kernel necessary equivalence into the rest gives:
$\exists k(k \wedge ((L'p) \leftrightarrow ([k]p)) \wedge ((k \equiv p) \vee (k \equiv \#t)))$
which distributes to:
$\exists k(k \wedge ((L'p) \leftrightarrow ([k]p)) \wedge (k \equiv p)) \vee \exists k(k \wedge ((L'p) \leftrightarrow ([k]p)) \wedge (k \equiv \#t))$
which is equivalent to: $(p \wedge ((L'p) \leftrightarrow ([p]p))) \vee (\#t \wedge ((L'p) \leftrightarrow ([\#t]p)))$
which is equivalent to: $(p \wedge ((L'p) \leftrightarrow \#t)) \vee (\#t \wedge ((L'p) \leftrightarrow \#f))$
which simplifies to be: $(p \wedge (L'p)) \vee (\neg(L'p))$
which is just: $(L'p) \rightarrow p$

FOL Example: ($Rigid$-$worlds$($\forall x$(($L'(p\ x)$) $\rightarrow$ ($p\ x$))))
where L has a second unwritten argument which is an association list binding
$'x$ to x.
By ML1 this is:
$\exists k(k \wedge (k \equiv \forall x((L'(p\ x)) \rightarrow (p\ x)) \wedge \forall x((L'(p\ x)) \leftrightarrow ([k](p\ x)))))$
By ML2 this is:
$\exists k(k \wedge (\forall x((L'(p\ x)) \leftrightarrow ([k]((p\ x)\{L'/[k]\}))))$
$\qquad \wedge (k \equiv (\forall x((L'(p\ x)) \rightarrow (p\ x))\{L'/[k]\})))$
which is: $\exists k(k \wedge (\forall x((L'(p\ x)) \leftrightarrow ([k](p\ x)))) \wedge (k \equiv \forall x(([k](p\ x)) \rightarrow (p\ x))))$
By the following Lemma we solve the kernel necessary equivalence getting:
$\exists k(k \wedge (\forall x((L'(p\ x)) \leftrightarrow ([k](p\ x)))) \wedge \exists S(k \equiv \forall x(([\ ](S\ x)) \rightarrow (p\ x))))$
Plugging in the kernel solutions gives:
$\exists S((\forall x(([\ ](S\ x)) \rightarrow (p\ x))) \wedge \forall x((L'(p\ x)) \leftrightarrow ([\forall x(([\ ](S\ x)) \rightarrow (p\ x))](p\ x))))$
which is equivalent to: $\exists S((\forall x(([\ ](S\ x)) \rightarrow (p\ x))) \wedge \forall x((L'(p\ x)) \leftrightarrow ([\ ](S\ x))))$
which is equivalent to: $(\forall x((L'(p\ x)) \rightarrow (p\ x))) \wedge \exists S \forall x((L'(p\ x)) \leftrightarrow ([\ ](S\ x)))$
Since $[\ ](\exists S \forall x((L'(p\ x)) \leftrightarrow ([\ ](S\ x))))$ is true we get just: $\forall x((L'(p\ x)) \rightarrow (p\ x))$.

Lemma: $(k \equiv \forall x(([k](p\ x)) \rightarrow (p\ x))) \leftrightarrow \exists S(k \equiv \forall x(([\ ](S\ x)) \rightarrow (p\ x)))$

Proof: The proof divides into two parts:

1. $(k \equiv \forall x(([k](p\ x)) \rightarrow (p\ x))) \rightarrow \exists S(k \equiv \forall x(([\ ](S\ x)) \rightarrow (p\ x)))$
   Letting S be $\lambda x(k \rightarrow (p\ x))$ gives:
   $(k \equiv \forall x(([k](p\ x)) \rightarrow (p\ x))) \rightarrow (k \equiv \forall x(([k](p\ x)) \rightarrow (p\ x)))$ which holds.

2. $\exists S(k \equiv \forall x(([\ ](S\ x)) \rightarrow (p\ x))) \rightarrow (k \equiv \forall x(([k](p\ x)) \rightarrow (p\ x)))$

Using the hypothesis to replace k in the conclusion it suffices to prove:
$(\forall x(([\ ](S\ x)) \rightarrow (p\ x))) \equiv \forall x(([\forall x(([\ ](S\ x)) \rightarrow (p\ x))](p\ x)) \rightarrow (p\ x))$ which
is true.

[5] discusses the propositional example:

$$(Rigid\text{-}worlds(((\neg(L'\alpha)) \rightarrow \pi) \wedge \bigwedge_{i=1,f} ((L'\rho_i) \leftrightarrow \rho_i))) \rightarrow \pi$$

where α is a randomly produced expression constructed from the fixed zero
arity predicates ρ_i and non fixed zero arity predicates π_i. By ML2* the modal
representation is:

$\exists k(k \wedge ((L'\alpha) \leftrightarrow ([k]\alpha)) \wedge (\bigwedge_{i=1,f}((L'\rho_i) \leftrightarrow ([k]\rho_i))) \wedge (k \equiv (((\neg([k]\alpha)) \rightarrow$
$\pi) \wedge \bigwedge_{i=1,f}(([k]\rho_i) \leftrightarrow \rho_i))))$

which is of size $\Theta(c+f)$ if α is of size $\Theta(c)$. Ignoring the conservative extention
L, a Tableaux Sequent Calculus for Propositional Logic is then used to try to
prove π from the disjunction of all the solutions to the kernel.

4   Results and Comparison of SOL and Z Modal Logic ATPs

To test the effects of the SOL and the Z Modal Logic representations of Rigid-
worlds we applied our automatic theorem provers for solving propositional problems in these representations to prove two classes of test theorems discussed in
[5]. These theorems were of the form:

$$(Rigid\text{-}worlds(((\neg(L'\alpha)) \rightarrow \pi) \wedge \bigwedge_{i=1,f} ((L'\rho_i) \leftrightarrow \rho_i))) \rightarrow \pi$$

where α is a sentence in clausal form not containing the L predicate. Each clause
has 3 randomly chosen literals. Two tests were made. Test 1 involved a constant
number of clauses and a linear increasing number f of fixed predicates. Test 2
involved the case where $f = 4$ and an exponentially (i.e. a power of 2) increasing
number of clauses. The results of Test 1 and Test 2 are given in Table 1 and
Table 2 respectively.

Table 1

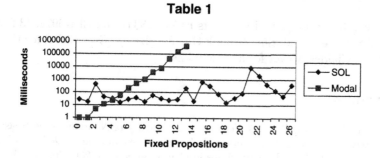

Table 1 shows that the execution time of the Modal ATP appears to increase exponentially as the number of fixed predicates increases; whereas the SOL ATP appears to increase at an asymptotically slower exponential rate. Conversely, Table 2 shows that the execution time of the SOL ATP appears to increase linearly as the number of clauses increases; whereas the Modal ATP appears to take constant time.

Table 2

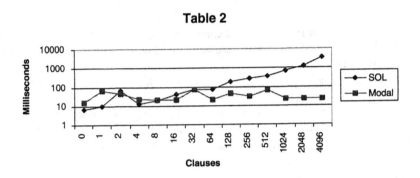

5   Conclusion

The existence of three different representations (i.e. Rigid-worlds, SOL and Z Modal Logic) for a nonmonotonic system allows different automatic theorem proving approaches to be developed. The results given herein suggest that different representations are useful for solving different classes of problems and, therefore, that a better automatic deduction system may be constructed by using multiple representations. One such approach would be to determine the class of a problem and then use the representation that was best for that class. Alternatively, the apparent asymptotic differences in the deductive behavior of different representations suggests that one could also attempt to solve a problem (whose class was unknown) with multiple representations at the same time without wasting much effort.

Acknowledgements

This research was supported by grants EIA-9818341 and EIA-9972843 from the National Science Foundation. I thank Guy Jacobs and Greg Siebel for their help and Marta Cialdea Mayer for additional help.

References

[1] Brown, Frank M., 1986 "A Commonsense Theory of Nonmonotonic Reasoning", *Proceedings of the 8th International Conference on Automated Deduction'*, Oxford England, July 1986, Lecture Notes in Computer Science 230, Springer-Verlag.

[2] Brown, Frank M. 1987. "The Modal Logic Z", The Frame Problem in AI"; *Proc. of the 1987 AAAI Workshop, Morgan Kaufmann*, Los Altos, CA.

[3] Brown, Frank M. 1989. "The Modal Quantificational Logic Z Applied to the Frame Problem", *advanced paper First International Workshop on Human & Machine Cognition*, May 1989 Pensacola, Florida. Abbreviated version published in *International Journal of Expert Systems Research and Applications, Special Issue: The Frame Problem. Part A.* eds. Keneth Ford and Patrick Hayes, vol. 3 number 3, pp169-206, JAI Press 1990. Reprinted in *Reasoning Agents in a Dynamic World: The Frame problem*, editors: Kenneth M. Ford, Patrick J. Hayes, JAI Press 1991.

[4] Brown, Frank M., 2003. "Solving Modal Equivalences", 2003, *2003 IEEE International Symposium on Intelligent Control*, Omni press, Oct 5-8 2003.

[5] Eiter, Thomas & Klotz, Volker & Tompits, Hans and Woltran, Stefan, 2002, "Modal Nonmonotonic Logics Revisited: Efficient Encodings for the Basic Reasoning Tasks", *Automated Reasoning with Analytic Tableaux and Related Method: Tableaux 2002*, LNAI 2381, Springer Verlag.

[6] Konolige, Kurt 1989. "On the Relation between Autoepistemic Logic and Circumscription Preliminary Report", *IJCAI89*.

[7] Moore, R. C. 1985. "Semantical Considerations on Nonmonotonic Logic", *Artificial Intelligence, 25.*

Logistica 2.0: A Technology for Implementing Automatic Deduction Systems

Frank M. Brown

University of Kansas
Lawrence, Kansas 66045
brown@ku.edu

Abstract. The Logistica 2.0 Deduction System Implementation Technology is a programming language extension to R5RS Scheme which automatically computes all possible combinations of values of multiply valued subexpressions. Multiple values are generated by multiple definitions of a symbol and by allowing Second Order patterns such as segment variables, which may match in different ways, as the parameters of lambda abstractions. This technology is briefly illustrated with an extensible deduction system involving the derivation of an axiom schema.

1 Introduction

Logistica 2.0 is a system for implementing automatic deduction systems and for embedding them into larger Artificial Intelligence reasoning applications. The basic idea of Logistica is to supplement the language facilities of R5RS Scheme [14], a particularly elegant dialect of Lisp [17] with additional capabilities including allowing symbols to be symbolic, variables to be multiply defined, and complex patterns such as segment variables to occur in the formal parameter list of lambda abstractions. A simple example illustrating some elementary capabilities of Logistica is given in section 2. Some conclusions are drawn in section 3.

2 A Simple Example

A fundamental problem in building Automatic Deduction Systems (and in building those applications such as automatic program verification systems [2,11], automatic complexity analysis systems [7], automatic design verification systems [10], and automatic design change systems [8] which depend on underlying automatic theorem proving technology) is how to allow them to increase their capabilities by incorporating and using previously proven rules of inference in proving subsequent theorems [5,6]. Although at first glance, one might think that this problem is solved by simply proving a lemma from the primitive axioms of the formal theory and then using it as a hypothesis in proving the theorem, this simplistic description ignores the fact that usually it is not actually a lemma which is used to prove the theorem, but instead it is an inference rule expressed in some metalanguage of that formal theory. Such inference rules are important

M. Cialdea Mayer and F. Pirri (Eds.): TABLEAUX 2003, LNAI 2796, pp. 246–251, 2003.

because they package up into the rule a significant amount of control information for the deductive process. For example, in trying to prove that p+q=r+p+r we would like to cancel the p's leaving the subgoal q=r+r to prove later. If we try to make this inference using the cancellation lemma of addition, namely: x+y=x+z iff y=z. Letting x:=p and y:=q we unfortunately find that r+p+r cannot in general be an instance of x+z. Thus this cancellation lemma cannot be used in a direct manner to simplify the above equation. However, we could achieve this simplification if this cancellation lemma of addition were replaced by a general cancellation rule of inference expressed in the English metalanguage as:

> "A sum containing a number equals another sum containing the same number if and only if the first sum with one occurrence of that number removed is equal to the second sum with one occurrence of that number removed."

In this case, letting the term in this inference rule be p, the equation p+q=r+p+r would then be seen to be equivalent to q=r+r since it is that equation with the two occurrences of p deleted. Logistica can represent this inference rule in visually appealing manner as a single line of code:

```
(define (=(+ _a x _b)(+ _c x _d))
        (=(+ _a _b)(+ _c _d)))
```

When =, +, p, q, and r are symbolic (i.e. defined to return themselves if no other rule is applicable) the application of this definition results in the appropriate cancellation:

$$(=(+\ p\ q)(+\ r\ p\ r))\ \Longrightarrow\ (=(+\ q)(+\ r\ r))$$

This Logistica program can be automatically proven to be a derived rule of inference of Number Theory [9] by simply defining some of the Peano Axioms of Number Theory and Varyadic structures with more primitive basic laws of recursion and mathematical induction as follows: First, varyadic functions, such as the N-ary plus written as: +, are defined in terms of binary functions, such as binary plus which is written as: ++, as follows:

```
(define (+ x _L) (++ x(+ _L)))
(define (+) 0)
(define (make-ind(^p L))
        (and (^p '())
             (->(^p L)(^p(cons(gensym)L)))))
```

The first definition, which will be called +list, says that + of one or more arguments is binary sum (i.e. ++) of the first argument with the sum of the remaining arguments. The second law, which is called +() says that the sum of no arguments is zero. The third law is the induction law for lists. It is used to prove general statements about segment variables. In this context it states that a property is true of zero or more arguments if and only if it is true of no arguments and if it is true of n arguments then it is true of n+1 arguments. For example, this law transforms: (foo _L) into:

```
(and(foo _'())
     (->(foo _L)(foo _(cons g1 L))))
```

which splices in as `(and(foo)(->(foo _L)(foo g1 _L))))`. Second, we define some axioms dealing with binary plus (i.e. ++) which are easily proven from the Peano axioms of number theory [Skolem]. The following Logistica definitions define two such axioms.

```
(define (++ x(++(!test y(<< y x)) z))
        (++ y(++ x z)))
(define (=(++ x y)(++ x z)) (= y z))
```

The first definition is the commutativity of binary plus. This definition re-orders terms across a level of ++ expressions where << is an alphabetic lexicographic well ordering on such terms. !test succeeds in matching if its first argument matches and its second argument is true. The second law says that an two binary sums with identical first arguments are equal if and only if their second arguments are equal.

Using these definitions, some elementary laws of equality and logic and supporting definitions such as lexicographic ordering the cancellation rule of inference can be proven as by evaluating a sequence of such rules of inference. As each such rule of inference in the sequence is proven, it is incorporated as new Logistica definition by orienting them, translating them to Logistica, and defining them. Orienting rules of inference is required since we must be able to apply them unidirectionally. This is done by choosing the left and right sides of each rule such that the left side of the rule is greater than the right side with respect to some well-founded complexity ordering. Here is an example call to a Logistica program which does all this:

```
(justifyAndCompile
     (=(+ _a x _b)(+ x _a _b))
     (iff(=(+ _a x _b)(+ _c x _d))
         (=(+ _a _b)(+ _c _d))) )
```

The result of this call is to prove the rule of inference in the input list in the order given and to add each such rule of inference as a new Logistica rule immediately after it is proven:

Metatheorem: VaryadicCommutativity of $+$: $+$ comm

Proof: $(= (+ \_a\ x\ \_b)\ (+\ x\ \_a\ \_b))$:$+$ list
$(= (+ \_a\ x\ \_b)\ (++\ x\ (+\ \_a\ \_b)))$:symmetry of $=$
$(= (++\ x\ (+\ \_a\ \_b))\ (+\ \_a\ x\ \_b))$:induction 1.1, 1.2

1.1 (= (++ x (+ _b)) (+ x _b))   :+ list
(= (++ x (+ _b)) (++ x (+ _b)))   :reflexitivity of =
#t
1.2 (->(=(++ x (+ _a _b)) (+ _a x _b))
    (= (++ x (+ w _a _b)) (+ w _a x _b)))   :+ list (twice)
(-> (= (++ x (+ _a _b)) (+ _a x _b))
    (= (++ x (++ w (+ _a _b))) (++ w (+ _a x _b))))   :++AC
(->(= (++ x (+ _a _b)) (+ _a x _b))
    (= (++ w (++ x (+ _a _b))) (++ w (+ _a x _b))))   := substitution
(= (++ w (+ _a x _b)) (++ w (+ _a x _b))))   :reflexitivity of =
#t

This rule is not orientable since the 2 sides are permutations of each other;
therefore, we compile extra code into it to apply the rule only if the left side is
greater than the right under our ordering.

```
(define(+ _a x _b)
    (if(and(not(equal?(+_a x _b)(+ x _a _b)))
       (<L (+ x _a _b)(+ _a x _b)))
       (cut!(+ x _a _b))
       #fail))
```

Metatheorem: +VaryadicCancellation for +:

Proof: (<-> (=(+ _a x _b)(+ _c x _d)) (=(+ _a _b)(+ _c _d)))
                                    :+Comm (twice)
(<-> (=(+ x _a _b)(+ x _c _d))(=(+ _a _b)(+ _c _d)))
                                    :+list (twice)
(<-> (=(++ x (+ _a _b))(++x (+_c _d)))(=(+ _a _b)(+ _c _d)))
                                    :+cancel
(<-> (=(+ _a _b)(+ _c _d))(=(+ _a _b)(+ _c _d)))
                                    :reflexitivity of =
#t

This inference rule is orientable according to our ordering and is thus added
to this environment:

```
(define (=(+ _a x _b)(+ _c x _d))
    (=(+ _a _b)(+ _c _d)))
```

The simplicity of this representation of inference rules and their justification
may be contrasted with metatheoretic approaches, such as the meaning function
approach [4] applied to this example in [3].

3   Conclusion

It is especially easy to represent environments of rule systems and axiom schemas in Logistica. In addition, because Logistica is essentially the addition of synergistic features to an already well-developed programming language based on higher order logic [12], it is also easy to embed such deduction systems into larger Artificial Intelligence programs. By embedding deduction into a system including Higher Order Logic we do not separate logic and control as suggested in [13]. Instead, we merge Logic and Control as suggested in [15, 16] but in a more sophisticated manner than is done in First Order Logic or in the Horn clause subset of First Order Logic. Instead, like [1] Logistica mixes logic and control in the framework of a Higher Order Logic.

Acknowledgements

This research was supported by grants EIA-9818341 and EIA-9972843 from the National Science Foundation. I thank Guy Jacobs for his comments and Marta Cialdea Mayer for additional help.

References

[1] Araya, C. and Brown, F.M, Schemata: A Language for Deduction, *9th. European Conference on Artificial Intelligence*, Stockholm, Sweden 1990.

[2] Boyer and J Strother Moore, *A Computational Logic*, Academic Press, New York, 1981.

[3] Boyer, R. S. and Moore, J Strother , "Metafunctions: proving them correct and using them efficiently as new proof procedures," *The Correctness Problem in Computer Science*, R. S. Boyer and J Strother Moore, eds., Academic Press, New York, 1981.

[4] Brown, F. M., "The Theory of Meaning", Department of Artificial Intelligence Research Report 35, University of Edinburgh, June 1977, University of Edinburgh, *Edinburgh Artificial Intelligence Research Papers 1973-1986*, Scientific Datalink microfische, 1989.

[5] Brown, F. M., "Towards the Automation of Set Theory and its Logic", *Artificial Intelligence*, Vol. 10, 1978.

[6] Brown, F. M., "An investigation into the goals of research in Automatic Theorem Proving as related to Mathematical Reasoning", Artificial Intelligence, 1980.

[7] Brown, F. M., "An Experimental Logic based on the Fundamental Deduction Principal", *Artificial Intelligence*, vol. 30 no. 2, 146 pages from page 117 to page 263, November 1986, North Holland.

[8] Brown, Frank M., "Prologue to a Theory of Design Change and its Automation", *Proceedings Mid-America Conference on Intelligent Systems*, 1994.

[9] Brown, Frank M. and David Leasure, "Proving Schematic Metatheorems by Induction, *Procdings of the Fourth Midwest Artificial Intelligence and Cognitive Science Society Conference*, 1992.

[10] Brown, F. M., Leasure, D. E. and Peterson, Niel, "Logistica: A High-Level Programming language for Implementing Design Verification Systems," June 1993 *AAAI workshop on Systems Engineering*.

[11] Brown, F. M., and Liu P., "A Logic Programming and Verification System for Recursive Quantificational Logic", *International Joint Conference on Artificial Intelligence*, 85. August 1985.

[12] Church, A. "The Calculi of Lambda-Conversion", *Annals of Mathematical Studies, N.6*, Princeton University Press, 1941.

[13] Hayes, P. J., Computation as Deduction, *Proceedings 2nd MFCS Symposium, Czechoslovakia Academy of Sciences*, Prague, Czechoslovakia, 1973.

[14] Kelsey, K., et. Al.   "Revised[5] Report on the Algorithmic Language Scheme" 20 February, 1988. http://download.plt-scheme.org/doc/203/html/r5rs/index.htm.

[15] Kowalski R., Predicate Logic as Programming Language, *Proceedings of IFIP74*, North-Holland, Amsterdam, 1974.

[16] Kowalski R., Algorithm = Logic + Control, *CACM*, August 1979.

[17] McCarthy, J. et.al., *LISP1.5 Programmers' Manual*, MIT Press 1962.

Fair Constraint Merging Tableaux in Lazy Functional Programming Style

Reiner Hähnle and Niklas Sörensson

Chalmers University of Technology & Göteborg University
Department of Computing Science, S-41296 Gothenburg, Sweden
{reiner,nik}@cs.chalmers.se

Abstract. Constraint merging tableaux maintain a system of all closing substitutions of all subtableau up to a certain depth, which is incrementally increased. This avoids backtracking as necessary in destructive first order free variable tableaux. The first successful implementation of this paradigm was given in an object-oriented style. We analyse the reasons why lazy functional implementations so far were problematic (although appealing), and we give a solution. The resulting implementation in Haskell is compact and modular.

1 Introduction

Until recently, implementations of free variable tableau proof procedures suffered from the necessity to backtrack over branch closing substitutions, if completeness was not to be sacrificed [3]. The central problem is that substitutions destructively change the rigid variables occurring in tableaux, which leads to complex dependencies between substitution and extension steps. Although by now there are ways to "repair" a tableau after a destructive closure step, the resulting proof procedures have unusual and relatively complicated rules, and a serious implementation of these ideas was not tried so far.

A fundamentally different way to cope with destructive closing substitutions is to simply enumerate *all* possible closing substitutions of a tableau in parallel. The fact that substitutions can be seen as term constraints suggests the phrase *constraint tableaux* [4] for a tableau procedure along these lines. Traditionally, this was considered too expensive in order to be viable and, if one uses naive breadth-first search, it certainly is. The breakthrough came with [1, 2], where a "lazy" stream of closing substitutions is associated with the subtableau below each tableau node. This requires to *merge* streams of closing sustitutions, so the resulting calculi are called *constraint merging tableaux*. In the presence of refinements such as pruning, subsumption, and simplification, they are a basis for a competitive implementation.

The implementation [1, 2] is object-oriented, but the term "lazy" suggests to use a programming language supporting lazy evaluation. Such implementations in Haskell were given in [5] and [4]. Our approach is related to the latter. Both, however, suffer from drawbacks, which we were able to remedy. A lazy

M. Cialdea Mayer and F. Pirri (Eds.): TABLEAUX 2003, LNAI 2796, pp. 252–256, 2003.

functional implementation is less straightforward than it seems at first. At its heart is the *merging* of streams of substitutions at branching nodes: if `refute` yields a stream of closing substitutions for a set of formulas, then the code for disjunctive formulas looks like this:

$$\texttt{refute } \{\phi \vee \psi\} \cup \texttt{B = merge (refute } \{\phi\} \cup \texttt{B) (refute } \{\psi\} \cup \texttt{B)}$$

The merger drives lazy evaluation of subtableaux. As we show below, one must design it very carefully to ensure fairness (and, hence, completeness). This was not properly addressed in functional implementations so far. Another problem is that any competitive implementation of a tableau-based proof procedure needs to incorporate refinements such as pruning and subsumption. Giese [2, p 45] reports that an attempt to combine fairness and refinements in a functional style resulted in a merger of overwhelming complexity. Both [5, 4] are not fair (as we show below) and do not feature refinements.

We implemented a constraint merging tableaux procedure in lazy functional style including the following features: (a) we identify and describe the fairness problems present in previous approaches and we solve them cleanly; (b) our implementation is compact (less than 100 lines), making it suitable for experimentation; (c) the input formula language may contain arbitrary Haskell functions; (d) basic refinements to improve efficiency. The source code is available at http://www.cs.chalmers.se/~nik/lazy.

2   The Implementation

In [5], explicit data structures are built up to represent tableaux, while the approach in [4] dispenses with them and keeps only the system of term constraints that would result from applying the rules. In our approach the notion of a formula is central. We exploit that certain kinds of first order formulas completely determine a tableaux for them up to the substitutions applied. Formally, call a free variable tableau for a formula ϕ, to which no closure rule has been applied, a **tableau template**. Note that a tableau template, in general, is an infinite tree.

Now, a formula is identified by such a tableau template, which in turn is represented by a Haskell function that produces a stream of closing substitutions for this formula and a given tableau branch. Expressions of type TT can be constructed using the following functions:

```
fresh :: (Term -> TT) -> TT
pLit  :: Atom -> TT              (<|>) :: TT -> TT -> TT
nLit  :: Atom -> TT              (<&>) :: TT -> TT -> TT
```

For example, the formula $\phi = (\neg p \vee q) \wedge \neg q$ could be constructed as follows: "phi = (nLit p <|> pLit q) <&> nLit q". We allow disjunction, conjunction, plus negation at the literal level. Universal quantification is discussed below. Functions have a string identifier, variables are identified by a unique label generated by the system. For simplicity, atoms are typed as terms. Here we declare a constant `zero`, a one-place function `suc`, and a one-place predicate `nat`:

```
zero = Fun "zero" []    suc x = Fun "suc" [x]    nat x = Fun "nat" [x]
```

Now we can build formulas containing variables. Consider, for example, the definition "sucNat x = nLit (nat x) <|> pLit (nat (suc x))". Here, we view the formula sucNat(x) with free variable x as a Haskell function sucNat with formal parameter x.

At the heart of first order theorem proving is the capability to obtain unlimited numbers of fresh instances of universally quantified formulas. We provide directly an operation called fresh that takes a formula with a free variable and produces an instance, where the free variable has been replaced with a new unique (ie, "fresh") name. The following example demonstrates how to emulate universal quantification, using fresh and recursive definitions.

```
zeroIsNat   = pLit (nat zero)
twoIsNotNat = nLit (nat (suc (suc zero)))
uSucNat     = fresh sucNat <&> uSucNat
countToTwo  = zeroIsNat <&> twoIsNotNat <&> uSucNat
```

The first order formula $nat(0) \wedge \neg nat(s(s(0))) \wedge (\forall x)(\neg nat(x) \vee nat(s(x)))$ is represented by this code, but the tableau building functions of type TT do much more than representing a formula: they build an infinite *tableau template*. There are infinitely many possible tableau templates for a given formula, however, a tableau template is completely determined by the *particular* definition of countToTwo. We happened to arrange the constituents of countToTwo in a fair manner, hence, tableau completeness guarantees that the corresponding tableau template can be completed to a proof.

Our formula input language can be mixed with arbitrary Haskell code to make it more expressive. It is easy, for example, to code a resource-bounded quantifier that can use at most n instances of its scope. The language of tableau templates allows even to control the *shape* of proofs. The following definition, for example, forces tableaux for the counting example to become linear:

```
linSucNat x = nLit (nat x) <|> (pLit (nat (suc x)) <&> fresh linSucNat)
countToTwo' = zeroIsNat <&> twoIsNotNat <&> fresh linSucNat
```

The trick is that recursion is only done in the right part of the disjunction, which leads to linear trees.

3   The Difficulty of Merging Substitutions

Consider the problem countToTwo from above. Below is the initial part of the tableau template for countToTwo, annotated with closing substitutions for each subtableau. At each node we provide the following information: in the first line, a node identifier, followed by the formula the node is labelled with. In the remaining lines (if any), the *new* closing substitutions that are possible at this node. The notation @N refers to the node label that is used to close the tableau (besides the current node).

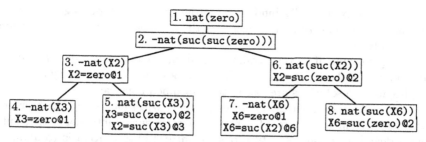

We consider the following strategy for merging substitutions: at each node, look at the combination of each pair of substitutions, where one substitution is from the left and one from the right subtableau. Wlog start with the first substitution in each subtableau, followed by the second substitution in the left and the first in the right. This weak requirement is enough to cause non-termination independently of how the remaining Cartesian product is enumerated.

The first pair of substitutions in the tableau above are the incompatible ones of node 3 (X2=zero) and node 6 (X2=suc(zero)). Then our enumeration looks for the next solution for node 3. This second substitution at node 3 must come from the combination of substitutions in nodes 4 and 5. Again, the first substitutions at this level are combined: X3=zero at node 4 and X3=suc(zero) at node 5. Again, they are incompatible, leading yet to another level of expansion. Note the similar situations in nodes 2, 3 and 4: all need the second element in their left subtree to compute the next pair of solutions. This is an invariant for the "leftmost" nodes of the constraint tree, which makes the whole computation non-terminating.

In the example, the mistake occurs first in the merging of the substitutions belonging to nodes 4 and 5: it would have been correct to compute the combination of *all new* substitutions at this level *at the same time*, before proceeding any further. This would result in the compatible pair (X3=zero,X2=suc(X3)), which quickly terminates the search.

Our example can be adapted to any merger built from any systematic enumeration of Cartesian products of *single* substitutions. This shows that any such approach is bound to be incomplete due to non-termination. For example, the implementations in [4, 5] suffer from exactly this problem.

Our solution combines, at each level of the search, *all new* substitutions with each other and passes them up, before proceeding further. To ensure this, one needs to record which substitutions were generated at the same node. The most straightforward way is to use doubly nested lists of substitutions, where the "inner" lists comprise exactly those substitutions belonging to the same node.

4   Refinements

For a subtableau starting with the introduction of a fresh variable, it suffices to know that it is closable *for some* value of this variable. In order to get rid of these *local* variables one introduces existential quantification in the constraint language, and existentially binds local variables as they "propagate upwards" past their scope. This technique decreases the number of variables in constraints considerably.

During merging, if two constraints c_1 and c_2, where c_2 is *subsumed* by c_1, are found in one of the input streams, then it is possible to throw away c_2. This is safe, because no solutions are lost in the subsumed constraint.

The vanilla implementation realizes a *branch selection* strategy with breadth first effect. Often it is more efficient to prioritize such subtableaux that suffer from a dearth of closing substitutions. A natural optimization is to allow the mergers to focus on only one branch until at least one solution is found in both of their branches.

5 Conclusion and Future Work

Our implementation is still naive. We were more interested in expressing the core algorithm in a concise way than producing a competitive theorem prover, so we paid little attention to efficiency of data structures. However, from the SYN category of TPTP 2.5.0, which contains about 1000 problems, 448 problems could be proven with a time limit of 5 min. Most of the successfully proved problems where classified as simple in TPTP, but 28 had a rating between 0.12 and 0.67.

In the future we would like to add more refinements, such as simplification [2], and equality handling. We made some experiments with pruning [3], which indeed gives a significant performance boost, but makes the implementation much less elegant. Hyper tableaux proved to be an effective refinement, and should be implemented as well. Our prover has a facility for graphical output which could not be described here for lack of space. There should be a library of formula constructors, for example, for various kinds of quantifiers or abstract data types.

References

[1] M. Giese. Incremental closure of free variable tableaux. In R. Goré, A. Leitsch, and T. Nipkow, editors, *Proc. Intl. Joint Conf. on Automated Reasoning IJCAR, Siena, Italy*, volume 2083 of *LNCS*, pages 545–560. Springer-Verlag, 2001.

[2] M. Giese. *Proof Search without Backtracking for Free Variable Tableaux*. PhD thesis, Fakultät für Informatik, Universität Karlsruhe, July 2002.

[3] R. Hähnle. Tableaux and related methods. In A. Robinson and A. Voronkov, editors, *Handbook of Automated Reasoning*, volume I, chapter 3, pages 101–178. Elsevier Science B. V., 2001.

[4] B. Ó Nualláin. Constraint tableaux. In *Position Papers presented at International Conference on Analytic Tableaux and Related Methods, Copenhagen, Denmark*, 2002.

[5] J. van Eijck. Constrained hyper tableaux. In L. Fribourg, editor, *Proc. Computer Science Logic, Paris, France*, volume 2142 of *LNCS*, pages 232–246. Springer-Verlag, Sept. 2001.

SOLAR: A Consequence Finding System for Advanced Reasoning

Hidetomo Nabeshima[1], Koji Iwanuma[1], and Katsumi Inoue[2]

[1] University of Yamanashi
4-3-11 Takeda, Kofu-shi 400-8511, Japan
{nabesima,iwanuma}@iw.media.yamanashi.ac.jp
[2] Kobe University
Rokkodai-cho, Nada-ku, Kobe 657-8501, Japan
inoue@eedept.kboe-u.ac.jp

1 Introduction

SOLAR is an efficient first-order consequence finding system based on a connection tableau format with Skip operation. Consequence finding [1, 2, 3, 4] is a generalization of refutation finding or theorem proving, and is useful for many reasoning tasks such as knowledge compilation, inductive logic programming, abduction. One of the most significant calculus of consequence finding is SOL [2]. SOL is complete for consequence finding and can find all minimal-length consequences with respect to subsumption. SOLAR (SOL for Advanced Reasoning) is an efficient implementation of SOL and can avoid producing non-minimal/redundant consequences due to various state of the art pruning methods, such as skip-regularity, local failure caching, folding-up (see [5, 6]).

SOLAR also achieves a good performance as a theorem prover. For 1,921 problems in TPTP v2.5.0 library which do not contain the equality, the experimental results show that SOLAR can solve 52% problems within 300 CPU seconds for each problem, whereas 50% are solved by OTTER 3.2. SOLAR is written in Java, and thus has the desirable features of high programmability, extensibility, reusability, and platform independence. Hence SOLAR can easily be incorporated into many AI programs. According to our knowledge, SOLAR is the first sophisticated implementation of first-order consequence finding calculus in the world.

2 Consequence Finding Procedure SOL

Consequence finding [1, 2, 3, 4] is a computation problem for finding important consequences from an axiom set, and is a generalization of refutation finding or theorem proving. However, in practice, the set of theorems derivable from an axiom set might be infinite, even if it is restricted to containing only the consequences that are minimal with respect to subsumption. Toward more practical automated consequence finding, Inoue [2] reformulated and restricted the attention to the problem for finding only "interesting" consequence formulas, called

M. Cialdea Mayer and F. Pirri (Eds.): TABLEAUX 2003, LNAI 2796, pp. 257–263, 2003.

characteristic clauses. The concept of characteristic clauses is useful for various reasoning problems of interest to AI, such as nonmonotonic reasoning, abduction, knowledge compilation (see [2, 1] for details), inductive logic programming [7, 8], multi-agent systems [6], bioinformatics [9] and distributed knowledge bases [10].

Inoue [2] proposed *SOL-resolution* for mechanically finding characteristic clauses within first-order logic, which can be viewed as either an extension of Loveland's model-elimination-like calculus [11] with *Skip* operation or a generalization of Siegel's propositional production algorithm [12]. Compared with other calculi, SOL-resolution can focus on generating only the characteristic clauses rather than all logical consequences. SOL-resolution is one of the most advanced and significant calculi for the consequence finding problem.

The original SOL-resolution [2] was given in a model-elimination-like chain format [11]. Iwanuma et al. [5] reformulated SOL-resolution within the framework of connection tableaux [13, 14] and proposed various complete pruning methods [5, 6] for enhancing the efficiency of SOL tableaux such as skip-regularity, local failure caching, folding-up.

We give a brief view of SOL tableaux. A *production field* $\mathcal{P}$ is a pair, $\langle L, Cond \rangle$, where L is a set of literals and is closed under instantiation, and *Cond* is a certain condition to be satisfied. When *Cond* is not specified, $\mathcal{P}$ is denoted as $\langle L \rangle$. A clause C *belongs to* $\mathcal{P} = \langle L, Cond \rangle$ if every literal in C belongs to L and C satisfies *Cond*. When Σ is a set of clauses, the set of logical consequences of Σ belonging to $\mathcal{P}$ is denoted as $\mathrm{Th}_{\mathcal{P}}(\Sigma)$. A production field $\mathcal{P}$ is *stable* if, for any two clauses C and D such that C subsumes D, D belongs to $\mathcal{P}$ only if C belongs to $\mathcal{P}$. The stability of a production field is important in practice [2], and we assume in this paper that production fields are stable.

Example 1. Let $\mathcal{L} = \mathcal{L}^+ \cup \mathcal{L}^-$ be the set of all literals in the first-order language, where $\mathcal{L}^+$ and $\mathcal{L}^-$ are the positive and negative literals in the language, respectively. The following are examples of stable production fields.

1. $\mathcal{P}_1 = \langle \mathcal{L} \rangle$: $\mathrm{Th}_{\mathcal{P}_1}(\Sigma)$ is the set of logical consequences of Σ.
2. $\mathcal{P}_2 = \langle \mathcal{L}^+ \rangle$: $\mathrm{Th}_{\mathcal{P}_2}(\Sigma)$ is the set of all positive clauses derivable from Σ.
3. $\mathcal{P}_3 = \langle \mathcal{L}^-,$ length is fewer than $k \rangle$: $\mathrm{Th}_{\mathcal{P}_3}(\Sigma)$ is the set of negative clauses implied by Σ consisting of fewer than k literals.

On the contrary, $\mathcal{P}_4 = \langle L,$ length is more than $k \rangle$ is not a stable production field. For example, if $k = 2$ and $L = \{\neg P, Q, R\}$, then $C = \neg P \vee Q$ subsumes $D = \neg P \vee Q \vee R$, and D belongs to $\mathcal{P}_4$ while C does not.

Given a set of clauses Σ, a newly added clause C and a production field $\mathcal{P}$, an *SOL-deduction from* $\Sigma + C$ *and* $\mathcal{P}$ satisfies the following:

Theorem 1. (Soundness and Completeness of SOL-Deduction). *[5]*

1. *Soundness: If a clause S is derived by an SOL-deduction from $\Sigma + C$ and $\mathcal{P}$, then S belongs to $\mathrm{Th}_{\mathcal{P}}(\Sigma \cup \{C\})$.*
2. *Completeness: If a clause F does not belong to $\mathrm{Th}_{\mathcal{P}}(\Sigma)$ but belongs to $\mathrm{Th}_{\mathcal{P}}(\Sigma \cup \{C\})$, then there is an SOL deduction of a clause S from $\Sigma + C$ and $\mathcal{P}$ such that S subsumes F.*

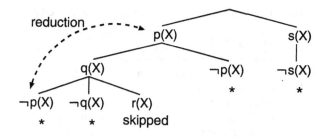

Fig. 1. Example of a tableau

In the above, S is an interesting clause newly obtained by adding C to Σ. We briefly explain SOL resolution by the next example.

Example 2.

$$C = p(X) \vee s(X),$$
$$\Sigma = \{q(X) \vee \neg p(X),\ \neg s(Y),\ \neg p(Z) \vee \neg q(Z) \vee r(Z)\},$$
$$\mathcal{P} = \langle \mathcal{L}^{+},\ \text{length is less than } 2\rangle.$$

Figure 1 is one of tableaux that is derived by an SOL-deduction from $\Sigma + C$ and $\mathcal{P}$. An SOL-deduction uses three inference rules — *Skip, Extension and Reduction.* **Skip** is a rule that skips a literal belonging to the production field. Skipped literals constitute a consequence clause. In Fig. 1, the node $r(X)$ is skipped. **Extension** rule expands a node with a clause of an axiom set in the same ways as the resolution principle does. The node $p(X)$ is extended with the clause $q(X) \vee \neg p(X)$. The node $\neg p(X)$ at the bottom is closed by **Reduction** because it has an ancestor $p(X)$ which is unifiable with the complement of $\neg p(X)$. A tableau is said to be *solved* if all leaf nodes of the tableau are marked. Figure 1 is a solved tableau, thus we can get a new consequence clause $r(X)$ which consists of all skipped literals.

3   SOLAR

SOLAR is an efficient implementation of consequence finding procedure SOL, and has various sophisticated pruning methods shown in Tab. 1. "full" denotes that a pruning method is fully implemented, while "partial" means that a pruning method is partially implemented since its full checking requires high computational cost. Merge with skipped literals and skip-regularity are native to SOL. Identical goal pruning and unit subsumption are specialization of regularity and TCS-freeness, respectively. Additionally, SOLAR uses a discrimination tree [15] for representing a set of clauses in order to enhance term indexing and retrieval.

The input of SOLAR is a description of a consequence finding problem that is compatible with the TPTP [16] format. Example 2 can be described as follows:

Table 1. Implemented pruning methods in SOLAR

Pruning method	Checking
Merge with skipped literals [5]	full
Unit Axiom/Lemma matching [5]	full
Identical reduction [5]	full
Folding-up [5]	full
Regularity [6]	partial
Tautology-freeness [6]	partial
Complement-freeness [6]	partial
Skip-regularity [6]	partial
TCS-freeness [6]	partial
Identical goal pruning	full
Unit subsumption [5]	partial
Order preserving reduction [5]	full
Local failure caching [5]	full

```
input_clause(clause1, top_clause, [p(X), s(X)]).
input_clause(clause2, axiom, [q(X),-p(X)]).
input_clause(clause3, axiom, [-s(Y)]).
input_clause(clause4, axiom, [-p(Z), -q(Z), r(Z)]).
production_field([predicates(pos_all), length < 2]).
```

top_clause means that the clause is newly added. production_field indicates the production field, and in this example, it allows to generate consequences that consist of less than 2 positive literals. If there is no top_clause and production_field, then the format is equivalent to TPTP's one and SOLAR tries to find a refutation as a theorem prover.

The output of SOLAR is the set of minimal-length consequences:

```
% ./solar example.sol
real time : 100 msec (0.1 sec)
3 consequences
[p(_0)]
[q(_0)]
[r(_0)]
```

SOLAR has many command-line options that enable/disable each pruning method, display several derived tableaux, and so on.

4 Performance of SOLAR

We show that merge, skip-regularity and local failure caching have a great ability for accelerating consequence finding computation. Table 2 shows the experimental results that compare these methods. We use an axiom set of TPTP library as

Table 2. Performance of consequence finding

Problem	params		none	merge	skip-reg	local fail	all
PUZ001-0.ax	dep ≤ 8	time [sec]	21.61	8.43	2.30	11.47	1.15
	len ≤ 3	infs	7,964,928	2,861,086	514,089	3,552,867	94,660
		concs	50	50	53	50	52
HWV001-2.ax	dep ≤ 4	time [sec]	32.94	23.77	20.56	16.65	10.05
	len ≤ 5	infs	74,110	57,374	63,532	48,804	34,855
		concs	1122	984	961	857	716
GEO001-0.ax	dep ≤ 4	time [sec]	33.99	31.97	32.96	3.79	3.94
	len ≤ 1	infs	5,431,218	4,910,903	5,089,578	417,915	415,070
		concs	34	34	34	9	9
BOO001-0.ax	dep ≤ 11	time [sec]	39.44	39.14	39.30	39.50	39.32
	len ≤ 1	infs	3,765,128	3,765,128	3,765,128	3,765,128	3,765,128
	TD ≤ 4	concs	4	4	4	4	4

Table 3. Performance of refutation finding

	5,181 problems possibly containing equality			1,921 problems containing no equality		
	Solved	Failed	Rate	Solved	Failed	Rate
SOLAR	1644	3537	31.7%	999	922	52.0%
Otter3.2	2047	3134	39.5%	960	961	50.0%

a consequence finding problem. SOLAR calculates minimal-length consequences derivable from the axiom set. "Problem" denotes an axiom set of TPTP library. "dep", "len" and "TD" in the column "params" mean the maximum search-depth, the maximum length of consequences, and the maximum term-depth of each literal in consequences, respectively. "none" represents that SOLAR does not use these pruning methods at all, and oppositely "all" uses all of them. "merge", "skip-reg" and "local fail" mean that SOLAR uses the corresponding pruning method without any other methods. "infs" is the total number of rules applied to tableaux, and "concs" is the number of the computed consequences. All experiments were conducted on a Pentium4 (2.53GHz) machine running JDK1.4.1_01 on Turbolinux Workstation 8.0 with 1GB memory.

Table 2 shows that all pruning methods reduce the search space and improve the speed. In particular, skip-regularity and local failure caching have great effects for PUZ001-0.ax and GEO001-0.ax, respectively. The result of BOO001-0.ax indicates that, although there is no effect of pruning, there is almost no overhead of these methods.

SOLAR also achieves a good performance as a theorem prover. We compared with OTTER 3.2 [17] for TPTP problem library v2.5.0 [16][1]. We experimented on 5,181 problems. Table 3 shows the experimental results as a theorem prover. "Solved" is the number of the proved problems within 300 CPU seconds and "Failed" shows the number of problems that could not be proved . SOLAR is especially superior to OTTER for 1,921 problems that do not contain the equality.

5   Conclusion

Consequence finding is an important technique for advanced reasoning such as nonmonotonic reasoning, abduction, multi-agent systems, bioinformatics. In such reasoning tasks, SOLAR is the first sophisticated implementation of first-order consequence finding calculus in the world, and can find out important consequences efficiently due to various state of the art pruning methods. SOLAR also achieves a good performance as a theorem prover.

del Val [1] defines a variant of SOL resolution called *SFK resolution* for finding characteristic clauses based on ordered resolution. A propositional version of SFK resolution has recently been implemented using ZBDDs (Zero-suppressed Binary Decision Diagrams) [18]. It is reported that the ZBDD implementation can handle problems with more than 10^{70} propositional clauses.

Acknowledgement

This research is partially supported by Grant-in-Aid for Scientific Research (No.13358004) from The Ministry of Education, Culture, Sports, Science and Technology of Japan, and Japan-Australia Collaboration sponsored by Japan Society for the Promotion of Science. We thank Ken Satoh, John Slaney, Abdul Sattar and Rajeev Goré for useful discussions provided through the Japan-Australia Advance Reasoning Systems Workshop.

References

[1] del Val, A.: A new method for consequence finding and compilation in restricted languages. In: Proceedings of AAAI-99. (1999) 259–264
[2] Inoue, K.: Linear resolution for consequence finding. Artificial Intelligence **56** (1992) 301–353
[3] Minicozzi, E., Reiter, R.: A note on linear resolution strategies in consequence-finding. Artificial Intelligence **3** (1972) 175–180
[4] Slagle, J. R., Chang, C., Lee, R. C.: Completeness theorems for semantic resolution in consequence-finding. In: Proceedings of IJCAI-69. (1969) 281–285
[5] Iwanuma, K., Inoue, K., Satoh, K.: Completeness of pruning methods for consequence finding procedure SOL. In: Proceedings of FTP-2000. (2000) 89–100

---

[1] This library consists of 6,672 problems. Since SOLAR can not interpret the input_formula() command currently, we excluded 1,491 problems including the command.

[6] Iwanuma, K., Inoue, K.: Conditional answer computation in SOL as speculative computation in multi-agent environments. In: Proceedings of CLIMA-02. (2002) 149–162 (It also appears in *Electronic Notes on Theoretical Computer Science* 70(5), 2002)

[7] Yamamoto, A.: Representing inductive inference with SOLD-resolution. In: Proceedings of IJCAI-97 Workshop on Abduction and Induction in AI. (1997) 59–63

[8] Inoue, K.: Induction, abduction, and consequence-finding. In: Proceedings of ILP-01. Volume 2157 of LNAI., Springer (2001) 65–79

[9] Reiser, P., King, R. D., Kell, D. B., Muggleton, S. H., Bryant, C. H., Oliver, S. G.: Developing a logical model of yeast metabolism. Electronic Transactions on Artificial Intelligence 5 (2001) 223–244

[10] McIlraith, S., Amir, E.: Theorem proving with structured theories. In: Proceedings of IJCAI-01. (2001) 624–631

[11] Loveland, D. W.: Automated Theorem Proving: a logical basis. North-Holland Publishing Company, Amsterdam (1978)

[12] Siegel, P.: Représentation et utilization de la connaissance en calcul propositionnel. Thèse d'État, Université d'Aix-Marseille II, Luminy, France (1987) (in French)

[13] Letz, R.: Clausal tableaux. In Bibel, W., Schmitt, P. H., eds.: Automated Deduction - A Basis for Applications. Volume I: Foundations. Kluwer, Dordrecht (1998) 39–68

[14] Letz, R., Goller, C., Mayr, K.: Controlled integration of the cut rule into connection tableau calculi. Journal of Automated Reasoning 13 (1994) 297–338

[15] McCune, W.: Experiments with discrimination-tree indexing and path indexing for term retrieval. Journal of Automated Reasoning 9 (1992) 147–167

[16] Sutcliffe, G., Suttner, C.: The TPTP problem library for automated theorem proving v2.5.0. http://www.tptp.org/ (2002)

[17] McCune, W.: OTTER 3.0 reference manual and guide. Technical Report ANL-94/6, Argonne National Laboratory (1994)

[18] Simon, L., del Val, A.: Efficient consequence finding. In: Proceedings of IJCAI-01. (2001) 359–370

CondLean:
A Theorem Prover for Conditional Logics

Nicola Olivetti and Gian Luca Pozzato

Dipartimento di Informatica, Università di Torino, Italy

Abstract. In this paper we present CondLean, a theorem prover for propositional conditional logics CK, CK+ID, CK+MP and CK+MP+ID. The theorem prover implements some recently introduced sequent calculi for these logics. CondLean is developed following the methodology of leanTAP and it is implemented in SICStus Prolog. It also comprises a graphical user interface implemented in JAVA. CondLean can be downloaded at the site www.di.unito.it/~olivetti/CONDLEAN/

1 Introduction

Conditional logics have found interesting applications in several areas of computer science and artificial intelligence. We just mention: knowledge representation, non-monotonic reasoning, deductive databases, belief revision and natural language semantics. In spite of their significance, very few proof systems have been proposed for these logics.

In [1] labelled sequent calculi SeqS[1] are introduced for minimal normal conditional logic **CK** and for three extensions of it, namely **CK+ID, CK+MP** and **CK+MP+ID**.

In this work we describe an implementation of SeqS calculi in SICStus Prolog. The program, called CondLean, gives a decision procedure for these logics; as far as we know this is the first theorem prover for these logics. For each system, we introduce three different versions:

1. a simple version, where Prolog *constants* are used to represent SeqS's labels;
2. a more efficient one, where labels are represented by Prolog *variables*, inspired by the free-variable tableaux presented in [2];
3. a "two-phase" theorem prover, which first attempts to prove a sequent by using an incomplete, but fast, proof procedure (phase 1), and then it calls the free-variable proof procedure (phase 2) in case of failure.

CondLean also comprises a **graphical interface** implemented in *Java*, using the se.sics.jasper package to link the graphical user interface to the SICStus Prolog kernel.

---

[1] S stands for CK, ID, MP or ID+MP.

M. Cialdea Mayer and F. Pirri (Eds.): TABLEAUX 2003, LNAI 2796, pp. 264–270, 2003.

2   Conditional Logics and Their Sequent Calculi

We consider a propositional language $\mathcal{L}$ over a set ATM of propositional variables. Formulas of $\mathcal{L}$ are built from propositional variables by means of the boolean operators $\rightarrow, \bot$ and the conditional operator $\Rightarrow$. We adopt the so-called propositional selection function semantics [3]. A selection function model for $\mathcal{L}$ is a triple $\mathcal{M} = \langle \mathcal{W}, f, [\ ] \rangle$, where $\mathcal{W}$ is a non-empty set of items called *worlds*, f is a function of type $f : \mathcal{W} \times 2^{\mathcal{W}} \longrightarrow 2^{\mathcal{W}}$, called the *selection function* and $[\ ]$ is an evaluation function of type $ATM \longrightarrow 2^{\mathcal{W}}$. $[\ ]$ assigns to an atom p the set of worlds where p is true. The evaluation function $[\ ]$ can be extended to every formula by means of the following inductive clauses: 1. $[\bot] = \emptyset$; 2. $[A \rightarrow B] = (\mathcal{W} - [A]) \cup [B]$; 3. $[A \Rightarrow B] = \{w \in \mathcal{W} \mid f(w, [A]) \subseteq [B]\}$. We say that a formula A is *valid* in a model $\mathcal{M}$ as above if $[A] = \mathcal{W}$. A formula A is *valid* (denoted by $\models A$) if it is valid in every model $\mathcal{M}$.

The above one is the semantics of the basic conditional logic **CK**, where no specific properties of the selection function f are assumed. Moreover, we consider the following extensions of **CK**: **CK+ID**, **CK+MP** and **CK+MP+ID**, obtained by postulating the semantic conditions (ID) and (MP) where (ID) is $f(x, [A]) \subseteq [A]$ and (MP) is $w \in [A] \rightarrow w \in f(w, [A])$. The two semantic conditions correspond respectively to the axiom schemata:

$$A \Rightarrow A \text{ and } (A \Rightarrow B) \rightarrow (A \rightarrow B)$$

In Figure 1 we present the calculi for CK and its mentioned extensions introduced in [1]; the calculi make use of labelled formulas, where the labels are drawn from a denumerable set $\mathcal{A}$; there are two kinds of formulas: 1. *labelled formulas*, denoted by $x: A$, where $x \in \mathcal{A}$ and $A \in \mathcal{L}$; 2. *transition formulas*, denoted by $x \xrightarrow{A} y$, where $x, y \in \mathcal{A}$ and $A \in \mathcal{L}$. A transition formula $x \xrightarrow{A} y$ represents that $y \in f(x, [A])$.

Definition (Sequent Validity). Given a model $\mathcal{M} = \langle \mathcal{W}, f, [\ ] \rangle$ for $\mathcal{L}$, and a label alphabet $\mathcal{A}$, we consider any *mapping* $I : \mathcal{A} \rightarrow \mathcal{W}$. Let F be a labelled formula, we define $\mathcal{M} \models_I F$ as follows: $\mathcal{M} \models_I x: A$ iff $I(x) \in [A]$ and $\mathcal{M} \models_I x \xrightarrow{A} y$ iff $I(y) \in f(I(x), [A])$. We say that $\Gamma \vdash \Delta$ is *valid* in $\mathcal{M}$ if for every mapping $I : \mathcal{A} \rightarrow \mathcal{W}$, if $\mathcal{M} \models_I F$ for every $F \in \Gamma$, then $\mathcal{M} \models_I G$ for some $G \in \Delta$. We say that $\Gamma \vdash \Delta$ is valid in a system (CK or one of its extensions) if it is valid in every $\mathcal{M}$ satisfying the specific conditions for that system (if any).

Theorem 1 (Soundness and Completeness [1]). A sequent $\Gamma \vdash \Delta$ is valid if and only if $\Gamma \vdash \Delta$ is derivable in SeqS.

As usual, in order to obtain a decision procedure we have to control the application of the contraction rules in a backward proof search of a sequent derivation. To this regard we have the following results:

Theorem 2 ([1], [4]). If $\Gamma \vdash \Delta$ is derivable using SeqCK (resp. SeqID), it has a derivation where there are no applications of (Contr L) and (Contr R).

$$(\mathbf{AX})\ \Gamma, F \vdash \Delta, F \qquad\qquad (\mathbf{A}\bot)\ \Gamma, x : \bot \vdash \Delta$$

$$(\mathbf{ContrL})\ \dfrac{\Gamma, F, F \vdash \Delta}{\Gamma, F \vdash \Delta} \qquad\qquad (\mathbf{ContrR})\ \dfrac{\Gamma \vdash \Delta, F, F}{\Gamma \vdash \Delta, F}$$

$$(\rightarrow \mathbf{R})\ \dfrac{\Gamma, x : A \vdash x : B, \Delta}{\Gamma \vdash x : A \rightarrow B, \Delta} \qquad\qquad (\rightarrow \mathbf{L})\ \dfrac{\Gamma \vdash x : A, \Delta \quad \Gamma, x : B \vdash \Delta}{\Gamma, x : A \rightarrow B \vdash \Delta}$$

$$(\mathbf{EQ})\ \dfrac{u : A \vdash u : B \quad u : B \vdash u : A}{\Gamma, x \xrightarrow{A} y \vdash x \xrightarrow{B} y, \Delta}$$

$$(\Rightarrow \mathbf{L})\ \dfrac{\Gamma \vdash x \xrightarrow{A} y, \Delta \quad \Gamma, y : B \vdash \Delta}{\Gamma, x : A \Rightarrow B \vdash \Delta} \qquad (\Rightarrow \mathbf{R})\ \dfrac{\Gamma, x \xrightarrow{A} y \vdash y : B, \Delta}{\Gamma \vdash x : A \Rightarrow B, \Delta}\ (y \notin \Gamma, \Delta)$$

$$(\mathbf{ID})\ \dfrac{\Gamma, y : A \vdash \Delta}{\Gamma, x \xrightarrow{A} y \vdash \Delta} \qquad\qquad (\mathbf{MP})\ \dfrac{\Gamma \vdash x : A, \Delta}{\Gamma \vdash x \xrightarrow{A} x, \Delta}$$

Fig. 1. Sequent calculi SeqS; the (ID) rule is for SeqID and SeqID+MP only; the (MP) rule is for SeqMP and SeqID+MP only

In contrast, *we cannot eliminate contractions in* SeqMP *and* SeqID+MP; more precisely, in these calculi we might need to use (Contr L) on conditional formulas $x: A \Rightarrow B$[2]. However, we can limit the application of left contraction on conditional formulas, as follows:

Theorem 3 ([4]). *If* $\Gamma \vdash \Delta$ *is derivable in* SeqMP (*resp.* SeqID+MP), *then it has a derivation where there are no applications of* (Contr R) *and there is at most one application of* (Contr L) *with constituent* $x: A \Rightarrow B$ *in each branch of the proof tree, for each conditional formula* $x: A \Rightarrow B$.

These results give a constructive proof of decidability of the respective systems, alternative to the semantic proof based on the finite model property.

3 Design of CondLean

In this section we present an implementation of the sequent calculi SeqS; it is a SICStus Prolog program inspired by leanTAP [5]. The program comprises a set of clauses, each one of them represents a sequent rule or axiom. The proof search is provided for free by the mere depth-first search mechanism of Prolog, without any additional ad hoc mechanism.

We represent each component of a sequent (antecedent and consequent) by a **list** of formulas, partitioned into three sub-lists: atomic formulas, transitions

---

[2] For example, we need (Contr L) to prove $x: \top \Rightarrow ((B \rightarrow (\top \Rightarrow B)) \rightarrow \bot) \vdash$ in CK+MP.

and complex formulas. Atomic and complex formulas are represented by a list like [x,a], where x is a Prolog constant and a is a formula. A transition $x \xrightarrow{A} y$ is represented by [x,a,y].

As we explained above, we present three different implementations. The first one, called **constant labels**, makes use of Prolog constants to represent SeqS's labels. The sequent calculi are implemented by the predicate

prove(Sigma, Delta, Labels).

This predicate succeeds if and only if $\Sigma \vdash \Delta$ is derivable in SeqS, where Sigma and Delta are the lists representing the multisets Σ and Δ, respectively and Labels is the list of labels introduced in that branch. For example, to prove x: $A \Rightarrow (B \wedge C)^3 \vdash x$: $A \Rightarrow B$, x: C in CK, one queries CondLean with the goal:

prove([[],[],[[x, a=>(b and c)]]], [[[x,c],[],[[x, a=>b]]], [x]).

Each clause of prove implements one axiom or rule of SeqS, except for contraction[4]; for example, the clause implementing ($\Rightarrow$ L) is:

prove([LitSigma,TransSigma,ComplexSigma],[LitDelta,TransDelta,
    ComplexDelta], Labels):-
  select([X,A=>B],ComplexSigma,ResComplexSigma), member(Y,Labels),
  put([Y,B],LitSigma,ResComplexSigma,NewLitSigma,NewComplexSigma),
  prove([LitSigma,TransSigma,ResComplexSigma],
    [LitDelta,[[X,A,Y]|TransDelta],ComplexDelta],Labels),
  prove([NewLitSigma,TransSigma,NewComplexSigma],
    [LitDelta,TransDelta,ComplexDelta],Labels).

The predicate select removes [X,A=>B] from ComplexSigma returning ResComplexSigma as result. The predicate put is used to put [Y,B] in the proper sub-list of the antecedent.

To search a derivation of a sequent $\Sigma \vdash \Delta$, CondLean proceeds as follows. First of all, if $\Sigma \vdash \Delta$ is an axiom, the goal will succeed immediately by using the clauses for the axioms. If it is not, then the first applicable rule will be chosen, e.g. if ComplexSigma contains a formula [X,A and B], then the clause for ($\wedge$ L) rule will be used, invoking prove on the unique premise of ($\wedge$ L). CondLean proceeds in a similar way for the other rules. The ordering of the clauses is such that the application of the branching rules is postponed as much as possible. When the ($\Rightarrow$ L) clause is used to prove $\Sigma \vdash \Delta$, a backtracking point is introduced by the choice of a label Y occurring in the two premises of the rule; in case of failure, Prolog's backtracking tries every instance of the rule with every available label (if more than one). Choosing, sooner or later, the right label to apply ($\Rightarrow$ L) may strongly affect the theorem prover's efficiency: if there are n

---

[3] CondLean extends the sequent calculi to formulas containing also $\neg$, $\wedge$, $\vee$ and $\top$.

[4] In SeqMP and SeqID+MP (ContrL) is "embedded" in ($\Rightarrow$ L), although in a controlled way in light of Theorem 3 (see above).

labels to choose for an application of ($\Rightarrow$ L) the computation might succeed only after n-1 backtracking steps, with a significant loss of efficiency.

Our second implementation, called **free-variables**, makes use of *Prolog variables* to represent all the labels that can be used in a single application of the ($\Rightarrow$ L) rule. This version represents labels by integers starting from 1; by using integers we can easily express constraints on the range of the variable-labels. To this regard the library `clpfd` is used to manage free-variable domains. As an example, in order to prove Σ', 1: $A \Rightarrow B \vdash \Delta$ the theorem prover will call prove on the following premises: $\Sigma' \vdash \Delta$, $1 \xrightarrow{A} V$ and V: B, $\Sigma' \vdash \Delta$, where V is a Prolog variable. This variable will be then instantiated by Prolog's pattern matching to apply either the (EQ) rule, or to *close a branch with an axiom*. Here below is the clause implementing the ($\Rightarrow$ L) rule:

```
prove([LitSigma,TransSigma,ComplexSigma],[LitDelta,
    TransDelta,ComplexDelta],Max):-
  select([X,A => B],ComplexSigma,ResComplexSigma),
  domain([Y],1,Max), Y#>X,
  put([Y,B],LitSigma,ResComplexSigma,NewLitSigma,NewComplexSigma),
  prove([NewLitSigma,TransSigma,NewComplexSigma],
    [LitDelta,TransDelta,ComplexDelta],Max),
  prove([LitSigma,TransSigma,ResComplexSigma],
    [LitDelta,[[X,A,Y]|TransDelta],ComplexDelta],Max).
```

The atom `Y#>X` adds the constraint `Y>X` to the constraint store: the constraints solver will verify the consistency of it during the computation. In SeqCK and SeqID we can only use labels introduced *after* the label `X`, thus we introduce the previous constraint. In SeqMP and SeqID+MP we can also use `X` itself, thus we shall add the constraint `Y#>=X`.

On a sequent with 65 labels on the antecedent this version succeeds in 460 mseconds, whereas the constant labels version takes 4326 mseconds.

We have also developed a third version, called **heuristic version**, that performs a "two-phase" computation: in "Phase 1" an *incomplete* theorem prover searches a derivation exploring a *reduced search space*; in case of failure, the free-variables version is called ("Phase 2"). Intuitively, the reduction of the search space in Phase 1 is obtained by committing the choice of the label to instantiate a free variable, whereby blocking the backtracking.

For SeqMP and SeqID+MP, the theorem prover can also apply (Contr L), although it needs to *at most once on each formula* $x : A \Rightarrow B$ *occurring in every derivation branch*. To implement this limited use of (Contr L) we allow to duplicate once the conditional formulas to which ($\Rightarrow$ L) is being applied. To this aim we introduce another argument `CondContr` to the `prove` predicate that now becomes:

prove(Sigma, Delta, Labels, CondContr).

The list `CondContr` stores the conditional formulas of the antecedent that have been duplicated so far. When ($\Rightarrow$ L) is applied to a formula x: $A \Rightarrow B$ in the antecedent, the formula is duplicated at the same time into the `CondContr` list; when ($\Rightarrow$ L) is applied to a formula in `CondContr`, in contrast, the formula is *no longer* duplicated. Thus the ($\Rightarrow$ L) rule is split in two rules, one taking care of "unused" conditionals of the antecedent, the other taking care of "used" (or duplicated) conditionals.

4   The Program CondLean

CondLean has also a graphical interface (GUI) implemented in Java. The GUI interacts with the SICStus Prolog implementation by means of the package `se.sics.jasper`. Thanks to the GUI, one does not need to know how to call the predicate `prove`. One just introduces a sequent in a text box and searches a derivation by clicking a button; moreover, one can choose the version of the theorem prover (constant labels, free-variables, heuristic version) and the intended system of conditional logic. If the sequent that has been introduced is valid, the program offers these options: display a proof tree of the sequent in a special window, build a latex file containing the proof tree, and view some statistics of the proof.

5   Statistics and Conclusions

The performances of the three versions of the theorem prover are promising even on a small machine[5]. We have tested CondLean obtaining the following results: in less than 2 seconds, the constant labels version succeeds in 79 tests over 90, the free-variables one in 73 (but 67 in less than 10 mseconds), the heuristic version in 78 (70 in less than 500 mseconds). The test samples have been generated by modifying the samples from [2]. Considering the sequent-degree (defined as the maximum level of nesting of the $\Rightarrow$ operator) as a parameter, the free-variables version succeeds in less than 1 second for sequents of degree 11 and in less than 2 seconds for sequents of degree 15.

In future research we intend to extend CondLean to other systems of conditional logics and to experiment standard refinements and heuristics to increase the efficiency of the theorem prover.

Acknowledgements

We are grateful to Matteo Baldoni for his precious help in improving the layout of this document.

---

[5] These results are obtained running SICStus Prolog 3.10.0 on a Pentium 166 MMX, 96 MB RAM machine.

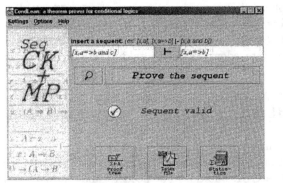
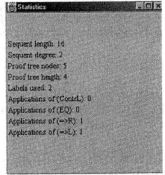

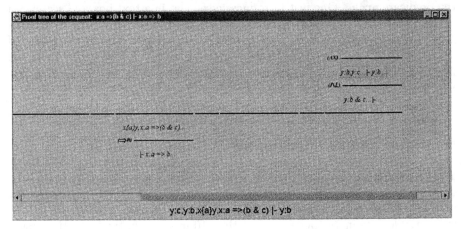

Fig. 2. Some pictures of CondLean

References

[1] N. Olivetti, C. B. Schwind. *A calculus and complexity bound for minimal conditional logic.* Proc. ICTCS01, LNCS 2202, 2001.

[2] B. Beckert, R. Gorè. *Free Variable Tableaux for Propositional Modal Logics.* Tableaux-97, LNCS 1227, Springer, 1997.

[3] D. Nute. *Topics in conditional logic.* Reidel, 1980.

[4] G. L. Pozzato. *Deduzione automatica per logiche condizionali: analisi e sviluppo di un theorem prover.* Tesi di laurea, Informatica, Università di Torino, 2003.

[5] B. Beckert, J. Posegga. *leanTAP: Lean tableau-based deduction.* Journal of Logic Programming, 28, 1996.

Author Index

Lecture Notes in Artificial Intelligence (LNAI)

Lecture Notes in Computer Science